Geomechanics of Marine Anchors

Geomechanics of Marine Anchors

Charles Aubeny

CRC Press
Taylor & Francis Group
Boca Raton London New York

CRC Press is an imprint of the
Taylor & Francis Group, an **informa** business

CRC Press
Taylor & Francis Group
6000 Broken Sound Parkway NW, Suite 300
Boca Raton, FL 33487-2742

First issued in paperback 2019

ISBN-13: 978-1-4987-2877-5 (hbk)
ISBN-13: 978-0-367-87341-7 (pbk)

Library of Congress Cataloging-in-Publication Data

Names: Aubeny, Charles, author.
Title: Geomechanics of marine anchors / by Charles Aubeny.
Description: Boca Raton, FL : CRC Press, Taylor & Francis Group, [2018] | Includes bibliographical references and index.
Identifiers: LCCN 2017017150| ISBN 9781498728775 (hardback : alk. paper) | ISBN 9781498728782 (ebook : alk. paper)
Subjects: LCSH: Offshore structures--Anchorage. | Anchors--Design and construction. | Marine geotechnics.
Classification: LCC TC1665 .A83 2018 | DDC 623.8/62--dc23
LC record available at https://lccn.loc.gov/2017017150

Visit the Taylor & Francis Web site at
http://www.taylorandfrancis.com

and the CRC Press Web site at
http://www.crcpress.com

To My Wife, Ulker,

for her support throughout the development of this book.

Contents

Preface

Geomechanics of Marine Anchors is intended for students and practitioners with an interest in the design of anchorage for floating structures. Much of the material presented in this book derives from experience with offshore civil infrastructure, which to date has largely been dominated by oil–gas exploration and production. More recently, various renewable energy systems are being proposed and developed that involve floating systems moored to the seabed. A need for efficient, reliable anchors is likely to be particularly important for such systems, owing to the relatively high capital costs of their support structures compared to traditional oil–gas systems. The basic mechanics of anchors used for offshore energy infrastructure also apply to anchors for ships and naval vessels. In fact, a chapter of this book devoted to drag embedment anchors is largely applicable to both offshore civil structures and ships. However, the constraints peculiar to ships, such as a need for rapid deployment and extraction, rule out a number of other types of anchors covered in this book that are often considered for civil infrastructure, such as suction caissons and piles.

Understanding anchor behavior requires a thorough grasp of the medium in which the anchors are embedded. A full chapter of this book is therefore devoted to a review of soil behavior to provide a foundation for later chapters. The presentation on soil behavior presumes that the reader has taken an introductory course in soil mechanics covering soil index properties and classification, seepage, consolidation, elasticity theory, and soil strength principles. Additionally, much of the analytical framework for understanding and predicting anchor performance is cast in terms of ordinary and partial differential equations; therefore, some familiarity with analytical and numerical solutions to differential equations is needed. Overall, the preparation from a typical undergraduate curriculum in civil engineering should provide sufficient background for utilizing this book. Additionally, this book should be considered suitable for specialists in coastal, ocean and mechanical engineering, although some advance preparation on soil mechanics may be needed. Most of the material presented in this book is suited to elective undergraduate and graduate students as well as practicing engineers.

A first objective of this book is to provide the engineer with an organized analytical framework for understanding and predicting anchor behavior. A number of analytical methods appear in the anchor literature for predicting anchor performance, each with their own range of applicability. Chapter 3 presents the available analytical tools, including plastic limit analysis, elastoplastic soil constitutive models, cavity expansion methods and finite element analyses. An effort has been made in the organization of the text to highlight the relative strengths and limitations of the various methods to assist the engineer in selecting the most appropriate suite of analysis methods for a given problem.

In regard to the second objective, a comprehensive presentation of the various anchor systems, the author's first impulse was to devote an entire chapter to each anchor type. Owing to overlapping analysis methodologies and many common features amongst the various anchors, this was found to generate needless redundancy. Thus, the anchors are grouped into two basic categories comprising cylindrical versus plate geometries. The former include suction caissons, driven tubular piles and most gravity installed (torpedo) piles. The major issues associated with cylindrical anchors—installation, ultimate pullout capacity, and elastic effects—are presented in Chapters 6, 7, and 8, respectively. Five types of anchors are considered in the plate category, drag embedment, suction installed, dynamically installed, pile driven, and helical. Drag embedment anchors largely fall into a category of their own, owing to complex interactions with the chain that govern their installation. Therefore, Chapter 9 is devoted exclusively to drag embedment anchors. The remaining plate anchor types are considered collectively as direct embedment anchors. Chapter 10 systematically addresses three major issues associated with direct embedment anchors: installation, loss of embedment during keying, and ultimate pullout capacity.

In writing a text of this type, a problem always arises in selecting an appropriate depth of coverage for the various topics. In the case of mature technologies, such as driven piles, a vast body of knowledge already appears in the published literature. In these instances, this book presents the basic principles for the sake of completeness, but does not attempt to fully incorporate the wide body of published information already in existence. By contrast, the literature on certain topics, such as a number of the plate anchor systems, is relatively sparse and often not easily accessible to many readers. In these cases, the text is organized largely with a view toward providing comprehensive coverage.

Finally I would like to acknowledge the various collaborations and associations from which I draw my experience with anchors and geomechanics. These include Dr. J.D. Murff and my other colleagues at Texas A&M University, Professor Robert Gilbert and his colleagues at the University of Texas Austin, Professor Mark Randolph and his colleagues at the University of Western Australia, Carl Erbrich and his colleagues at Fugro Advanced Geomechanics, Knut Andersen and his colleagues at the Norwegian Geotechnical Institute, Ed Clukey from the Jukes Group and Arash Zakeri from BP America Inc.

Author

Dr. Charles Aubeny is a professor at Texas A&M University where he is coordinator of the geotechnical group. He is an expert in offshore foundations and serves on the Offshore Site Investigation and Geotechnics Committee for the Houston Branch of the Society of Underwater Technology. He is the recipient of the ASCE Thomas A. Middlebrooks Award and a Fellow of the American Society of Civil Engineers.

Chapter 1

Introduction

Anchors are a critical component of floating and compliant structures used in coastal and ocean engineering applications. This book presents the application of geomechanics for understanding and predicting anchor performance. Geomechanics provides the analytical framework for characterizing the interaction of the anchor with the soil in which it is embedded. It is central to two aspects of anchor behavior: embedment of the anchor into the seabed to a sufficient depth to ensure that it can fulfill its intended function and the resistance of the anchor to pullout under applied working loads. Other engineering specialties contribute to anchor design, each of which could occupy a volume of its own; these include fatigue and corrosion, mooring line system design in the water column, the dynamics of floating structures, the mechanical hardware for anchor installation, instrumentation for monitoring anchor performance, and the vessels and logistics for transporting and installing anchors. These topics are largely outside of the scope of this book, aside from occasional mention of their impact on geotechnical behavior. Thus, anchor geomechanics is one of a number of specialties contributing to the design of mooring systems, but one of central importance.

The focus of the coverage in this book is largely directed toward anchors for offshore energy infrastructure, historically dominated by oil–gas exploration and production, but with an emerging focus on floating renewable energy systems. Many of the basic principles guiding the analysis of anchors for offshore energy infrastructure apply equally to anchors for ships and naval vessels, and findings from naval research laboratories has often been applied to anchors used for floating energy facilities. However, design constraints can tend to push the development of anchors for these applications in diverging directions. A main driver guiding the design of anchors for offshore energy development is maximizing the pullout load capacity of anchors. Pullout capacity is also a consideration for ship anchors, but other factors constrain their design, such as rapid installation and extraction, and the necessity of avoiding excessively deep embedment to reduce the potential for snagging cables and pipelines. A glance at an anchor load capacity chart will show commonly used anchors for shipping to have an order of magnitude lower capacity than the high-capacity anchors favored by the energy industry. The apparent "inefficiency" of commonly used anchors for ships is driven by constraints that do not normally occur in energy applications.

In regard to organization of this book, this introductory chapter provides an overview of the common types of floating and compliant structures secured by anchors, different categories of anchors, the physical mechanisms by which anchors resist pullout, how the nature of loading affects anchor performance, and criteria for evaluating anchor performance. This introductory chapter contains few quantitative formulations and is intended to provide the reader with an intuitive grasp of these topics; rigorous development of theoretical formulations and analytical methods is saved for later chapters. The remainder of this book comprises three major segments:

- Chapters 2 through 5 cover basic principles relevant to all anchor types, in successive order these include soil mechanics, relevant analytical methods from geomechanics, fundamental solutions that have been applied to all categories of anchors, and anchor line mechanics.
- Chapters 6 through 8 cover cylindrical anchors, with Chapter 6 covering caisson and pile installation, Chapter 7 ultimate load capacity, and Chapter 8 elastic response and soil–pile interaction.
- The final major segment is plate anchors, with Chapter 9 covering drag embedded anchors (DEAs) and Chapter 10 covering various types of directly embedded plate anchors.

1.1 FLOATING AND COMPLIANT STRUCTURES

As development progresses to greater water depths, the cost of fixed structures increases exponentially, driving the use of floating or compliant structures secured to the seabed by anchors. In deep water, it is also more difficult and expensive to hold the natural periods well below the dominant sea state period. The more common of these structures are described below.

1.1.1 Semisubmersible

A semisubmersible (Randall, 1997) is a floating structure comprising large vertical columns supported by large pontoons at the bottom. It floats high in the water while being towed to the site. The pontoons are then flooded to partially submerge the structure, the majority of which is below water. Only the columns are exposed to environmental loading, which contributes to the stability of the structure. A semisubmersible can be moored to the seabed using catenary or taut systems, as illustrated in Figures 1.1 and 1.2. Catenary moorings apply essentially horizontal loads on anchors, while taut moorings are inclined up to approximately 30–35° from horizontal. The mooring lines can comprise chain or wire rope, the latter are lighter but more prone to damage and corrosion. In deep waters, catenary moorings require excessively large offsets, resulting in an increased likelihood of encroaching upon adjacent tracts, crossing mooring lines from adjacent facilities, or crossing undersea pipelines. Hence, for these situations, taut mooring systems provide a more attractive method of mooring floating platforms. In deep water, the weight of the mooring line itself becomes a significant design consideration, a problem which has been addressed through the introduction of lightweight synthetic mooring lines (Vryhof, 2015).

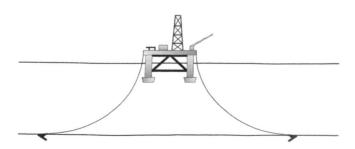

Figure 1.1 Semisubmersible with catenary mooring system.

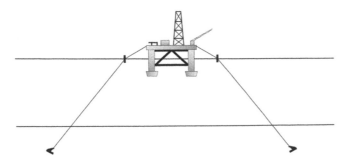

Figure 1.2 Semisubmersible with taut mooring system.

1.1.2 Tension leg platform

A tension leg platform (TLP) consists of a semisubmersible vessel moored by vertical tendons connected to the seafloor, as illustrated in Figure 1.3. The excess buoyancy of the structure—typically on the order of 15%–25% of the platform displacement—maintains the tendons in tension even under the worst storm loading conditions. Conoco constructed the first TLP in the British sector of the North Sea in the early 1980s in approximately 150 m of water. This structure was selected for relatively shallow water as a test of the concept prior to use in deep waters. Since then, several deep water TLPs (>450 m) have been installed. The foundations for TLPs usually consist of large diameter pipe piles that provide resistance to uplift through skin friction, although suction caissons (Section 1.2.1) can be used. Loading on the foundation consists of a tensile mooring force and cyclic loading owing to wave loading on the superstructure. Loading is predominantly vertical with lateral forces estimated at less than 10% of vertical forces. A critical loading condition most likely occurs when a change in mean sea level owing to a storm surge occurs in conjunction with

Figure 1.3 Tension leg platform. (a) Schematic and (b) View from surface.

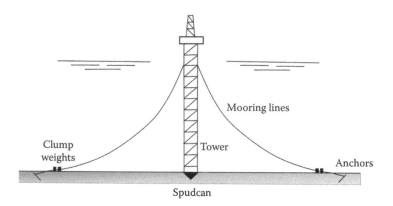

Figure 1.4 Compliant guyed tower.

large wave loading. TLPs can provide a feasible platform alternative to water depths from approximately 450 to at least 1000 m.

1.1.3 Guyed tower

A guyed tower (Figure 1.4) comprises a slender steel truss supported by a compression spud-can foundation and held upright by guy wires secured by anchors. Heavy clump weights are inserted between the anchors and the tower. Under normal conditions, the guy restrains the tower without lifting the clump weights. However, under extreme weather conditions, the clump weights lift from the seabed to provide additional restoring force to resist the wave loading. This concept was first employed for the Lena tower in the Gulf of Mexico in 305 m of water. The tower was secured by 20 wire rope mooring lines and 179 clump weights.

1.1.4 Spar

The Spar platform (Figure 1.5) is an old idea that emerged as a popular deepwater alternative because of various advantages over other solutions. It is essentially a deep-draft, truncated cylinder, with a moonpool that supports a platform by means of excess buoyancy. The hull is a steel cylinder, which encloses soft tanks in the bottom and hard tanks in the top part to provide the required buoyancy. Variable ballast allows adjustments to be made to maintain a constant draft when the topside loads change. The spar can be tied to the sea bottom by vertical tethers but more commonly by catenary or taut mooring lines. In oil-gas applications drilling and production risers run down a centerwell, which shields them from wave and current loading. As a result, the spar is very stable, relatively insensitive to deck loads and easy to transport. All natural periods of moored spars are much longer than that of waves; therefore, their linear dynamic response to wave action is small. There are, however, nonlinear effects, owing to different frequencies, which can cause significant responses in the natural frequencies of the platform, and which should be considered.

1.1.5 Arrays of floating units

Renewable energy can be harvested from sources that include offshore wind, tidal, and wave energy. Much of this energy is located at sites where floating units are likely to be

(a)

(b)

Figure 1.5 Spar. (a) Schematic and (b) View from surface.

most feasible. For example, over 60% of US offshore wind energy potential is in water depths greater than 60 m where floating offshore wind turbines are more economical than fixed bottom systems. The industry is in a very early stage of development, and the early models for floating offshore wind turbines tend to build on the concepts described above for exploiting offshore hydrocarbon resources. It is uncertain how these will evolve, but a prominent feature of renewable energy development that distinguishes it from oil–gas applications is that floating units are deployed in arrays, rather than isolated single units. This creates an opportunity for increased economy in anchors by attaching more than one mooring line to a single anchor, a concept that is in the very early stages of development (Fontana et al., 2016). Aside from an obvious benefit of cost reductions, multiline attachments provide a promising method for providing redundancy in design at little added cost. In principle, three mooring lines can secure a floating wind turbine, but this leads to a fragile system, as a single damaged mooring line can lead to loss of the unit. Distributing load to additional anchors can also permit smaller, more economical mooring lines. Noting that even temporary moorings for semisubmersibles for oil–gas exploration typically employ 8–12 moorings, multiline moorings can facilitate welcome resilience to renewable energy systems. Conceptually, anchors that possess a vertical axis of symmetry—caissons and piles are discussed later in this chapter—readily lend themselves to multiline moorings, as depicted in Figure 1.6. Nevertheless, plate anchors having a preferential direction of load resistance can still be configured into a multiline arrangement by transmitting mooring line loads through an intermediate attachment ring down to the anchors, although the potential gains in economy will likely be less for plate rather than cylindrical shapes. Since caissons and piles are typically more expensive than plate anchors, multiline mooring systems may

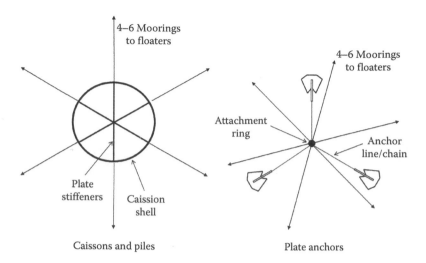

Figure 1.6 Possible multiline attachment configurations.

well level the playing field for selecting the most suitable anchor alternative for floating offshore wind turbines.

1.2 ANCHOR TYPES

This book considers seven types of anchor types. Figure 1.7 shows several example anchors, including driven piles, a suction embedded plate anchor (SEPLA), a suction caisson, a DEA, and a vertically loaded anchor (VLA), alternatively termed a drag embedded plate anchor. Of the seven types considered in this book, three can be considered as relatively mature technologies—suction caissons, piles, and DEAs. Two are relatively recent innovations, but have experienced a sufficient number of field deployments based on which they may reasonably be considered economically viable—dynamically installed piles and SEPLAs. Finally, two anchor types are limited in the presence of commercial offshore deployments. However, ongoing research indicates that they are technically feasible and have promises for economic viability—dynamically embedded plate anchors (DEPLAs) and helical piles. It is emphasized that the term "mature" refers not so much to their level of technological development as to the scale of their deployments to date. For example, piles and DEAs are among the oldest of anchor alternatives and have experienced widespread deployment, but there is plenty of room for innovation and improvement of predictive models of performance for these anchors.

The various anchors have advantages and limitations in regard to the soil types and loading conditions for which they are most suitable, which will be the focus of most of the discussion that follows. Although this book focuses primarily on geotechnical aspects of anchor performance, some mention in cost differences is essential to guide the discussion. In this regard, DEAs may be considered as the least expensive of the mature technologies, piles most expensive, with suction caissons somewhere in between. In terms of scale, a factor of five between least and most expensive is not unrealistic; however, it is emphasized that costs are highly sensitive to project-specific factors such as equipment availability and vessel costs. Moreover, cost comparisons should be considered in light of differences in reliability amongst the various anchor types.

(a)   (b)

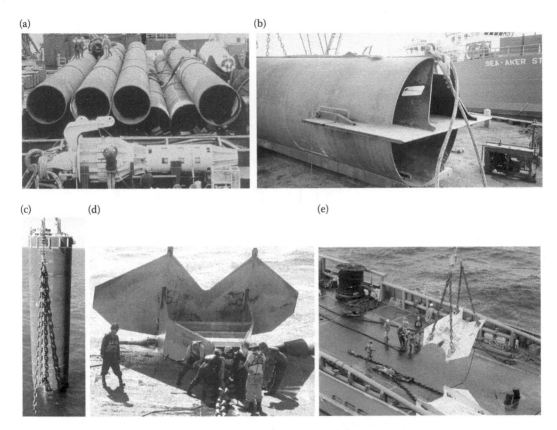

(c)   (d)   (e)

Figure 1.7 Example anchors. (a) Driven piles, (b) suction embedded plate anchor (SEPLA, courtesy of Intermoor), (c) suction caisson, courtesy of E. Clukey (d) drag embedment anchor (DEA, courtesy of Delmar), and (e) VLA, vertically loaded anchor, courtesy of Delmar.

1.2.1 Suction caissons

Suction caissons (Andersen et al., 2005) are large cylindrical anchors that are installed partially by self-weight penetration, with penetration to full installation depth accomplished by the application of "suction," which is actually a differential pressure induced by pumping through a valve in the top cap (Figure 1.8). Suction caissons have advantages over piles in that they avoid the requirement for heavy underwater hammers. Moreover, they do not require auxiliary platforms such as jack-up rigs to support the installation operation. Suction installation is feasible in a range of soil profiles, including soft clays, stiff clays, and sand. Suction installation in clays is a matter of applying sufficient underpressure to overcome the various sources of soil resistance that are described in detail in Chapter 6. The maximum underpressure is limited by considerations of stability of the plug of soil inside the caisson—excessive underpressure can cause the soil plug to heave into the interior of the caisson. Installation in sands also relies on underpressure to advance the caisson, but the underpressure also induces upward seepage flow inside the caisson which dramatically reduces the resistance to penetration. Caisson aspect ratio (ratio of length to diameter) is largely dictated by considerations for safely installing the caisson. Optimal aspect ratios are typically in the range of 1.5–3 in stiff clays, less than 1.5 for dense sands, and greater than 5 for soft clays. Although suction caissons are versatile from the standpoint that they can be installed in a wide range of soil profiles, heterogeneous soil profiles are problematic.

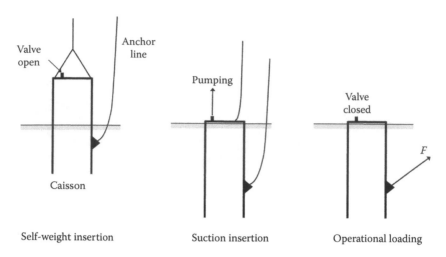

Figure 1.8 Suction caisson installation.

For example, interbedded clays can interrupt the upward seepage that is essential for reducing resistance to penetrate in sands. Recent developments in suction installation (Tjelta, 2015) show that suction installation is possible in heterogeneous soil profiles by cycling or jetting, but the process is more complex—and requires considerably more expertise—than in the more benign homogeneous soil profiles. Proof testing to verify load capacity is not normally required for suction caissons (Andersen et al., 2005), since their positioning is well controlled, their design is supported by numerical studies calibrated to model tests, and their design is based on the data of site-specific soils.

In general, suction caissons can resist both horizontal and vertical loads, and are therefore suitable anchors for either catenary or taut mooring systems. For reasons to be discussed subsequently, suction caissons experience substantial reductions in load capacity when subjected to sustained vertical loads. For this reason, they are not particularly attractive for anchorages involving this type of loading, such as TLPs. Horizontal load capacity of suction caissons is sensitive to the depth at which the load is applied; therefore, an optimal load attachment (pad eye) depth that produces maximum capacity should be selected. Methods for estimating the optimal load attachment depth as well as predicting ultimate load capacity are presented in Chapter 7. Suction caissons are normally designed with relatively thin walls; typical diameter-to-wall-thickness ratios are in the range of 125–160, in contrast to 10–40 for piles. Various internal plate stiffeners and ring stiffeners are typically required for preventing buckling failure during installation (Section 6.6) and structural failure during operational loading. Wall-thickness selection and stiffener design is within the province of structural engineers, but geotechnical engineers are typically called for providing soil reaction pressures to support the structural analyses, a topic addressed in Chapter 8.

1.2.2 Piles

In spite of the many attractive features of suction caissons, piles remain a viable anchor alternative in many instances. Piles (Figure 1.9) can be installed in a wide range of geologic profiles, including profiles containing calcarenite layers and rock (Vijayvergiya et al., 1977) that would normally not be considered feasible for suction caissons. Additionally, heterogeneous soil profiles that, as described above, present a challenge for suction caisson installation, are

(a) (b)

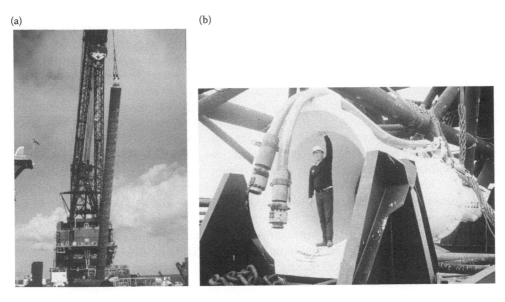

Figure 1.9 Driven pile. (a) Tubular pile and (b) Hammer.

not particularly problematic for driven pile installation. In regard to load resistance capabilities, the shortcomings of suction caissons in resisting pure vertical load were noted earlier.
For this reason, piles remain the anchor of choice for TLP systems, which involve nearly pure
vertical loading on anchors. Piles can also resist horizontal loads. Short piles—piles with
length-to-diameter ratios less than 6—experience negligible bending when subjected to horizontal and moment loading. Thus, their mechanical behavior under applied loads is similar
to suction caissons, and the limit analysis methods presented in Chapter 7 for caissons also
apply to short piles. Longer piles behave as axially–laterally loaded slender structural mem-
bers, and prevailing practice models piles as beam columns supported by "soil springs" representing the continuum in which they are embedded. While elastic response can be potentially
significant for any anchor type, slender piles are the only anchor type considered in this book
for which elastic behavior can be said to truly dominate their design. Chapter 8 presents
methods for evaluating the elastic response of piles to mooring line loads.

While a wide variety of pile types, materials, and methods of installation have evolved in
general geotechnical practice, much of offshore pile practice uses driven open-ended tubular piles, with diameters frequently in excess of 2 m. Onshore methods for driving piles
include impact hammers powered by steam, air, or diesel; and, in cohesionless soils, vibratory hammers that advance the pile by liquefying the soil while applying a downward load.
Historically, onshore technology was adapted to offshore pile installations using a "follower" that was attached to the pile to transmit the stress wave from the water surface to the
pile. This approach was technically feasible to water depths up to approximately 300 m. In
the mid-1970s, development at greater water depths spurred the development of underwater hydraulic hammers. Various developments in this technology are described by Cox and
Christie (1976), Jansz (1977), Heerema (1980), Aurora (1984), and van Luipen (1987). Wave
equation analyses are used for estimating driving stresses, hammer selection, and penetration rates. They are also used in a reverse sense to back-analyze driving records to estimate
pile load capacity.

An alternative to hammer driven installation is jetting, which is commonly used for conductor installation in oil–gas applications (Zakeri et al., 2014). Jetting involves extraction

of soil through the interior of the pile to reduce resistance to penetration. This mode of penetration is in contrast to conventional pile driving by which the volume of the penetrating pile is accommodated by displacing the soil. Although jetted piles will likely have less than half the axial capacity of driven piles, they have some appeal for offshore renewable energy applications. Objectionable acoustic emissions associated with pile driving are relatively limited for oil–gas projects involving some 10–20 piles; however, environmental concerns become more serious in renewable energy projects potentially involving hundreds or even thousands of piles. As a much quieter installation method, jetting may well be a favorable alternative to driving from the standpoint of acoustic emissions. Since much of pile design is empirical, largely based on load tests on driven piles, adoption of jetted piles would likely require a recalibration of much of the current design methodology. A second alternative to mitigating acoustic emissions is a "bubble curtain"—a ring of bubbles streaming upward from the seabed to interrupt sound wave propagation (Love and Arndt, 1948). However, the effectiveness of bubble curtains may be limited owing to sound transmission through the seafloor (Lee et al., 2011). Further, difficulties arise in attenuation of low frequencies, which requires large stable freely rising bubbles. Lee et al. (2011) investigate the use of tethered latex balloons to overcome this difficulty.

1.2.3 Drag embedded anchors

A DEA is a bearing plate (termed the "fluke") inserted into the seabed by dragging with wire rope or chain (Figure 1.10). The fluke is attached to the anchor line by a "shank" comprising one or more plates. Self-embedment of the anchor is achieved by controlling the line of action of the anchor line force—by setting the fluke–shank angle—such that "failure" occurs roughly parallel to the fluke so that the anchor will dive downward when dragged. DEAs are normally designed for several possible fluke–shank angle settings according to soil type. In stiff clays and sands, the fluke–shank angle is typically on the order of 30°, while 50° is typical for soft clays. DEAs are less expensive to install relative to caissons and driven piles and, as will be discussed subsequently in more detail, plate anchors are also generally more efficient than caissons and piles in terms of the ratio of load capacity to anchor weight. However, DEAs do not have the advantage enjoyed by caissons and piles in terms of their ability to be precisely positioned. Further, the load capacity of the anchor depends on the depth of anchor penetration, which cannot be predicted with a high degree of uncertainty. Nevertheless, the uncertainty in DEA load capacity can be mitigated substantially by

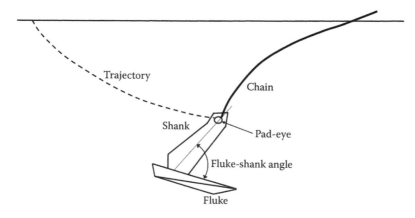

Figure 1.10 Drag embedded anchors.

proof load testing of anchors following installation, which is usually required for permanent installations. It should be noted that obtaining a vessel capable of proof testing is not a trivial matter, as the load capacity of some anchors may exceed the available bollard pull of many anchor handling vessels (AHVs). Winch-driven tensioners may alleviate this obstacle in cases where another anchor is available to provide the necessary reaction force (Vryhof Anchors, 2015).

DEAs in sands and stiff clays experience minimal penetration into the seabed, typically less than 1–2 fluke lengths. Consequently, these anchors have minimal vertical load capacity and their suitability is restricted to catenary systems. Effectively, the anchor provides resistance to horizontal loads, while the dead weight of the anchor chain resists vertical loads. The extensive lengths of anchor chain required for this arrangement may easily exceed the costs of the anchors; thus, in comparing anchor alternatives, it is important to compare the cost of the entire mooring system, rather than considering only the cost of the anchors. Although caissons and piles are more expensive than DEAs, they typically have vertical load capacities suitable for taut moorings. The reduced mooring line costs for such systems may well offset the differences in anchor costs.

In contrast to the case of sands and stiff clays, DEAs in soft clays will penetrate to substantial depths, on the order of tens of meters in some cases. Since soft clay profiles typically exhibit increasing strength with depth, increased penetration leads to increased DEA load capacity, including a substantial capability for resisting vertical loads. Thus, DEAs in soft clay can provide suitable anchorage for both catenary and taut mooring systems. To enhance the pullout resistance of DEAs in soft clays, so-called VLAs were developed. A VLA is installed in the same manner as a DEA, but it features a releasable shank that can be opened after drag installation. Most VLA designs employ a shear pin that ruptures when the mooring line force exceeds a certain level, although more recently some mechanical release mechanisms have been developed. Two types of shank have evolved for VLAs: a rigid bar, and a bridle arrangement comprised of chains (Figure 1.11). Since chain mooring lines tend to inhibit anchor penetration, VLAs are usually used in conjunction with wire rope. As noted earlier, the fluke–shank angle of a DEA in soft clay is typically set at approximately 50° to promote diving of the anchor during drag embedment, but maximum pullout capacity of the anchor is developed when the shank is normal to the fluke. Thus, increasing the fluke–shank angle at the end of drag installation provides added pullout capacity. A down side of increasing the fluke–shank angle is that it leads to a failure mechanism involving upward movement of the anchor when the anchor is overloaded. This behavior is undesirable because, as the anchor migrates upward into weaker soil, it rapidly loses load capacity

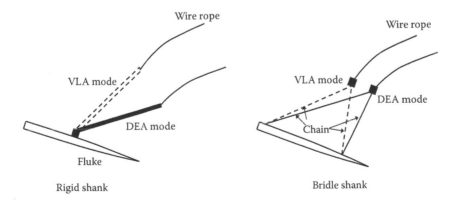

Figure 1.11 Vertically loaded anchors.

leading to a brittle failure mode. To reduce the potential for a brittle failure mode, most VLAs now feature a "near-normal" orientation for the released shank, with the opened shank oriented to slightly less than 90° from the fluke—78° is the largest angle known to the author. At this shank orientation, additional load capacity is achieved when the shank is released. However, if overloaded, the anchor retains a tendency to dive downward. While the near-normal concept definitely leads to reduced tendency for upward movement, the downward motion of the anchor is conditional on the orientation of the anchor line and fluke at the time of failure. Therefore, it is not advisable to simply assume *a priori* that an overloaded VLA will continue to dive downward. Chapter 9 provides further details on this topic.

As both DEA and VLA load capacity depend on embedment depth, the load capacity prediction requires two key calculations: (1) prediction of the anchor trajectory and (2) prediction of load capacity for the depth of embedment achieved. The former calculation depends on the interaction between the anchor chain or wire line. Historically, predictions of drag anchor performance were based on model tests at the prototype and laboratory scale (e.g., Naval Civil Engineering Laboratory, 1987). More recent trends have experienced the application of analytical methods, including the introduction of limit equilibrium analyses (Stewart, 1992; Neubecker and Randolph, 1995a; Dahlberg, 1998), finite element studies and plastic limit analyses of soil–anchor interactions (Rowe and Davis, 1982a; O'Neill et al., 2003; Murff et al., 2005), and analytical formulations for mooring line behavior and interaction between the mooring line and the anchor (Vivatrat et al., 1982; Neubecker and Randolph, 1995b; Aubeny and Chi, 2010).

1.2.4 Suction embedded plate anchors

The SEPLA conceived and developed by Dove et al. (1998) and Wilde et al. (2001) is a plate anchor installed by attaching it to the tip of a suction caisson (Figure 1.12). The suction caisson is installed in the conventional manner, but retracted by overpressure to leave the plate behind. The vertically oriented plate is then "keyed" to turn into the direction of the applied mooring line load. Some loss of embedment occurs during the keying process, with a concomitant reduction in load capacity (Gaudin et al., 2006); consequently, considerable research attention has been devoted to analyzing the loss of embedment. Song et al. (2009) indicate that, for typical loading conditions, keying results in a loss of embedment of

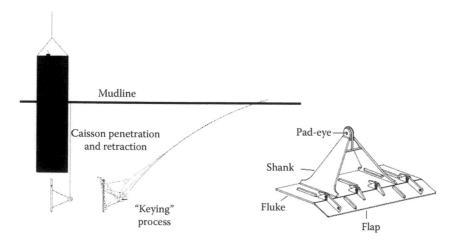

Figure 1.12 Suction embedded plate anchor.

approximately 0.4–0.6 times the height of the anchor. Some SEPLA design concepts incorporate a flap designed to minimize the loss of embedment during keying.

SEPLAs have the appeal that they combine the precise vertical and horizontal positioning of a suction caisson with the lightness and efficiency of a plate anchor. Deployment is largely limited to soft clay soil profiles. Since the anchor is deeply embedded, it has capabilities for resisting vertical loads and is a suitable anchorage alternative for taut mooring systems. Installation time is lengthy—on the order of 20% greater than for a conventional suction caisson installation (typically 1 day)—since both suction installation and overpressure retraction are required. Accordingly, installation costs are greater than that for drag embedded plate anchors, and the benefit of precise positioning of the plate is obtained at the price of higher installation costs. Moreover, transport costs should be factored into any cost comparisons. Owing to their size, suction caissons require large transport vessels, which can be a major cost driver. By contrast, SEPLA deployments require the transport of only single suction caisson in conjunction with a large number of compact plate anchors that can easily be arranged on the deck of a transport vessel. Overall, for soft clay soil profiles where capabilities for resisting vertical/inclined loads and precise anchor positioning are a significant concern, SEPLAs can be considered a competitive alternative.

1.2.5 Dynamically installed piles

Dynamically installed piles, sometimes termed "torpedo piles" penetrate into the seabed by the kinetic energy obtained from free fall through the water column and their own self-weight. This type of anchor (Figure 1.13a) typically is rocket shaped or cylindrical with a nose cone, with up to four stabilizing fins at the top of the pile. The Delmar Omni-Max anchor is somewhat of an exception in that it comprises three plates oriented at 120° from one another rather than a cylindrical body (Figure 1.13b). In this sense, it may be classified as a plate anchor. However, since its mode of installation more closely follows other gravity installed piles, it is included in this discussion. The piles are typically 12–15 m long and 0.8–1.2 m in diameter, with dry weights in the range of 500–1000 kN (Randolph et al., 2011). Reported free fall heights are as high as 170 m (Ehlers et al., 2004), but are more commonly in the range of 50–75 m (Zimmermann et al., 2009; Lieng et al., 2010). Prediction of impact velocity as the pile falls through the water column is a critical part of the overall prediction of penetration depth into the seabed and load capacity; guidelines on the analysis and

(a)

(b)

Figure 1.13 Dynamically installed piles.

selection of a drag coefficient are given by Hasanloo et al. (2012). Impact velocities are up to 30 m/s (Randolph et al., 2011). Penetration in soft clays is typically 2–3 pile lengths.

Dynamically installed piles offer the advantage of quick, economical installation. They can be installed with a single AHV and require minimal mechanical equipment—in contrast to the heavy underwater hammers required for piles or pumps for suction caissons. Ehlers et al. (2004) report successful deployments in soil profiles that include normally consolidated clay, overconsolidated clay, sand and calcareous deposits. Considering one comparative study of transportability, Zimmermann (2009) reported that the AHV under consideration could transport in a single trip either 4 suction caissons, 8 torpedo anchors, or 12 VLAs. Thus, economies in transport can be realized for torpedo piles, but not perhaps as much as for plate anchors. Since they embed relatively deeply, dynamically installed piles can resist both horizontal and vertical loads and are thus suitable for both catenary and taut mooring systems.

Prediction of load capacity requires accurate prediction of penetration depth which, in turn, requires accurate prediction of the impact velocity as well as accurate accounting of the effects of rate-dependent soil shearing resistance and inertial resistance. Rate-dependent soil shearing resistance is a particular challenge, since the strain rates occurring during dynamic penetration lie well above the range for which most experimental evaluations of soil strength are conducted. Performance can be enhanced through judicious positioning of the pad eye. Shelton (2007) and Zimmermann et al. (2009) report that placing the pad eye of the Omni-Max anchor near the pile tip promotes diving behavior when the anchor is overloaded.

1.2.6 Dynamically installed plate anchors

A DEPLA is a plate anchor that is installed by free falling through the water column in a manner similar to dynamically installed piles. This anchor concept has the advantage of the lightness and efficiency of a plate anchor combined with the low installation cost of a dynamically installed pile. Two approaches have evolved in the development of dynamic installation of plate anchors. The first approach, developed by O'Loughlin et al. (2014), consists of attaching the plate anchor to a detachable torpedo pile—a "follower"—as depicted in Figure 1.14. The follower can be detached from the plate anchor by triggering a shear pin in a manner similar to the shank release mechanisms for VLAs. The anchor itself comprises two crossed circular plates to form a cruciform configuration. After installation, the pile is retracted and the plate is keyed into position in a manner similar to the keying process for SEPLAs. DEPLA installation using a follower is essentially a dynamic pile installation; therefore, the analytical framework for modeling installation of the two anchor types is similar. The DEPLA concept has been validated through centrifuge model tests, numerical and analytical studies, and reduced-scale field trials, based on which a rigorous design procedure has been developed by O'Loughlin et al. (2016). While this anchor alternative is novel, the technology is promising. O'Loughlin et al. (2015) estimate that the reductions in fabrication costs alone for dynamically installed plate anchors can lead to cost savings of 70%–80%, and this does not allow for the potentially far greater cost savings associated with reduced AHV costs. The studies to date have focused on DEPLA installations in soft clay soil profiles.

The second DEPLA concept involves direct embedment of the plate anchor from free fall without the use of a follower to enhance penetration. A "Flying Wing" anchor based on this concept is currently under investigation. Preliminary single gravity model tests indicate the potential for deployment in both soft clays and sands. As with the other DEPLA approach, the plate is oriented vertically following installation and must then be keyed into position.

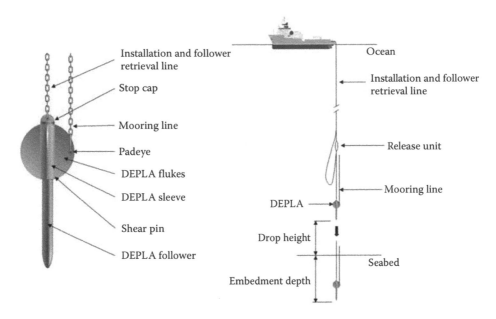

Figure 1.14 Dynamically embedded plate anchor.

This anchor concept is a novel approach. Therefore, continuous research is required for establishing fundamental questions in regard to penetration depth, loss of embedment owing to keying, and bearing resistance of the anchor.

1.2.7 Other direct embedment plate anchors

In addition to the SEPLA described earlier, other forms of directly embedded plate anchors include helical anchors and driven or jetted plate anchors. These types of anchor can be deeply embedded in clays or sands, providing the capability for developing vertical load capacity in sands. Helical anchors have received relatively little attention for offshore floating structures in the energy industry, although they have long been in use for anchoring vessels and submarine pipelines. A helical anchor (Figure 1.15) comprises one or more helical plates having a carefully controlled pitch welded to a central shaft. The central shaft functions to transmit torque during installation and transmit axial loads to the helical plates. Installation torque correlates to anchor load capacity, thereby providing a basis for confirming design load capacity predictions. In addition to their versatility from the standpoint of soil type, helical anchors have an advantage of relatively rapid, inexpensive installation. Small diameter helical anchors can be economically installed by remotely operated underwater vessels (ROVs). Helical anchors have the capability to resist lateral as well as axial loads (Prasad and Narasimha Rao, 1996), but the helices appear to enhance the lateral efficiency compared to a single shaft pile. Current models are relatively small compared to many marine anchors. Elkasabgy and El Naggar (2015) characterize a 0.601-m helix as being "large diameter," which is not inaccurate for land-based applications, would be considered relatively small in marine applications. However, if their value can be shown for marine applications, it is likely that they can be scaled upward to achieve larger capacities.

For estimating load capacity of single-helix anchors in clay, early studies include the investigation of Vesic (1971), Meyerhof and Adams (1968), Meyerhof (1973), and Das (1978, 1980), from which semiempirical design equations were developed. More recently, rigorous

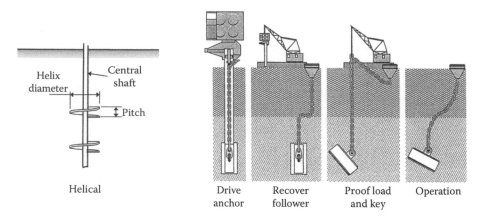

Figure 1.15 Direct embedment plate anchors.

theoretical solutions based on plasticity theory were developed for flat (i.e., the effect of pitch is neglected) circular plates in cohesive soil (Martin and Randolph, 2001; Merifield et al., 2003), which are described in Chapter 4. For multihelix anchors, interference and overlapping stress zones between adjacent plates complicate the analysis. Merifield (2011) presents theoretical solutions and proposed design charts for multihelix anchors in clays having up to three plates. A somewhat similar trajectory of development evolved for sands, with early empirical methods for estimating load followed by successively more rigorous limit equilibrium, cavity expansion, plasticity, and finite element solutions. Giampa et al. (2017) evaluate the accuracy of these capacity prediction methods through comparisons to intermediate-scale anchor tests. A finite element plane strain solution by Rowe and Davis (1982b), adjusted for three dimensional effects, proved to be the method that compared most favorably to measurements.

The solutions described above apply to monotonic loading, with no consideration of cyclic load capacity or upward displacements under cyclic loading. Model tests on cyclically loaded single-plate helical anchors in medium dense sand by Newgard et al. (2015) indicate that the anchor experiences upward ratcheting movements, which lead to significant reduction in anchor capacity under cyclic loading relative to monotonic load capacity. As discussed by Newgard et al. (2015), improved performance in regard to cumulative cyclic displacements may be possible through anchor redesign or using additional anchor plates. Additionally, rather than considering measures purely associated with anchor design, operation and maintenance measures may be employed. For example, damaged moorings may be reinstalled following storm loading, or retorqueing to recover embedment loss may be possible.

Aside from the helical concept for direct embedment, plate anchors can be installed by attaching a plate to a pile tip, which is driven or jetted into the seabed (Forrest et al., 1995; NAVFAC, 2011). The pile is subsequently extracted for reuse. The installation largely employs existing pile installation technology. A subcategory of this type of anchor is the "umbrella" anchor having vertically oriented plates during installation that rotate to a horizontal position during keying. Similar to helical anchors, these anchors enjoy the advantage of precise positioning. Moreover, they can be installed in a broad range of soil deposits including soft clays, stiff clays, dense sands, glaciated soils, and corals. They can be keyed to orient them to the anticipated direction of loading; hence, they are capable of resisting inclined and vertical loads. The low skin friction associated with jetting installation, normally considered a liability for piles, is actually beneficial, as it facilitates extraction of the pile at the end of plate installation.

1.2.8 Dead weight anchors

In certain situations, such as rock seabeds, penetration into the seabed is not possible, rendering the anchor systems described above impractical. In these cases, consideration may be given to dead weight anchors. Reinforced concrete or scrap steel may be used for this purpose, but are likely to be costly. As a less expensive alternative, Flory et al. (2016) describe a dead weight anchor comprising a ballast-filled flexible bag. The ballast can be any pourable materials, such as sand, drilling mud, or steel pellets. The bags are then grouped within a containment net, as shown in Figure 1.16.

Vertical capacity is simply the submerged weight of the anchor. Load capacity for the system is relatively small—on the order 5 tonnes for the Flory et al. system. In regard to horizontal load capacity, Flory et al. (2016) report a friction coefficient of 0.45 for this

(a)

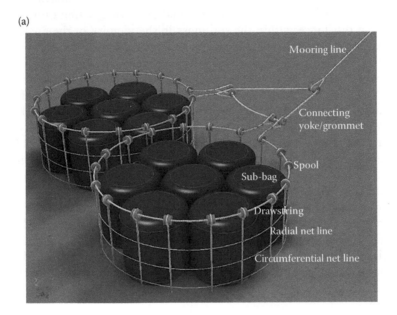

(b)

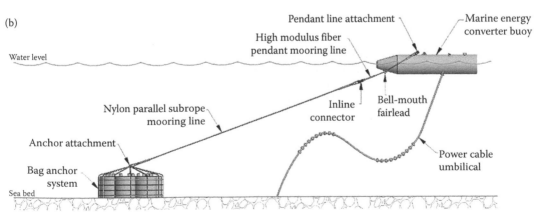

Figure 1.16 Dead weight anchors. (a) Anchor bag system and (b) Attachment to buoy.

system. These capacities are a small fraction of what can be achieved by pile/caissons and plate anchors. Nevertheless, in applications where the load demand is small and seabed conditions are difficult, a simple dead weight system may merit consideration.

1.3 MECHANISMS OF RESISTANCE

The physical mechanisms of resisting forces acting on an anchor should be understood to analyze both anchor installation and anchor pullout capacity. Methods for evaluating resistance comprise much of the content of this book. These methods can be easily misapplied if an intuitive understanding of the basic mechanics is lacking. The following discussion focuses on intuitive concepts, with more rigorous treatment left for later chapters.

1.3.1 Soil shearing resistance

Soil shearing resistance is often the primary source of pullout capacity for an anchor. Shearing resistance may be in the form of frictional resistance acting on the surface of the anchor or bearing resistance within the soil mass in the vicinity of the anchor. Figure 1.17 illustrates the issues associated with calculating skin resistance for the case of piles and caissons in clay. Firstly, pile installation disturbance weakens the soil through the process of remolding as well as the generation of large excess pore pressures that reduce the effective stress, $\sigma'_{rii} < \sigma'_{h0}$. Over time "setup" occurs, primarily owing to the process of consolidation; i.e., the dissipation of excess pore pressures induced during installation. Thixotropy—the time-dependent increase in remolded clay at constant water content—can also contribute the strength gain during the early stages of consolidation (Jeanjean, 2006). Subsequent axial loading induces changes in total stress $\Delta\sigma_{rr}$ as well as generation of additional excess pore pressures Δu. Depending on the rate of axial loading, dissipation of excess pore pressures can occur, such that partially drained or fully drained conditions prevail. Although the process is often described as "side friction," actual slippage can occur either at the soil–pile interface or entirely within the soil adjacent to the pile. Strain-softening behavior at the soil–pile interface can lead to progressive failure as described by Randolph (2003) and presented in detail in Chapter 8. Analytical methods that have been developed for simulating pile installation disturbance and setup, discussed in Chapter 3, provide useful insights into

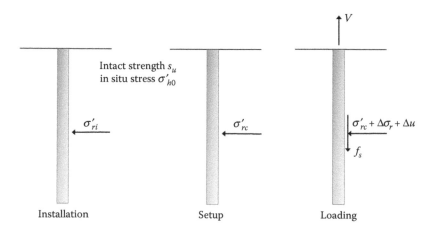

Figure 1.17 Skin resistance on piles in clay.

the factors that can affect side resistance. However, current prevalent practice employs the "α -method," where measured skin resistance is simply correlated to the intact soil undrained shear strength, $f_s = \alpha \, s_u$. The parameter α characterizes the net effect of a number of complex processes and, not unexpectedly, wide variations in back-calculated α occur for similar site conditions. Values of α can vary from nearly unity in normally consolidated clays to less than one-half in over-consolidated clays. Side resistance calculations for plate anchors generally parallel the α-method approach for piles, although the database of supporting physical measurements is less extensive.

Side resistance for piles in sand broadly parallels the discussion above in regard to clays, but differs in a number of details. Installation disturbs the soil but excess pore pressures rapidly dissipate. Therefore, a setup process dominated by dissipation of excess pore pressures does not occur. Nevertheless, time-dependent setup occurs for piles in sands, although the physical mechanisms are poorly understood. Observations of time-dependent increase in side resistance in sands date back to the 1970s; an excellent review of previous studies as well as presentation of recent studies of the effects of pile aging are given by Gavin et al. (2015). Although the mechanisms of aging are poorly understood, field measurements indicate that time-dependent increases in radial effective stresses and interface dilative stresses contribute to increased shaft capacity. Reported setup times are on the order of 1 year. As described in Chapter 7, shaft resistance is predicted based on correlations to *in situ* effective stress or, more recently, correlations to cone penetration test tip resistance.

Bearing resistance involves mobilization of soil shearing resistance outside of the immediate area of the soil–anchor interface, as depicted in Figure 1.18. As the resisting soil mass is largely undisturbed, the analytical estimates of bearing resistance using plasticity theory are often possible, at least for simple loading conditions. Chapter 3 describes plasticity theory and Chapter 4 presents some fundamental solutions that have wide applicability to anchor problems. Bearing resistance in clays is usually expressed as a multiple of the soil undrained shear strength, for example, $q_{ult} = N_c \, s_u$. Not surprisingly given that it involves mobilization of shearing resistance well outside the boundaries of the anchor, bearing resistance is much more effective than frictional. Chapters 3 and 4 present bearing factors for various shapes that vary from approximately $N_c = 7.5$–13, in contrast to the multiplier on soil strength for

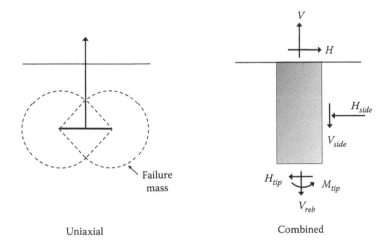

Uniaxial   Combined

Figure 1.18 Bearing resistance and combined loading.

side resistance, $\alpha < 1$. Thus, a plate anchor that functions by mobilizing bearing resistance can have the pullout capacity as a friction pile many times its size.

Despite the importance of plasticity theory, most analytical solutions apply to "uniaxial" load conditions such as pure translation relative to a given axis of the anchor or pure rotation. Many practical situations arise that deviate from simple uniaxial loading. For example, the sketch in the right-hand side of Figure 1.18 shows a caisson subjected to combined horizontal and vertical loading as would occur in a taut mooring system. In this case, the soil resistance on the side of the caisson is a combination of horizontal bearing resistance and axial side friction. The resistance at the tip is an even more complex combination of vertical reverse end bearing, horizontal sliding and moment resistance. With some unusual exceptions—Chapter 4 gives an analytical solution for combined shear-torsion loading on a plate—analytical solutions are not available for these types of load combinations. In these instances, recourse is generally made to numerical finite element or finite difference solutions to map out a "yield locus," defining the load combination at which yield occurs.

1.3.2 Gravity

The contribution of soil self-weight to anchor load capacity is a common source of confusion, because the contribution only occurs under certain circumstances. We can first consider the case of a horizontally embedded plate anchor shown in Figure 1.19. The sketch on the left considers the case in which no gap develops between the soil and the underside of the plate. This condition occurs when gravity forces are sufficiently large relative to soil shearing resistance such that the soil flows in to fill any gap that may tend to form beneath the plate. Quantitative assessment of this condition is discussed in Chapter 3. The absence of gapping is beneficial from the standpoint of mobilized bearing resistance, since shearing resistance is mobilized both above and below the plate. However, it turns out that soil weight makes no contribution to anchor capacity in this case. When described in terms of virtual work, the positive work performed by the upward-moving soil above and below the plate is exactly balanced by the negative work performed by the soil flowing down around the sides, resulting in zero net contribution of soil weight to anchor capacity. The contrasting condition of gap formation is shown on the right-hand sketch in Figure 1.19. In this case, the plate must overcome the weight of the overlying soil in addition to the soil shearing resistance. In general, gapping is detrimental to anchor capacity, since the added contribution of soil self-weight typically does not offset the loss of bearing resistance beneath the plate. The soil self-weight contribution is nevertheless not negligible and should be considered when gapping is presumed to occur.

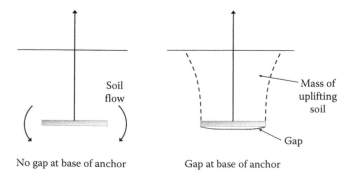

No gap at base of anchor   Gap at base of anchor

Figure 1.19 Soil self-weight contribution to plate anchor capacity.

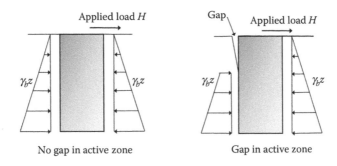

Figure 1.20 Soil self-weight contribution to horizontal load capacity of piles and caissons.

The effect of soil self-weight and gapping on horizontal load capacity of piles and caissons is similar to that described above for plates. As shown in the sketch on the left-hand side of Figure 1.20, in the absence of gapping, the self-weight stress from the soil acts equally on the active and passive sides of a horizontally loaded pile. Thus, the net contribution of soil self-weight to horizontal load capacity is zero. However, in the presence of gapping, the horizontal stresses in the active zone drop to zero as the soil separates from the pile wall. As is the case for the plate anchor, gapping results in a loss of shearing resistance in the region in the vicinity of the gap; however, this loss is partially offset by an added contribution from soil self-weight. Detailed analysis of this effect is given in Chapter 4.

Soil self-weight can also contribute to vertical pullout capacity of suction caissons, although not for the potential failure modes commonly considered in suction caisson design. Figure 1.21 shows the potential modes of vertical failure in caissons. The first assumes that reverse end bearing mobilizes at the base of the caisson. In this case, the virtual work argument stated above for plate anchors applies: the positive virtual work performed by soil inside the caisson is exactly offset by the negative virtual work of the downward moving soil outside the caisson; therefore, soil weight makes no contribution to pullout capacity. The second possible failure mode is a pure pullout failure, which is typically associated with sustained loading where suction cannot mobilize to provide reverse end bearing resistance. By definition, no soil movement occurs for this failure mechanism, so soil weight also does

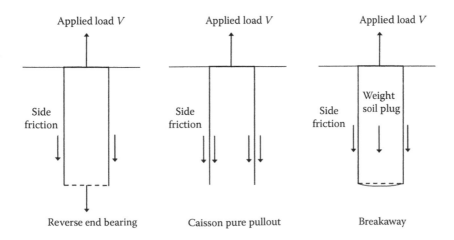

Figure 1.21 Possible vertical failure modes for suction caissons.

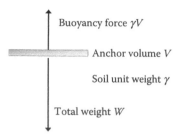

Figure 1.22 Buoyant weight of anchor.

not contribute to pullout capacity. The final possible mode involves loss of reverse end bearing resistance with the side friction on the interior of the caisson being sufficient to cause the caisson and interior soil plug to move as a unit. In this case, the weight of the soil plug clearly contributes to soil resistance.

The weight of the anchor itself will also contribute to vertical load capacity, although the contribution is typically relatively minor. When evaluating the contribution of anchor weight, the buoyant weight of the anchor should be used. In this case, the buoyant weight is the weight of the anchor minus the weight of the material that is displaced by the anchor (Figure 1.22). When the anchor is fully embedded in the soil, the calculation should be based on the total unit weight of the soil. For a steel anchor (specific gravity = 7.8), the buoyancy effect reduces the total anchor weight by approximately 20%–25% for typical soil densities. If part of the anchor protrudes above the mudline—for example, the top cap of a suction caisson—the weight of the protruding portion should be reduced by the displaced volume of water in this portion of the anchor.

1.3.3 Structural resistance

Anchor load capacity is largely synonymous with "geotechnical" capacity; that is, capacity derived from soil shearing resistance and, to a lesser extent, the weight of the soil. The most significant exception to this generalization is laterally loaded flexible piles (Figure 1.23). As will be discussed in Chapter 8, elastic deformation, specifically pile curvature, is so great that the pile fails structurally in bending well before the ultimate shearing resistance from the soil is fully mobilized; thus, the load capacity of the pile itself becomes a primary factor controlling ultimate load capacity. Analytical treatment of this behavior may follow the elastic analysis methods for piles described in Chapter 8. Alternatively, the plastic limit methods discussed in Chapter 3 can also be applied to flexible piles, provided that a plastic hinge is placed in the pile to account for structural yield (Murff and Hamilton, 1993; Chen et al., 2016). Suction caissons typically have aspect ratios less than 6, which is too short for significant beam bending action to occur; thus, they are typically treated as rigid bodies for which geotechnical resistance controls the ultimate load capacity.

1.3.4 Inertia

Rapidly applied loads can mobilize two forms of additional soil resistance: increased shearing resistance owing to the rate dependence of soil undrained shearing resistance, and resistance associated with particle acceleration within the moving soil mass. The former is a soil strength issue, which will be discussed separately. The latter is termed "inertial resistance," which is discussed here. Short duration impact loads for which inertial resistance may be

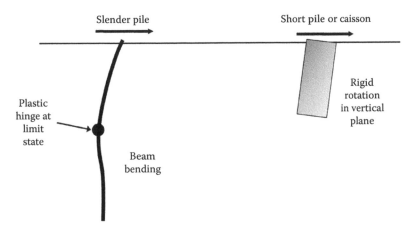

Figure 1.23 Flexible and rigid piles.

significant can arise from sources such as a ship impacting a mooring, various installation and construction operations, and blast loading. Hermann (1981) states that in situations where the load duration (Figure 1.24) is less than 0.02 seconds, the inertial effects may have to be considered. For load durations less than 0.0035 seconds, the Naval Civil Engineering Laboratory (NCEL) guidelines show a tripling of anchor resistance owing to inertial effects. This guidance applies to plate anchors used by naval vessels, it will not necessarily apply to all dimensions and anchor types considered in this book. Nevertheless, it does convey some sense of the time scale of impulse loading for which inertial effects may be significant.

Aside from short-duration impulse loading, significant inertial resistance can arise during continuous penetration processes involving sufficiently high velocities, such as those occurring during dynamic pile installation. True (1974) includes drag resistance (Figure 1.24) in the summation of forces resisting penetration during dynamic pile installation and characterizes the resistance in terms of a drag force similar to that from classical fluid mechanics. Subsequent studies on dynamic installation by Kim et al. (2015) found that inertial forces

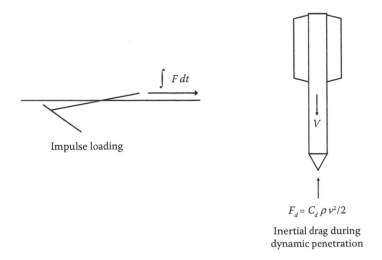

Figure 1.24 Inertial forces on anchors.

are most significant during the early stages of dynamic penetration, after which shearing resistance dominates.

1.4 LOADING CONDITIONS

Characteristics of anchor loading that can prove critical to anchor performance include load duration, cyclic loading, load rate, point of application, and orientation. The following paragraphs discuss how these characteristics affect anchor performance.

1.4.1 Sustained loading

Anchor loads can be of short duration, such as wind and wave loading, or sustained such as that can occur in various taut mooring systems. Load duration is particularly critical for anchors in clay, owing to the tendency for excess (transient) pore water pressures to develop, which in turn influences soil strength. Excess pore pressures in clays under applied loading arise from shearing of the soil as well as changes in mean total stresses in the soil mass. Both topics are beyond the scope of a summary discussion and are addressed in detail in Chapters 2 and 3. The present discussion is limited to some illustrative examples.

Although geotechnical engineers often expect dramatic differences between short-term (undrained) and long-term (drained) load capacity, the most significant differences in bearing resistance are actually limited to cases where asymmetric generation of excess pore pressure occurs. A case in point is the reverse end bearing capacity of a suction caisson (Figure 1.25). When subjected to vertical uplift, the soil mass beneath the caisson tip experiences extension, which generates negative excess pore pressures ("suction") that contribute to pullout resistance. Over time, the process of consolidation leads to water flow into this region, with a concomitant loss of suction and reduction in vertical load capacity of the anchor. Thus, the vertical load capacity of a suction caisson can be significantly less under sustained versus rapid loading. Pile anchors are also vulnerable to the same effect but, since reverse end bearing is typically a small fraction of total capacity, the reduction is of minor significance. This situation can be contrasted to two commonly occurring cases where anti-symmetric distributions of excess pore pressure develop, as shown in Figure 1.25. The first

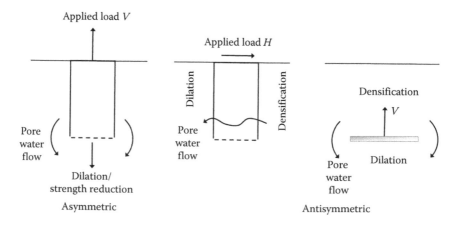

Figure 1.25 Effect of sustained loading on anchors.

case is a laterally loaded caisson. Upon loading, positive excess pore pressures develop ahead of the caisson, with negative pore pressures developing on the back side. During consolidation, water flows from front to back, with soil densification ahead of the caisson being offset by dilation on the back side. Secondary effects may occur, such as gapping behind the caisson. However, reductions comparable to those associated with a loss of reverse end bearing are generally not expected to occur. A second example of antisymmetric excess pore pressure generation is the case of a deeply embedded plate anchor. Negative and positive excess pore pressures develop, respectively, below and above the plate. As excess pore pressures dissipate, dilation beneath the plate is offset by densification above the plate. Recent numerical and model test studies on vertically loaded plate anchors (Han et al., 2016) indicate no reduction in capacity when sustained loads are less than 88% of ultimate monotonic undrained capacity. When sustained loads exceed 60% of ultimate monotonic capacity, breakaway does occur beneath the plate, but the loss of suction is offset by densification above the plate. Although antisymmetric excess pore pressure distributions are rather benign in regard to load capacity under sustained loading, other aspects of anchor performance may still have to be considered. For example, displacements associated with the consolidation process will still occur and will have to be considered if limitations on displacements are included in the design basis.

When side resistance is considered, the possible offsetting effects noted above for bearing resistance cannot be counted upon, and the load capacity reductions for sustained loading are potentially significant. In a discussion of sustained loading effects on axial load effects, Doyle (2011) cites the following: (1) laboratory creep tests indicate that sustained loads exceeding 80% of the soil undrained shear strength can induce failure, (2) long-term model tests by Edil and Mochtar (1988) show continuing postconsolidation displacements when sustained loading exceeds 30% of the undrained pile capacity, and (3) in a discussion of remolded clays, Terzaghi and Peck (1964) report that when a sustained shear stress exceeds 50% of the peak strength creep is likely to occur. In the absence of long-term creep data, Doyle recommends a conservative approach of not permitting sustained loads to exceed 30% of the monotonic undrained load capacity of a friction pile.

1.4.2 Cyclic loading

Analysis methods for anchors almost universally evaluate load capacity in terms of a quasistatic monotonically applied load. As shown in Figure 1.26, the actual loading on anchors is dominated by wave and wind loading, which is cyclic and episodic in nature. A full solution of the boundary value problem, accounting for the effects of cyclic loading on soil

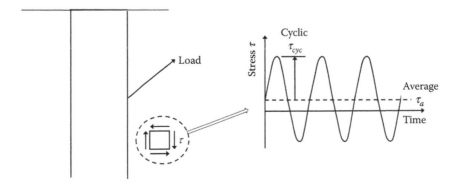

Figure 1.26 Cyclic loading.

stress–strain-strength properties, at all points within the soil mass surrounding the anchor, is seldom performed in practical design situations. Rather, a simplified approach is typically adopted in which the soil strength profile is modified to account for cyclic loading effects. Details of selecting cyclic strength for use in design are discussed in Chapter 2, but salient considerations include the following:

- Cyclic loading can reduce soil shear strength owing to breakdown of the soil structure and create a tendency for volumetric compression, which, under undrained conditions, can lead to accumulation of pore pressure (Andersen, 2015). The degree of degradation is highly sensitive to reversals in cyclic loading or, in other words, the magnitude of the average cyclic stress τ_a. For example, the most severe degradation occurs when complete reversal of the cyclic stress occurs, or when $\tau_a = 0$.
- The rate of cyclic loading also influences soil behavior and rates can vary widely for different sources of loading—for example earthquake, wave, and tidal. Wave loading is often a dominant concern for anchors; typical periods are 10–20 seconds, which are 2–3 orders of magnitude greater than load application rates in standard laboratory tests for measuring shear strength. As discussed in Chapter 2, the undrained strength of clays increases with strain rate, which tends to offset the strength loss owing to cyclic loading. Clukey et al. (2013) note that the cyclic soil strength can exceed the static strength when cyclic stresses are small relative to average stresses, and vice versa when stress reversals occur.
- Cyclic loading often involves random amplitudes. As it is not practical in design situations to attempt to simulate a large number of cycles of random loading in laboratory tests or in numerical simulations, simplified procedures have been developed where a random load series is transformed into equivalent packets of uniform amplitude cyclic loads (Andersen, 2015).
- Laboratory tests provide information on soil stress–strain-strength behavior at a single point in the soil mass, but load capacity of an anchor is governed by the response of the entire soil mass surrounding the anchor, within which stress histories and strain rates can be highly variable. Therefore, it is important that the assessment of cyclic strength reflects load histories in that portion of the soil mass judged to be governing load capacity.

1.4.3 Eccentric loading

Eccentric loading is defined here as any system of loads that causes the anchor to rotate. Eccentric loading is not simply a matter of the line of action of the resultant applied load not passing through the centroid of the anchor, because soil resisting forces also affect rotational response. It is more convenient to consider eccentric loading on anchors in terms of the center of rotational resistance, which is the point about which pure moment (zero net force) is generated when rotation occurs. The center of rotational resistance depends on both the anchor geometry and the distribution of soil resistance, with the center of rotational resistance of an anchor in a heterogeneous soil profile differing substantially from that in a uniform soil. The center of rotational resistance also has significance where the anchor experiences pure translation (generally desirable) when the line of action of an applied force passes through it. Chapter 7 presents detailed procedures for computing the load capacity reduction owing to eccentric loading, but an intuitive grasp of the source of the reduction is readily apparent from the sketch on the left side of Figure 1.27. Instead of resisting the applied load H, the soil below the center of rotation exerts a force that actually

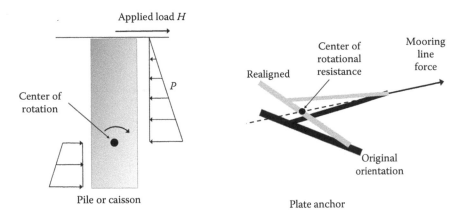

Figure 1.27 Eccentric loading on anchors.

acts in the same direction as *H*; thus, eccentric loading can produce sharp reductions in pile load capacity. When physically feasible, eccentric loading can be minimized by placing the applied load (the pad-eye depth) near the center of rotational resistance of the caisson, which typically occurs somewhere near 60% of the pile length. As the center of rotational resistance in practical problems depends on an imperfectly defined soil strength profile, the optimal load attachment depth for minimizing rotation and maximizing load capacity cannot be predicted with perfect accuracy. Therefore, estimates of lateral load capacity should be based on realistic assessments of the uncertainty in optimal load attachment depth. Inclined loading adds an additional complication to positioning the pad eye to minimize rotation, because the load inclination angle changes the depth at which the line of action of the anchor line force intersects the centerline of the pile and, therefore, the eccentricity of loading.

Eccentric loading of piles and caissons manifests itself primarily through a reduction in lateral load capacity. This is not necessarily the case for plate anchors. The moment capacity of plate anchors is often so low that, when subjected to an eccentric load, the plate simply realigns itself such that the line of action of the applied load passes through the center of rotational resistance of the plate, as depicted in the sketch on the right side of Figure 1.27. The tendency of plate anchors to realign themselves in this manner factors into many of the plate anchor trajectory prediction calculations in Chapters 9 and 10.

1.4.4 Out-of-plane loading

Anchors are normally designed for resisting loads acting within a plane of symmetry of the anchor; Figure 1.28 illustrates some examples. Forces acting outside the plane of intended loading can arise from a number of circumstances. Misalignment of the anchor can occur during installation; for example, the design of suction caissons typically considers a possible twist of 5–10° during installation. Additionally, the floating unit itself can move off station, generating an additional source of out-of-plane loading. More serious out-of-plane loading can occur during partial failure of a mooring system, with out-of-plane load angles approaching 90° (Ward et al., 2008) being estimated from back-analysis of some damaged systems during hurricanes.

In the case of piles and caissons, the out-of-plane load generates a torsion load on the anchor. For small out-of-plane angles, this imposes a load demand that should be resisted

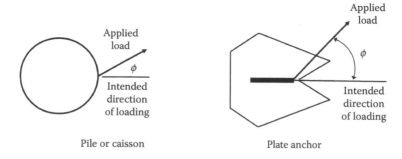

Figure 1.28 Out-of-plane loading on anchors.

by side resistance on the walls of the caisson. This demand reduces the capacity available to resist vertical and in-plane horizontal loads, thereby reducing the overall capacity of the anchor. The reduction in capacity is typically modest—on the order of a few percent—but it is typically considered in the capacity estimate, as described in Chapter 7. For the more extreme out-of-plane load angles that can occur with partial mooring system failure, the caisson can fail in torsion. However, the failure does not necessarily result in complete loss of the anchor. Following a torsion failure, the anchor will spin into the direction of the load. The skin resistance will be reduced as the soil is sheared to its remolded state, but back-analysis of partial mooring system failures indicates continued but impaired functioning of the pile (Ward et al., 2008).

Out-of-plane loading of plate anchors results in complex 6-degrees of freedom (DOF) loading conditions acting on the anchor (Aubeny et al., 2010), in contrast to the conventional in-plane case involving 3-DOF loading, an in-plane moment and forces normal and parallel to the plate. Chapter 3 presents studies on idealized rectangular plate anchors subjected to 6-DOF loading conditions. Studies are limited for plate anchors having complex geometry typical of DEAs, but some model tests have been performed for DEAs in soft clays (Aubeny et al., 2010), which are presented in Chapter 9.

1.5 ANCHOR PERFORMANCE

General factors to be considered in selection of the most appropriate anchor for a given situation include the ability to install the anchor in the soil profile under consideration, load resistance characteristics of the anchor, anchor efficiency and cost, and the ability of the anchor to maintain its embedment under sustained or cyclic loading. Recognizing that local project specific factors—equipment availability, vessel availability, environmental constraints, etc.—can be important if not decisive considerations, the following discussion provides an overview of general considerations guiding anchor selection.

1.5.1 Installation and load capacity characteristics

Table 1.1 summarizes the capabilities of the various anchor types in regard to the soil conditions for which they may be reliably installed, the type of loading (horizontal, inclined, and pure vertical) under consideration. Heterogeneous soil profiles can also limit the feasibility of deploying certain anchor types, because they can preclude or inhibit installation as well as adversely impact the load resisting capabilities of the anchor.

Table 1.1 Installation and load resistance capabilities for various anchor types

Anchor type	Soil conditions where installation is possible	Allowable/optimal load orientations	Problems in stratified soil profiles	Comments
Suction caissons	Soft clay, stiff clay, sand	Horizontal, inclined	Yes, but installation may be possible	Vertical capacity in clays reduced under sustained loading
Driven piles	Soft clay, stiff clay, sand, soft rock	Vertical, inclined, horizontal	No	Low efficiency in resisting horizontal loading
Drag embedded anchor (DEA)	Soft clay, stiff clay, sand	Horizontal	Yes, but installation may be possible	Can mobilize vertical capacity in soft clays
Vertically loaded anchor (VLA)	Soft clay	Inclined	Yes, but installation may be possible	Brittle failure in pure vertical loading
Dynamically installed pile (torpedo)	Soft clay, stiff clay, sand	Horizontal, inclined, vertical	Uncertain	Case history data in stratified profiles not identified
Suction embedded plate anchor (SEPLA)	Soft clay	Horizontal, inclined, vertical	Yes	Brittle failure in pure vertical loading
Dynamically installed plate anchor (DEPLA)	Soft clay, sand	Horizontal, inclined, vertical	Uncertain	Some models have potential for penetration in sand
Pile driven plate anchor (PDPA)	Soft clay, stiff clay, sand	Horizontal, inclined, vertical	No	

Suction caissons can be installed in a wide range of homogenous soil profiles (soft clay, stiff clay, and sand) with relative ease, but interbedded soils make installation difficult. Recent innovations in the installation technology make installation in stratified soils possible, but it is still difficult. Even if successfully installed, there may be issues with load capacity if the stratification is such that it compromises the mobilization of reverse end bearing resistance. A prime advantage of suction caissons is that their short aspect ratio leads to much more efficient mobilization of lateral resistance than driven piles. Accordingly, caissons perform best when horizontal loading is dominant. Suction caissons in clays can resist pure vertical loads, but since they can lose their reverse end bearing resistance under sustained vertical loading, which can comprise 25% or more of the vertical pullout capacity, they often lose their attractiveness in situation involving nearly vertical uplift, as occurs for TLP loading.

Driven piles can be installed in virtually any soil profile, and they can be designed to withstand virtually any load combinations. However, they are relatively inefficient in resisting lateral loads, so their competiveness is likely to be marginal, when another alternative (especially suction caissons) performs the same function. Piles often become attractive for heterogeneous soil profiles, for which installation of many of the other anchor alternatives becomes either difficult or impossible. Dynamically installed piles have been successfully installed in soft clays, stiff clays, and sands (Ehlers et al., 2004). Given this versatility, they may perform well in heterogeneous soil profiles but, given the lack of case histories, their performance is designated as uncertain in Table 1.1 for stratified profiles.

DEAs in sands and stiff clays cannot penetrate to depths great enough to generate significant vertical load capacity, which effectively rules them out for any mooring system requiring vertical anchor capacity, such as a taut mooring system. DEAs in soft clays embed deeply into the soil profile generating substantial vertical load capacity, which can be further enhanced by opening the anchor shank when VLAs are used. Although deeply embedded VLAs can resist pure vertical loading, their ideal niche is usually considered as anchorage for taut mooring systems. Similar to suction caissons, DEAs/VLAs can be installed in a broad range of homogeneous soil profiles, but deep embedment becomes questionable in stratified profiles. Even where heterogeneous soil conditions exist, successful deployment may be possible. For example, if a soft clay stratum overlies a sand or stiff clay stratum, the anchor will not penetrate the stiffer layer deeply, but can penetrate to a sufficient extent to mobilize substantial load capacity. DEAs can also penetrate localized stiff layers if the layer thickness is sufficiently small.

SEPLA deployments appear to have been limited to date to soft clay profiles. Similarly, development of DEPLAs to date has largely been directed toward soft clay profiles. Conceptually, a DEPLA could penetrate sand or stiff clay, and the wing anchor concept (dynamic plate penetration that does not employ a follower) is under evaluation for deployments in both clays and sands. Both SEPLAs and DEPLAs are suitable for resisting horizontal, inclined, or purely vertical loads. Heterogeneity does not appear to be a major impediment to pile driven plate anchor installation. They are similar to conventional piles in that they can resist vertical, horizontal, and inclined loads.

Compared to compression foundations, precise horizontal positioning of anchors is often not critical. Nevertheless, precise positioning could be a factor in anchor selection, an example being the positioning of pile anchors for TLPs, which typically have tight installation tolerances. Vertical penetration of the anchor is often of greater concern, as greater embedment depth normally leads to increased load capacity. Table 1.2 lists general uncertainties in

Table 1.2 Sources of uncertainty in position of installed anchors

Anchor type	Horizontal placement	Vertical embedment
Suction caissons	High precision	Slight uncertainty owing to internal plug heave
Driven piles	High precision	Minimal uncertainty
Drag embedded anchor (DEA)	Drag distance extends 10 s of meters	Shallow penetration usually presumed in sand and stiff clay, high uncertainty in soft clay
Vertically loaded anchor (VLA)	Drag distance extends 10 s of meters	High uncertainty in trajectory penetration
Dynamically installed pile	Uncertainty from free fall through water/soil and keying	Uncertainty in penetration depth and embedment loss during keying
SEPLA	Some uncertainty in anchor translation during keying	Some uncertainty in embedment loss during to keying
DEPLA	Uncertainty from free fall through water/soil and keying	Uncertainty in penetration depth and embedment loss during keying
Helical anchor	High precision	Minimal uncertainty
Hammer-driven, vibrated in-place and jetted plates	Some uncertainty in anchor translation during keying	Some uncertainty in embedment loss during keying

positioning associated with various anchor types, excluding consideration obvious sources of positioning problems associated with failed installations or refusal. Suction caissons, piles, and helical and other directly embedded plate anchors can be positioned to a high degree of precision. A small degree of uncertainty exists for vertical penetration of suction caissons owing to plug heave, possibly preventing full penetration, but the embedment loss is typically small and, as discussed in Chapter 6, rather predictable. Drag embedment anchors for all soil conditions require drag distances of tens of meters to mobilize the requisite load capacity. Therefore, precise horizontal positioning is not possible. In stiff clays and sands, shallow embedment is typically taken as a given, so the issue is not so much uncertainty in penetration depth as uncertainty in load capacity. In spite of recent improvements in the understanding of drag anchor behavior in soft clays, accurate prediction of embedment depth is difficult and recourse to proof load tests is required for confirming that adequate penetration has been achieved. SEPLAs and hammer-driven plate anchors can be installed at a vertical orientation to a high degree of horizontal and vertical precision, but the keying process will lead to both a modest loss of embedment as well as a shift in the horizontal position of the plate. The order of magnitude of these movements is typically less than a fluke length. Therefore, these uncertainties are considerably less than the uncertainties noted above for DEAs. Dynamically installed piles and DEPLAs both involve free falling through the water column, impact penetration into the soil, and keying—all of which introduce uncertainty into the horizontal position and vertical penetration depth. To provide a sense of the degree of magnitude the latter, Zimmerman et al. (2009) report that in a series of 54 Omni-Max anchor installations in soft Gulf of Mexico clay, measured tip penetration depths varied from 10.7 to 20.1 m. It should be noted that, from the standpoint of its effect on anchor capacity, uncertainty in dynamically installed anchor penetration depth could be self-compensating, because a higher soil strength will tend to reduce penetration depth but with a concomitant increase in anchor pullout resistance.

1.5.2 Efficiency

A simple measure of anchor efficiency may be taken as the ratio of its ultimate load capacity to its weight; in fact, many charts of anchor capacity are expressed in this form. As an example of the comparative efficiency of piles/caissons to plate anchors, Figure 1.29 compares efficiencies of a typical suction caisson to a plate anchor in a soft clay profile for pure vertical loading. The caisson has a load capacity of a little more than eight times of its weight, in contrast to the plate anchor having an efficiency of approximately 40. The difference arises from two advantages enjoyed by plate anchors. First, the plate anchor derives its load capacity almost entirely from bearing resistance, in contrast to the caisson that derives about half its capacity from reverse end bearing. As discussed earlier, bearing resistance is an order of magnitude more effective in mobilizing soil shearing resistance than side friction. Secondly, the mass of the plate anchor is concentrated at a depth, where the soil is strongest and the greatest contribution to load capacity is generated. By contrast, much of the mass of piles and caissons is distributed through weak soils that make only a minor contribution to load capacity—much of the mass of a pile or caisson simply provides a load path to transmit resistance mobilized near the bottom of the pile to the mooring line. It is noted that caissons will be somewhat more efficient when horizontal loads are considered—the horizontal load capacity of the caisson in this example is approximately twice the vertical load capacity. Nevertheless, the efficiency of a pile or caisson will seldom approach that of a plate anchor. As discussed earlier, flexible piles are even less efficient than short piles or caissons, since structural yield occurs well before the full geotechnical resistance is mobilized.

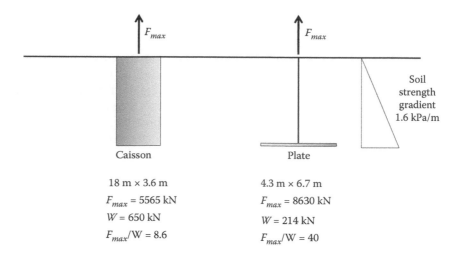

Figure 1.29 Geotechnical efficiency.

Material quantities obviously affect overall costs, but perhaps a more significant driver of costs is the transport volume of an anchor. As depicted in Figure 1.30, plate anchors also have an edge over caissons, as they can be stored on the deck of an AHV in greater quantities than caissons. Some drag anchor designs have detachable shanks, which can permit further economy in transport. Dynamically installed anchors are an example of

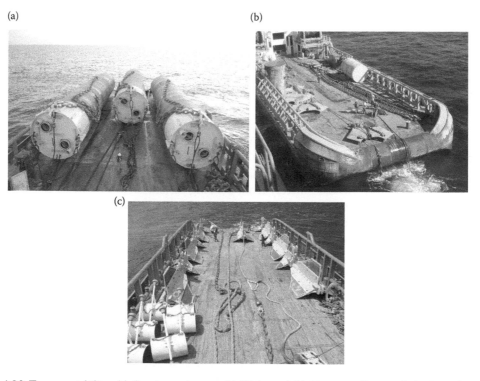

Figure 1.30 Transportability. (a) Suction caissons, (b) VLAs and (c) Dynamically installed piles. (Courtesy of Delmar.)

when transportability can be a more significant concern than simple weight efficiency. They necessarily have a high mass to ensure penetration, but they are more compact than caissons. Therefore, reduced costs in the form of smaller AHVs and fewer trips are possible. Useful studies on the implications of weight and volume differences among various anchor types include comparisons of suction caissons, SEPLAs, torpedo piles, and DEPLAs by O'Loughlin et al. (2015) and comparisons of suction caissons, torpedo piles, and VLAs by Zimmermann et al. (2009).

1.5.3 Maintaining embedment

Beyond simple consideration of load capacity is the kinematic response of the anchor if the load capacity is exceeded. Of particular concern is the potential for a reduction in the anchor embedment depth, which typically leads to a reduction in anchor capacity and an undesirable brittle failure mode for anchor pullout. Analysis of anchor failures (Ward et al., 2008) show suction caissons to fail structurally when overloaded—typically at the pad eye before the anchor pulls out—so a brittle pullout mechanism does not appear to be a significant issue for caissons—at least for long caissons in soft clay. However, plate anchors have been observed to experience pullout failures. Aside from not performing their intended function of providing anchorage, a failed anchor can skip along the seabed, generating a hazard to pipelines, umbilicals, and transmission lines in the area. Therefore, the postyield kinematic behavior of a plate anchor requires consideration in their design.

As shown in Figure 1.31, an overloaded anchor can either experience a loss of embedment or continue to dive downward, according to the system of forces acting on the anchor, specifically the angle of the mooring line at the mudline, the fluke angle, and the fluke–shank angle. As discussed earlier in the discussion of VLAs, orienting the anchor shank normal to the fluke maximizes anchor load capacity, but it also promotes a brittle failure mode. The "near-normal" design concept for VLAs typically limits the fluke–shank angle to less than 75–80° to reduce the tendency for upward motion of the anchor. This practice is consistent with plasticity-based analysis by Aubeny and Chi (2014) indicating a brittle failure mode becomes more likely for fluke–shank angles greater than 80°. Furthermore, influencing the kinematic behavior of the anchor is the mooring line angle at the mudline and the anchor fluke angle. For mooring line angles greater than 45°, a brittle failure mode is likely to occur irrespective of the anchor design. As taut mooring lines are frequently less than this angle, it is therefore not unrealistic to expect that a properly designed VLA can dive when overloaded. As shown in Figure 1.31, the fluke angle also influences whether an anchor will rise or dive. Theoretical analyses and observed behavior of drag embedment

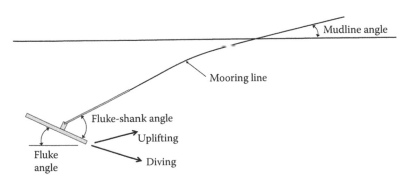

Figure 1.31 Anchor kinematics.

(discussed in Chapter 9) show that the anchor fluke angle decreases with increasing drag distance and anchor embedment depth. Thus, increased embedment gives the benefit of increased load capacity, but also increases the potential for a brittle failure mechanism. Similar to any plate anchor, helical anchors can experience brittle failure under vertical loading, but retorqueing to restore lost embedment is conceptually possible.

1.6 INDUSTRY GUIDELINES

A number of organizations—American Bureau of Shipping (ABS), American Petroleum Institute (API), Det Norski Veritas (DNV), Naval Facilities Engineering Command (NAVFAC)—have developed guidance documents for designing various types of anchors (Table 1.3). The ABS manual is a general manual for offshore floating wind installations, including specific guidance on anchor design. The API RP 2A LRFD (load and resistance factor design) and API RP 2A WSD (working stress design) guidelines are for fixed offshore platforms, but the pile design methodology applies equally to anchor applications. API RP 2SK is devoted to floating structures, with specific guidance on piles, suction caissons, VLAs, and SEPLAs. DNV publishes a large number of standards. Those directly relevant to anchors are listed in Table 1.3, but others may be of interest to anchor designers. The standards are well referenced, so additional available standards are easy to identify, particularly in the general guidance documents such as DNV-OS-E301 and DNV-OS-J101. NAVFAC SP-2209-OCN is a general manual for marine geotechnical engineering. Aside from guidance on background soil mechanics and geotechnical site characterization, it provides coverage for helical, vibrated-in, and driven plate anchors that historically have not received much usage in offshore energy industry applications.

Table 1.3 Industry guidelines for anchor design

Organization	Publication	Description/Comment
American Bureau of Shipping (ABS)	Guide for Building and Classing Floating Offshore Wind Turbine Installations	General guidance for offshore floating wind turbines, coverage of anchors in Chapter 8.5
American Petroleum Institute (API)	API RP 2A LRFD (1993) API RP 2A WSD (2000)	For fixed structures, pile design guidelines relevant to anchor piles
	API RP 2SK (2005)	Stationkeeping for floating structures, including guidance on piles, suction caissons, SEPLAs, and drag anchors
Det Norski Veritas (DNV)	DNV-OS-E301 (DNV, 2008)	General guidance on mooring systems
	DNV-OS-J101 (DNV, 2007)	General guidance for offshore wind turbines
	DNV-RP-E301 (DNV, 2012)	Design and installation of fluke anchors
	DNV-RP-E302 (DNV, 2002)	Design and installation of plate anchors in clay
	DNV-RP-E303 (DNV, 2005)	Design and installation of suction anchors in clay
Naval Facilities Engineering Command (NAVFAC)	SP-2209-OCN	General manual on marine engineering, guidance on anchors in Chapters 6 and 7

REFERENCES

ABS, 2013, *Guide for Building and Classing Floating Offshore Wind Turbine Installations*, American Bureau of Shipping, Houston, 119p.

Andersen KH, 2015, Cyclic soil parameters for offshore foundation design, *Proceedings of the International Symposium on Frontiers in Offshore Geotechnics ISFOG15*, Oslo, Taylor & Francis Group, London, pp 5–82.

Andersen KH, Murff JD, Randolph MF, Clukey E, Jostad HP, Hansen B, Aubeny C, Sharma P, Erbich C, and Supachawarote C, 2005, Suction anchors for deepwater applications, *Keynote lecture, International Symposium on Frontiers in Offshore Geotechnics*, Perth, Australia, September 2005, pp 3–30.

API RP 2A LRFD, 1993, *Recommended Practice for Designing and Constructing Fixed Offshore Platforms*, American Petroleum Institute, API Publishing Services, Washington, DC, 224p.

API RP 2A WSD, 2000, *Recommended Practice for Designing and Constructing Fixed Offshore Platforms*, American Petroleum Institute, API Publishing Services, Washington, DC, 226p.

API RP 2SK, 2005, *Recommended Practice for Design and Analysis of Stationkeeping Systems for Floating Structures*, American Petroleum Institute, API Publishing Services, Washington, DC, 184p.

Aubeny CP and Chi C-M, 2014, Analytical model for vertically loaded anchor performance, *ASCE J Geotech Geoenviron Eng*, 140(1), 14–24.

Aubeny C, Gilbert R, Randall R, Zimmerman E, McCarthy K, Chen C, Drake A, Yeh P, Chi C-M, and Beemer R, 2010, *The Performance of Drag Embedment Anchors (DEA)*, MMS, Offshore Technology Research Center, College Station, TX.

Aubeny CP and Chi C-M, 2010, Mechanics of drag embedment anchors in a soft seabed, *ASCE J Geotechn Geoenviron Eng*, 136(1), 57–68.

Aurora RP, 1984, Experience with driving 84 inch piles with underwater and above water hammers at the South Brae Platform, North Sea, *Proceedings of the 16th Annual Offshore Technology Conference*, Houston, Paper 4803, pp 237–242.

Chen J, Gilbert R, Choo YS, Marshall P, and Murff J, 2016, Two dimensional lower bound analysis of offshore pile foundation systems, *Inter J Numer Anal Meth Geomech*, 40(2), 1321–1338.

Clukey EC, Gilbert RB, Andersen KH, and Dahlberg R, 2013, Reliability of suction caissons for deep water floating facilities, *Foundation Engineering in the Face of Uncertainty: Honoring Fred H Kulhawy, GSP 229*, Withiam JL, Phoon K-K, and Hussein M (eds.), ASCE, 2013 GeoCongress, San Diego, pp 456–474.

Cox BE, and Christie WW, 1976, Underwater pile driving test offshore Louisiana, *Proceedings of the 8th Annual Offshore Technology Conference*, Houston, Paper 2478, pp 611–616.

Dahlberg R, 1998, Design procedures for deepwater anchors in clay, *Proceedings of the 30th Offshore Technology Conference*, Houston, OTC 8837, pp 559–567.

Das BM, 1978 Model tests for uplift capacity of foundations in clay, *Soils and Foundations*, 18(2), 17–24.

Das BM, 1980, A procedure for estimation of ultimate uplift capacity of foundations in clay, *Soils Found*, 20(1), 77–82.

de Aguiar CS, de Sousa JRM, Ellwanger GB, Porto EC, de Medeiros Junior CJ, and Foppa D, 2009, Undrained load capacity of torpedo anchor in cohesive soils, *Proceedings 28th International Conference on Offshore Mechanics and Arctic Engineering*, ASME, New York, OMAE2009-79465, pp 1–13 (electronic format).

DNV, 2002, *Offshore Standard DNV-RP-E302 Geotechnical Design and Installation of Plate Anchors in Clay*, Det Norsk Veritas, Hovik, Norway, 39p.

DNV, 2005, *Offshore Standard DNV-RP-E303 Geotechnical Design and Installation of Suction Anchors in Clay*, Det Norsk Veritas, Hovik, Norway, 24p.

DNV, 2007, *Offshore Standard DNV-OS-J101 Design of Offshore Wind Turbine Structures*, Det Norsk Veritas, Hovik, Norway, 142p.

DNV, 2008, *Offshore Standard DNV-OS-E301 Position Mooring*, Det Norsk Veritas, Hovik, Norway, 71p.

DNV, 2012, *Offshore Standard DNV-RP-E301 Geotechnical Design and Installation of Fluke Anchors in Clay*, Det Norsk Veritas, Hovik, Norway, 46p.

Dove P, Treu H, and Wilde B, 1998, Suction embedded plate anchor (SEPLA): A new anchoring solution for ultra-deep water mooring, *Proceedings of the DOT Conference*, New Orleans.

Doyle E, 2011, Driven pile design for tension leg platforms, *Deepwater Foundations and Pipeline Geomechanics*, McCarron WO, (ed.), J. Ross Publishing, Fort Lauderdale, FL.

Edil TB and Mochtar IB, 1988, Creep response of model piles in clay, *ASCE J Geotech Eng*, 114(11), 1245–1260.

Elkasabgy MA and El Naggar MH, 2015, *Lateral Performance of Large-Capacity Helical Piles, IFCEE 2015*, American Society of Civil Engineers, pp 868–877.

Ehlers CJ, Young AG, and Chen J-H, 2004, Technology assessment of deepwater anchors, *Offshore Technology Conference*, Houston, Texas, OTC 16840, pp 1–17 (electronic format), doi:10.4043/16840-MS.

Flory JF, Banfield SJ, Ridge IML, Yeats B, Mackay T, Wang P, Hunter T, Johanning L, Herduin M, and Foxton P, 2016, Mooring systems for marine energy converters, *MTS/EEE Oceans 16*, Monterey, CA, pp 1–13 (electronic format).

Fontana CM, Arwade SR, DeGroot DJ, Myers AT, Landon M, and Aubeny C, 2016, A multi-line concept for floating offshore wind turbines, *Proceedings of the 35th International Conference Ocean, Offshore and Arctic Engineering, OMAE2016*, ASME, Busan.

Forrest J, Taylor R, and Brown L, 1995, *Design Guide for Pile-driven Plate Anchors*, Technical Report TR-2039-OCN, Naval Facilities Engineering Service Center, Port Hueneme, CA, March 1995.

Gaudin C, O'Loughlin CD, Randolph MF, and Lowmass AC, 2006, Influence of the installation process on the performance of suction embedded plate anchors, *Geotechnique*, 56(6) 381–391.

Gavin K, Jardine R, Karlsrud K, and Lehane B, 2015, The effects of pile ageing on the shaft capacity of offshore piles in sand, *Proceedings of the International Symposium on Frontiers in Offshore Geotechnics ISFOG15, Oslo*, Taylor & Francis Group, London, pp 129–152.

Giampa JR, Bradshaw AS, and Schneider JA, 2017, Influence of dilation Angle on drained shallow circular anchor uplift capacity, *Int J Geomech*, 17(2), ISSN 1532-3641.

Han C, Wang D, Gaudin C, O'Loughlin C, and Cassidy MJ, 2016, Behavior of vertically loaded anchors under sustained uplift, *Geotechnique* 66(8), 681–693.

Hasanloo D, Pang H, and Yu G, 2012, On the estimation of the falling velocity and drag coefficient of torpedo anchor during acceleration, *Ocean Eng*, 42, 135–146.

Heerema EP, 1980, An evaluation of hydraulic vs. steam pile driving hammers, *Proceedings of the 12th Annual Offshore Technology Conference*, Houston, Paper 3829, pp 321–330.

Hermann HG, 1981, *Design Procedures for Embedment Anchors Subjected to Dynamic Loading Conditions*, Technical Report R-888, Naval Facilities Engineering Command & Naval Civil Engineering Laboratory, 70p.

Jansz JW, 1977, North Sea pile driving experience with a hydraulic hammer, *Proceedings of the 9th Annual Offshore Technology Conference*, Houston, Paper 2840, pp 267–274.

Jeanjean P, 2006, Setup characteristics of suction anchors for soft Gulf of Mexico clays: Experience from field installation and retrieval, *Offshore Technology Conference*, Houston, OTC 18005, pp 1–6 (electronic format).

Kim Y, Hossain MS, Wang D, and Randolph MF, 2015, Numerical investigation of dynamic installation of torpedo anchors in clay, *Ocean Eng*, 108, 820–832.

Lee KM, Argo TF, Wilson PS, and Mercier RS, 2011, Sound propagation in water containing large tethered spherical encapsulated gas bubbles with resonance frequencies in the 50 Hz to 100 Hz range, *J Acoust Soc Am*, 130(5), 3325–3332.

Lieng JT, Tjelta TI, and Skaugset K, 2010, Installation of two prototype deep penetrating anchors at the Gjoa Field in the North Sea, *Offshore Technology Conference*, Houston, OTC 20758, pp 1–9 (electronic format), doi:10.4043/20758-MS.

Love DP and Arndt WR, 1948, A sheet of air bubbles as an acoustic screen for underwater noise, *J Acoust Soc Am*, 20, 143–145.

Martin CM and Randolph MF, 2001, Applications of the lower and upper bound theorems of plasticity to collapse of circular foundations. *Proceedings of the 10th International Conference on the International Association of Computer Methods and Advances in Geomechnics*, Tucson, Vol. 2, pp 1417–1428.

Merifield RS, 2011, Ultimate uplift capacity of multiplate helical type anchors in clay, *J Geotech Geoenviron Eng*, 137(7), 704–716.

Merifield RS, Lyamin AV, Sloan SW, and Yu HS, 2003, Three dimensional lower bound solutions for stability of plate anchors in clay, *J Geotech Geoenviron Eng*, 129(3), 243–253.

Meyerhof GG, 1973, Uplift resistance of inclined anchors and piles, *Proceedings of the 8th International Conference on Soil Mechanics and Foundation Engineering*, Vol. 2:1, A. A. Balkema Publishers, Rotterdam, Netherlands, pp 167–172.

Meyerhof GG, and Adams JI, 1968, Ultimate uplift capacity of foundations, *Can Geotech J*, 5(4), 225–244.

Murff JD and Hamilton JM, 1993, P-Ultimate for undrained analysis of laterally loaded piles, *ASCE J Geotech Eng*, 119(1), 91–107.

Murff JD, Randolph MF, Elkhatib S, Kolk HJ, Ruionen RM, Strom PJ, and Thorne CP, 2005, Vertically loaded plate anchors for deepwater applications, *Proceedings of the International Symposium on Frontiers in Offshore Geotechnics, IS-FOG05*, Perth, pp 31–48.

Naval Civil Engineering Laboratory (NCEL), 1987, *Drag Embedment Anchors for Navy Moorings, Techdata Sheet 83-08R*, NCEL, Port Hueneme, CA.

NAVFAC, 2011, *SP-2209-OCN Handbook for Marine Geotechnical Engineering, Naval Facilities Engineering Command*, Engineering Service Center, Port Hueneme, USA.

Neubecker SR and Randolph MF, 1995a, The performance of embedded anchor chains systems and consequences for anchor design, *Proceedings of the 28th Offshore Technology Conference*, Houston, OTC 7712, pp 191–200.

Neubecker SR and Randolph MF, 1995b, Profile and frictional capacity of embedded anchor chain, *J Geotech Eng Div, ASCE*, 121(11), 787–803.

Newgard JT, Schneider JA, and Thompson DJ, 2015, Cyclic response of shallow helical anchors in a medium dense sand, *Frontiers in Offshore Geotechnics III*, Meyer (ed.), Taylor & Francis Group, London, 913–918.

O'Loughlin CD, Blake A, Richardson MD, Randolph MF, and Gaudin C, 2014, Installation and capacity of dynamically embedded plate anchors as assessed through centrifuge tests, *Ocean Eng*, 88, 204–213.

O'Loughlin CD, Blake AP, and Gaudin C, 2016, Towards a simple design procedure for dynamically embedded plate anchors, *Geotechnique*, 66(9), 741–753.

O'Loughlin CD, White DA, and Stanier SA, 2015, Novel anchoring solutions for FLNG – Opportunities driven by scale, *Proceedings of the Offshore Technology Conference*, Houston, OTC 26032-MS, pp 1–32 (electronic format).

O'Neill MP, Bransby MF, and Randolph MF, 2003, Drag anchor fluke–soil interaction in clays, *Can Geotech J*, 40, 78–94.

Prasad YVSN and Narasimha Rao S, 1996, Lateral capacity of helical piles in clays, *J Geotech Eng*, 122(11), 938–941.

Randall RE, 1997, *Elements of Ocean Engineering, Society of Naval Architects and Marine Engineers*, Society of Naval Architects and Marine Engineers, Alexandria, VA, 332p.

Randolph MF, 2003, Science and empiricism in pile foundation design, *Geotechnique* 53(10), 847–875.

Randolph MF, Gaudin C, Gourvenec SM, White DJ, Boylan N, and Cassidy MJ, 2011, Recent advances in offshore geotechnics for deep water oil and gas developments, *Ocean Eng*, 38(7), 818–834.

Rowe RK and Davis EH, 1982a, The behaviour of anchor plates in clay, *Geotechnique*, 32(1), 9–23.

Rowe RK and Davis EH, 1982b, The behaviour of anchor plates in sand, *Geotechnique*, 32(1), 24–41.

Shelton JT, 2007, OMNI-Max anchor development and technology, *Oceans 2007*, IEEE, Vancouver, Canada, 10 pages (electronic format).

Song Z, Hu Y, O'Loughlin CD, and Randolph MF, 2009, Loss in anchor embedment during plate anchor keying in clay, *ASCE J Geotech Geoenviron Eng*, 135(10), 1475–1485.

Stewart WP, 1992, Drag embedment anchor performance prediction in soft soils, *Proceedings of the 24th Offshore Technology Conference*, Houston, OTC 6970, pp 241–248.

Terzaghi K and Peck R, 1964, *Soil Mechanics in Engineering Practice*, John Wiley and Sons, New York.

Tjelta TI, 2015, The suction foundation technology, *Frontiers in Offshore Geotechnics III*, Meyer V (ed.), CRC Press, Taylor & Francis Group, London.

True DG, 1974, Rapid penetration into seafloor soils, *Proceedings of the 6th Annual Offshore Technology Conference*, Houston, Paper 2095, pp 607–618.

van Luipen P, 1987, The application of the hydraulic underwater hammer in slender and free riding mode with optional underwater power pack, *Proceedings of the 19th Annual Offshore Technology Conference*, Houston, Paper 5423, pp 561–568.

Vesic AS, 1971, Breakout resistance of objects embedded in ocean bottom, *J Soil Mech Found Div*, 97(9), 1183–1205.

Vijayvergiya VN, Cheng AP, and Kolk HJ, 1977, Design and installation of piles in chalk, *Proceedings of the 9th Offshore Technology Conference*, Houston, OTC 2938, pp 459–464.

Vivatrat V, Valent PJ, and Ponterio AA, 1982, The influence of chain friction on anchor pile design, *Proceedings of the 14th Annual Offshore Technology Conference*, Houston, Texas, OTC 4178, pp 153–163.

Vryhof Anchors, 2015, Anchor Manual 2015, the Guide to Anchoring, Global Maritime, Vryhof Anchors BV, 168p.

Ward EG, Mercier RS, Zhang J, Kim MH, Aubeny C, and Gilbert RB, 2008, No MODUs Adrift, Final Project Report prepared for the Minerals Management Service, Offshore Technology Research Center, Texas A&M University, OTRC Library Number: 3/08C188.

Whittle, AJ, 1992, Assessment of an effective stress analysis for predicting the performance of driven piles in clays, *Proceedings of the Conference on Offshore Site Investigation and Foundation Behaviour*, London, Vol. 28, pp 607–643.

Wilde B, Treu H, and Fulton T, 2001, Field testing of suction embedded plate anchors, *Proceedings of the 11th ISOPE Conference, Stavanger*, International Society of Offshore and Polar Engineers, Cupertino, California, pp 544–551.

Zakeri A, Liedke E, Clukey EC, and Jeanjean P, 2014, Long-term axial capacity of deepwater jetted piles, *Geotechnique*, 64(12), ICE Publishing, Thomas Telford, Ltd., London, pp 966–980.

Zimmerman EH, Smith MW, and Shelton JT, 2009, Efficient gravity installed anchor for deep water mooring, *Offshore Technology Conference*, Houston, Paper OTC 20117, pp 1–9 (electronic format), doi:10.4043/20117-MS.

Chapter 2

Soil behavior

Predicting anchor performance requires a thorough understanding of the soil mechanics controlling anchor behavior. Moreover, it is essential to have a working knowledge of soil sampling, laboratory testing, and *in situ* tests used for obtaining the soil data required to support the design of an anchor. This chapter provides an overview of the soil mechanics principles that are employed in the analysis methods discussed in later chapters. The presentation presumes familiarity with the basic concepts that are covered in a typical undergraduate course in soil mechanics, including soil index properties, phase relationships, and soil classification.

2.1 UNDRAINED BEHAVIOR

Undrained loading denotes a condition where the rate of loading exceeds the rate of pore water pressure dissipation to a sufficient degree that negligible changes in water content occur during applied loading. Typical examples of undrained conditions include anchor penetration into a clay seabed and rapidly applied loads, such as wave loading, on an anchor. Rapidly applied loads induce excess pore pressures in soil. The rate of dissipation of these pore pressures is governed by the coefficient of consolidation (discussed subsequently) of the soil. If uncertainty exists, strict determination as to whether a given loading condition is undrained is possible using analytical or numerical models of consolidation for establishing the rate of pore water pressure dissipation relative to the rate of applied loading. When pore pressures induced by loading are positive (typical of soft clays and compression foundations), soil shearing resistance increases as pore water pressures dissipate, with a concomitant increase in load capacity over time as drainage occurs. Owing to the tendency for zones of negative excess pore pressure to develop around anchors subjected to uplift loads, the consequences of drainage are less favorable for anchors than for compression foundations (Figure 1.24), since the soils can actually weaken over time as negative pore pressures dissipate. Thus, an assumption of undrained behavior does not necessarily produce conservative estimates of anchor capacity.

In principle, soil response to undrained loading can proceed on an effective stress basis, thereby explicitly considering changes in pore water pressure induced by the applied loads. While a number of constitutive models of soil behavior (discussed in Section 3.3) actually take this approach, most design methods adopt a total stress approach, whereby laboratory measurements of stress–strain strength under undrained loading conditions are applied directly to the design of the anchor or foundation, with the response of the pore water being implicit in the analysis. A condition of constant water content and volume implies that the shearing resistance of soil is independent of applied external loads, which justifies a total

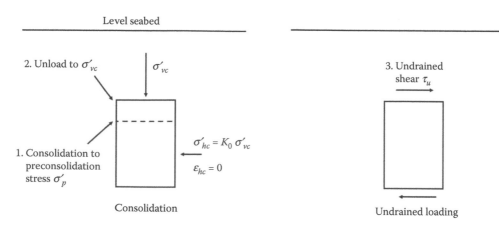

Figure 2.1 Framework for characterizing undrained behavior.

stress formulation. Relevant undrained material parameters can include the undrained shear strength s_u and the undrained elastic modulus E_u.

The design of laboratory tests for obtaining the data required for this purpose should consider three crucial factors affecting the undrained behavior: (1) the consolidated stress state of the soil prior to undrained loading, (2) the prior stress history of the soil, and (3) the conditions of undrained shearing. Figure 2.1 depicts the idealized framework that is commonly employed when developing a laboratory test program to characterize the soil stress–strain-strength behavior. Even with the simplifying idealization, the approach requires the determination of *in situ* vertical stress σ'_{vc}, the preconsolidation stress σ'_p, and the *in situ* horizontal stress σ'_{hc}. Once the *in situ* conditions of consolidation have been established, it remains to define the soil under conditions of undrained loading. Figure 2.1 shows undrained loading for conditions of direct simple shear (DSS), one of many possible conditions of undrained shear. Other factors that should be considered in the assessment of undrained soil behavior are the inevitable disturbance that occurs when sampling the soils to be tested and the sensitivity of undrained soil response to the rate of straining.

2.1.1 Consolidation stress state

Clay deposits generally consolidate under self-weight under conditions of anisotropic consolidation; that is, the horizontal effective stress σ'_{hc} differs from the vertical stress σ'_{vc}. For the common case of a level seabed, the soil experiences one-dimensional compression involving zero horizontal strain, and the principal stresses are aligned vertically and horizontally. This process is termed K_0-consolidation, where the coefficient of earth pressure at rest is defined as $K_0 = \sigma'_{hc}/\sigma'_{vc}$. Laboratory undrained strength tests are typically designed for replicating this condition in the laboratory. Additionally, laboratory testing requires a selection of an appropriate level of vertical stress σ'_{vc} to consolidate the soil specimen. A number of issues arise in connection with selecting an appropriate stress state to apply to a soil specimen during laboratory testing, including sample quality, overconsolidation ratio (OCR), and the degree to which cementation and other interparticle bonds influence the soil stress–strain-strength behavior; Section 2.1.13 addresses some of these issues.

Determination of σ'_{vc} is computationally simple, but requires accurate soil unit weight data as well as reliable characterization of *in situ* pore water pressures u_0, neither of which are particularly easy to acquire. Rapid deposition with consequent underconsolidation can produce vertical hydraulic gradients in marine clay profiles, which can significantly affect the magnitude of effective stress. Since the effective stress state affects virtually every aspect of the undrained stress–strain-strength behavior, reliable determination of steady state pore pressures is essential. Moreover, simply assuming a hydrostatic pore pressure distribution is a questionable practice. In principle, *in situ* tests such as the piezocone holding test used for measuring the horizontal coefficient of consolidation c_h (Section 2.2.4) can also be used for measuring the steady state pore pressure at a given elevation. However, the time for full dissipation of excess pore pressures surrounding a conventional piezocone can exceed 24 hours, which is not practical for most offshore site investigations.

Two alternatives exist for measuring u_0: piezoprobes and piezometers. A piezoprobe test advances a tapered shaft into the soil and halts penetration at a selected depth of interest. Pore pressures are continuously measured for monitoring their decay to u_0. Owing to the reduced diameter of the taper, dissipation initially occurs much more rapidly than for a conventional piezocone. However, Whittle et al. (2001) found that long-term dissipation is retarded by a second pulse of pore water pressure originating from above the taper, thereby resulting in times for full dissipation of excess pressures comparable to that for a conventional piezocone. They, therefore, developed a modified piezoprobe design having a second pore pressure measurement point above the taper. Using their two-point matching procedure, they show that reasonable estimates of u_0 are possible from incomplete dissipation records within an acceptable test time duration. The second alternative method of measuring u_0 is piezometers that can be either pushed in or installed in a prebored hole. Strout and Tjelta (2007) provide details on alternative piezometer designs along with their advantages and disadvantages. DeGroot et al. (2012) judge that, by either method for measuring u_0, one cannot estimate the vertical effective stress σ'_{vc} within an accuracy of greater than 10%.

2.1.2 Preconsolidation stress

The yield stress of a soil specimen under conditions of confined (K_0) loading is termed the preconsolidation pressure σ'_p. Loading at stress levels below σ'_p involves relatively small strains and (nearly) elastic behavior. While the preconsolidation stress is, of course, a key parameter in deformation calculations, it also turns out to be essential to understand and predict undrained shear strength. Laboratory measurement of σ'_p is covered in the discussion of consolidation, Section 2.2.1. The basic aspects of the preconsolidation stress critical to understand the undrained shear behavior are discussed below.

OCR is defined as the ratio of the preconsolidation stress σ'_p to the *in situ* stress σ'_{v0} as follows:

$$\text{OCR} = \frac{\sigma'_p}{\sigma'_{v0}} \tag{2.1}$$

Normally consolidated soils have an OCR = 1, and overconsolidated soils have OCR > 1.

A number of mechanisms can cause preconsolidation (Holtz and Kovacs, 1981). Most obvious are changes in total stress owing to processes such as glaciation or removal of overburden, and changes in effective stress associated with altered seepage conditions. In the case of removal of overburden (Figure 2.2a), the preconsolidation stress exceeds the effective overburden stress by a roughly constant amount. Desiccation—which can

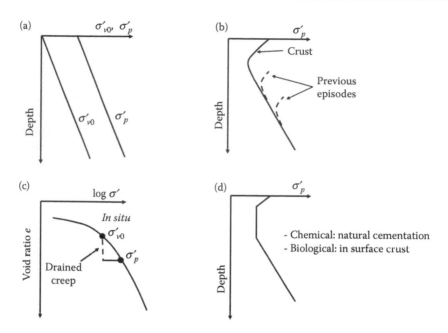

Figure 2.2 Mechanisms of preconsolidation. (a) Load removal, (b) desiccation, (c) aging, and (d) physico-chemical and biological.

be due to evaporation, vegetation, or freezing—produces a zone of elevated preconsolidation stress levels; these typically occur at the surface, but previous episodes of sea level fluctuation can lead to similar zones occurring at depth. Drained creep or "aging" (Figure 2.2c) involves settlement and densification of the soil under conditions of constant effective stress. The denser state of the soil causes it to behave as if it has experienced a stress level in the past that exceeds the current stress state. Physicochemical mechanisms such as cementation or other bonding mechanisms can also control the preconsolidation stress. In deepwater regions throughout the world, a surface crust (1–2 m deep) is sometimes encountered. Various investigators (e.g., Ehlers et al., 2005; Kuo and Bolton, 2009) suggest that the anomalously high strengths in this crust could be due to biological activity.

2.1.3 Coefficient of earth pressure at rest

The importance of the at-rest earth pressure coefficient K_0 in predicting the strength-deformation behavior and other problems in soil mechanics (e.g., the horizontal stress on underground structures) has motivated a great deal of research on this topic. The well-known equation of Jaky (1944) for normally consolidated cohesionless soils provides K_0 as follows:

$$K_{0NC} = 1 - \sin\phi' \tag{2.2}$$

Data for clays (e.g., Brooker and Ireland, 1965; Ladd et al., 1977; Mayne and Kulhawy, 1982) support the validity of this relationship for normally consolidated clays. For unloading, Schmidt (1966) proposed an exponential relationship between OCR and the earth pressure coefficient for conditions of unloading:

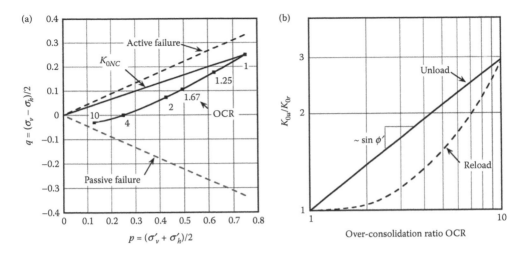

Figure 2.3 K_0 Consolidation behavior. (a) Plot in stress space and (b) unload-reload behavior.

$$\frac{K_{0u}}{K_{0NC}} = OCR^m \tag{2.3}$$

A statistical study (Mayne and Kulhawy, 1982) for both sands and clays indicate that the exponent in Equation 2.3 may be estimated as $m = \sin \phi'$. Owing to hysteresis (Campanella and Vaid, 1972), Equation 2.3 is not applicable under conditions of reloading.

Figure 2.3 illustrates the K_0-consolidation behavior in terms of a plot in $p - q$ stress space, where p' is the average of the vertical and horizontal effective stresses and q is half of the deviator stress. Under normally consolidated loading conditions, K_0 is essentially independent of consolidation stress level. Upon unloading, the effects of increased K_0 with increasing OCR are shown in the figure. In this instance, the consolidated stress state is isotropic ($K_0 = 1$) at OCR = 4. At elevated OCR, the horizontal stress actually exceeds the vertical stress. The active and passive failure envelopes are also superimposed on this plot. From classical soil mechanics theory, the earth pressure coefficients for the active and passive failure conditions are as follows:

$$K_a = \frac{1 - \sin\phi'}{1 + \sin\phi'} \tag{2.4}$$

$$K_p = \frac{1 + \sin\phi'}{1 - \sin\phi'} \tag{2.5}$$

The earth pressure coefficient for passive failure K_p represents an upper bound for the at-rest earth pressure coefficient under unloading, K_{0u}.

It is emphasized that, while the relationships presented above provide useful insights into horizontal stress, they apply to highly idealized conditions of one-dimensional loading. Little data are available on K_0 associated other mechanisms of preconsolidation such as desiccation, aging, and cementation.

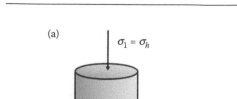
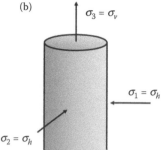
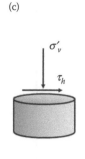

Figure 2.4 Advanced laboratory strength tests. (a) Triaxial compression CK_0UC, (b) triaxial extension CK_0UE, and (c) direct simple shear CK_0U-DSS.

2.1.4 Advanced laboratory strength tests

Laboratory tests for measuring the undrained stress–strain-strength behavior of silts and clays included the triaxial shear test and the DSS test (Figure 2.4). As discussed earlier, undrained behavior is highly sensitive to the conditions of consolidation prior to undrained shearing. Therefore, test procedures that attempt to capture the effects of anisotropic consolidation (typically K_0) are widely employed in offshore geotechnical practice. This text will use the widely adopted nomenclature in which the first two letters (CK_0) denote the conditions of consolidation, the third letter $(U = \text{undrained})$ denotes the condition of drainage during shear, and the remaining letters denote the mode of shearing $(C = \text{compression}, E = \text{extension, and } DSS = \text{direct simple shear}).$

The triaxial test (Bishop and Henkel, 1976) employs a pressurized fluid-filled chamber to control horizontal stress σ_h and a piston to control vertical stress σ_v. Conditions of drainage are controlled by a valve that is opened during the consolidation phase of the test and closed during undrained shearing. Automated systems can achieve the K_0-consolidation state that can be achieved by increasing and adjusting the vertical and horizontal stresses to ensure that axial strain equals volumetric strain. Alternatively, for manually controlled triaxial systems, K_0 can be estimated in advance as described in the previous section, and anisotropic consolidation can be performed at a predetermined ratio σ'_h/σ'_v. During the undrained stage of the test, the sample can be tested in either compression or extension (Figure 2.4a and b). Friction constrains the top and bottom of the specimen, thereby producing a barrel-shaped mode of deformation. Nevertheless, the stress and strain nonuniformities in the specimen are typically considered sufficiently minor to the test to be the representative of single-element behavior.

Test results are normally presented in terms of effective stress path $(p' - q)$ and stress–strain plots as shown in Figure 2.5a. Normally consolidated (OCR = 1) specimens generally generate positive pore pressures when subjected to shear, with concomitant reductions in effective stress occurring as the test progresses. This behavior is attributed to the volume change tendencies of the soil. Normally consolidated soils tend to contract during shear, thereby resulting in increases in pore pressure under undrained (zero volume change) conditions. The opposite tendency occurs for highly overconsolidated soils, which exhibit dilative behavior during shear, with consequent negative changes in pore pressure under undrained conditions. Soil response is highly dependent on the mode of shearing. A normally consolidated soil subjected to triaxial compression (TC) typically exhibits brittle behavior; that is, peak resistance mobilizes at small strains, followed

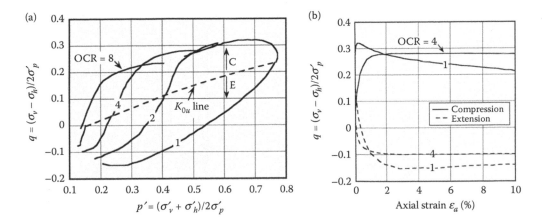

Figure 2.5 Typical triaxial compression and extension behavior. (a) Stress paths and (b) stress–strain curves.

by significant strain softening. By contrast, a normally consolidated soil subjected to triaxial extension (TE) has a ductile response. The tendency for brittle behavior in TC diminishes with increasing OCR. Undrained shear strength is typically considered as the peak value occurring during shear. Thus, the soil shown in the example has a strength $s_{uTC}/\sigma'_p = 0.31$ in TC and $s_{uTE}/\sigma'_p = 0.15$ in TE. Soil strength anisotropy will be taken up in more detail subsequently.

The DSS (Bjerrum and Landva, 1966; Dyvik et al., 1987) test employs a cylindrical soil specimen enclosed in a reinforced rubber ring that imposes a condition of zero lateral strain, $\varepsilon_h = 0$. Therefore, simple application of a vertical load achieves the K_0-consolidation state. After the desired consolidation stress is applied, a shear stress τ_h is applied. An undrained condition is achieved by continuously adjusting the vertical stress σ'_v to maintain a constant specimen height. Examination of the boundary conditions imposed on the soil specimen during shear shows that a nonuniform stress field will occur. Nevertheless, the test is normally interpreted based on the assumption of single-element behavior corresponding to conditions of simple shear. With some exceptions, this framework for test interpretation is satisfactory. Figure 2.6 shows typical trends in measured effective stress path and stress–strain behavior. In contrast to the triaxial test, the stress state in the DSS is not fully defined, so the effective stress path is plotted in terms of vertical effective stress σ'_v. Much of the behavior parallels that discussed above in regard to triaxial shear behavior; namely, the soil is contractive at low OCR and becomes progressively more dilative as OCR increases. The peak shear stress τ_{hmax} is considered as the undrained shear strength in the DSS shearing mode s_{uDSS}. Shear strength measured in the DSS test is typically intermediate between that measured in TC and TE.

2.1.5 Normalized behavior

Effective stress paths and stress–strain curves are frequently normalized by effective vertical consolidation stress (e.g., q/σ'_{vc}, p'/σ'_{vc}). In many cases, soil specimens consolidated to the same OCR but different σ'_{vc} exhibit similar stress path and stress–strain behavior when the test data are expressed in normalized form (Ladd et al., 1977). Figure 2.7a and b depicts an idealization of this behavior. When it occurs, normalizable soil behavior is advantageous, since data from a single test series can be extrapolated to other consolidation stress levels. It is also integral to the stress history and normalized soil engineering parameters

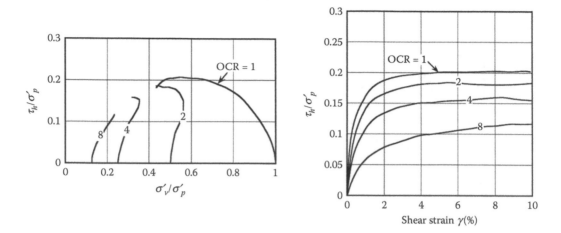

Figure 2.6 Typical direct simple shear data.

(SHANSEP) method for characterizing strength, which is discussed later in this chapter. Real soil behavior is seldom perfectly normalizable, but it frequently conforms sufficiently to this ideal for practical design purposes. The semi-logarithmic void ratio versus stress (e-log σ'_{vc}) plot from the one-dimensional consolidation tests provides some insight into normalizable soil behavior (Figure 2.7c). The virgin compression line (VCL) for insensitive clays is typically nearly linear, and effective and deviator stresses at failure (p'_f, q_f) plot parallel to the VCL. This implies a constant ratio between σ'_{vc} and p' and, consequently, normalizable behavior. An assumption of normalizable behavior is strictly valid only if the soil maintains

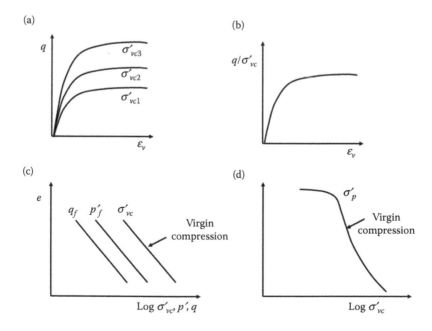

Figure 2.7 Normalized soil behavior. (a) Stress–strain, (b) normalized behavior, (c) normalized stress–strain, and (d) sensitive and cemented soils.

its basic behavior when it is loaded beyond the preconsolidation stress into the virgin compression region. For sensitive and naturally cemented soils (Figure 2.7d), loading beyond the preconsolidation stress breaks the interparticle bonds, fundamentally altering the behavior of the soil. In these cases, the soil behavior clearly depends on σ'_{vc}. Therefore, normalizable behavior cannot be assumed. Irrespective of whether or not the soil behavior is normalizable, to facilitate comparisons amongst various soil groups, it is frequently convenient to express the soil strength in the normalized form as follows:

$$\text{Undrained strength ratio} = USR = \frac{s_u}{\sigma'_{vc}} \tag{2.6}$$

2.1.6 Anisotropy

As is apparent from the above discussion, undrained strength s_u can be highly sensitive to the mode of shearing. The term "anisotropy" will be used in this discussion in connection with this behavior, thereby recognizing that the tests shown in Figure 2.4 differ in other respects beyond the direction of applied loading. For example, while TC and TE tests involve different orientations of the major principal stress σ_1, they also involve different intermediate principal stress conditions. In one case $\sigma_2 = \sigma_3$ and in the other $\sigma_2 = \sigma_1$. Thus, differences in compression and extension behavior cannot be attributed solely to load direction. An alternative framework is to consider a stress space of five independent shearing modes, with anisotropy in stress–strain-strength response linked to "direction" in stress space as opposed to a physical direction of loading. The discussion of yield loci and constitutive models in Section 3.1.1 further elaborate on this topic. For now, the discussion focuses on the empirical evidence of anisotropy, with the term "anisotropy" simply denoting differences in undrained behavior for different modes of shearing.

Analytical models of anchor performance can treat anisotropy either by selecting an appropriate "average" strength representative of all shearing modes or by explicitly incorporating anisotropy into the analysis, with the former approach being somewhat more common. In either event, familiarity with basic trends in anisotropic strength behavior as measured by conventional methods is useful. In a compilation of strength data on natural samples from various sources Ladd (1991) developed correlations to plasticity index (PI); Figure 2.8 summarizes the trends. In low plasticity soils, the DSS and TE strengths are, respectively, on the order of two-thirds and one-half of the TC strength. Undrained strength in TC is insensitive to PI, but both the DSS and TE strengths increase with increasing plasticity. It is emphasized that the trends shown in this figure are peak strengths, which occur at substantially different strain levels amongst the different shearing modes. Consideration of mobilized shearing resistance at comparable levels of strain should be considered in the actual assessments of strength anisotropy.

Actual three-dimensional stress states surrounding loaded anchors can generate shearing modes and load combinations that are not necessarily accurately represented by conventional triaxial and DSS tests. For example, plane strain conditions can develop around vertically oriented plates, thereby generating shear in the so-called plane-strain active (PSA) and plane-strain passive (PSP) modes, (Figure 2.9). While the PSA and PSP shear modes bear resemblance to TC and TE modes, specialized tests developed to measure soil strength in these shearing modes show consistent differences in comparison to triaxial test data. Ladd et al. (1977) assembled from a variety of sources for both undisturbed and remolded K_0-normally consolidated clays where comparisons of triaxial to plane strain measurements were possible. The data summary in Table 2.1 show that the strength in TC is approximately

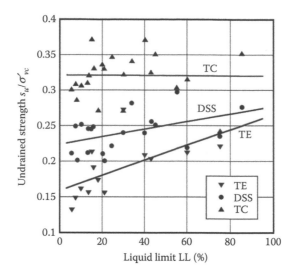

Figure 2.8 Undrained strength anisotropy in conventional CK_0U tests. (From Ladd CC, 1991, *J Geotech Eng*, ASCE, 117(4), 540–615, with permission from ASCE.)

90%–97% of that in the PSA mode. The TE-PSP comparisons showed even greater differences, with the extension strength being approximately 80%–85% of the PSP value.

Another shearing mode frequently of interest is cavity expansion, which is a prominent (but not the only) mode of shearing in processes such as pile penetration and pile translation. Since shearing in this mode requires special tests such as the true triaxial apparatus (Wood, 1981) or the directional shear cell (Arthur et al., 1977), most determinations of s_{uCE} have been made in a research setting. Albeit limited that this type of data can provide a useful guide for extrapolating data from conventional tests to general conditions of undrained

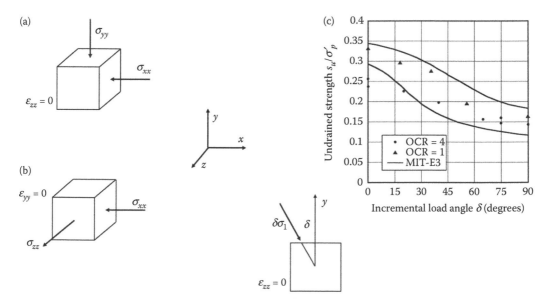

Figure 2.9 Other undrained CK_0U shearing modes. (a) Plane strain active/passive (PSA/PSP), (b) cavity expansion (CE), and (c) directional shear cell (DSC).

Table 2.1 Undrained strength K_0-normally consolidated clays triaxial versus plane strain shear

Soil (reference)	Specimen state	LL/PI	s_{uTC}/s_{uPSA}	s_{uTE}/s_{uPSP}
Haney Sensitive Clay Vaid and Campanella (1974)	Undisturbed	44/18	0.905	0.795
Boston Blue Clay Ladd et al. (1971)	Resedimented	41/21	0.97	0.815
Weald Clay Henkel and Wade (1966)	Remolded	46/26	0.93	–
Atlantic Generating Station (UBC (1975), unpublished from the University of British Columbia)	Undisturbed	71/40	0.87	0.83
San Francisco Bay Mud Duncan and Seed (1966)	Undisturbed	88/45	0.945	–
Conn. Valley Varved Clay (MIT (1976), unpublished data from the Massachusetts Institute of Technology)	Undisturbed	65/39 clay 35/12 silt	0.895	0.84

shear. Whittle and Aubeny (1992) compiled data from a variety of sources (Ladd et al., 1971; Wood, 1981; Seah, 1990) on undrained shear strength of resedimented Boston Blue Clay, including s_{uCE} data. Table 2.2 summarizes the comparisons for OCR = 1 and 4. Since the DSS strength is commonly considered as representative of all shearing modes in anchor load capacity calculations, strengths in the various shearing modes are also normalized by s_{uDSS} to facilitate comparisons. In this case, s_{uDSS} appears to provide a fair representation of s_{uCE}. This is not the case for the other shearing modes, but since s_{uDSS} is intermediate between the compression/extension and active/passive strengths, the implicit assumption is that the compensating errors will produce reasonable estimates of anchor capacity. This assumption is not unreasonable, but the reliance on compensating errors should be kept in mind when selecting strength parameters for specific situations.

A final example of undrained strength anisotropy considered here is that of the directional shear cell (Figure 2.9c). This test has the capability for applying incremental changes in the major principal stress $\delta\sigma_1$ at a fixed inclination angle δ. The test is conducted on K_0-consolidated specimens under plane strain conditions; thus, it can probe soil behavior between the limiting active (PSA, $\delta = 0$) and passive (PSP, $\delta = 90°$) undrained loading conditions. Figure 2.9c after Whittle et al. (1994) show the measured trends for normally and lightly overconsolidated resedimented Boston Blue Clay (BBC). Moreover, predictions from the MIT-E3 soil constitutive model (Whittle and Kavvadas, 1993), which is further discussed in Chapter 3, are shown. Noting that laboratory tests for measuring undrained stress–strain-strength behavior under arbitrary loading conditions can be difficult and expensive to execute, generalized constitutive soil models such as MIT-E3 and MIT-S1 (Pestana and Whittle, 1999) are powerful tools for extrapolating conventional undrained strength measurements to general conditions of loading.

Table 2.2 Undrained shear strength of K_0-consolidated resedimented BBC in various shearing modes

Overconsolidation ratio OCR	1		4	
Shearing mode	s_u/σ'_{vc}	s_u/s_{uDSS}	s_u/σ'_{vc}	s_u/s_{uDSS}
Triaxial compression CK$_0$UC	0.33	1.65	1.04	1.73
Triaxial extension CK$_0$UE	0.14	0.70	0.52–0.60	0.87–1.0
Plane strain active CK$_0$U-PSA	0.34	1.7	0.84–1.04	1.4–1.73
Plane strain passive CK$_0$U-PSP	0.16–0.19	0.8–0.95	0.52–0.60	0.87–1.0
Direct simple shear CK$_0$U-DSS	0.20	1.0	0.56–0.64	1.0
Cavity expansion CK$_0$U-CE	0.21	1.05	0.64–0.76	1.07–1.27

2.1.7 Influence of overconsolidation

Undrained shear strength increases with increasing OCR. Both critical state soil mechanics (Roscoe and Burland, 1968) and laboratory measurements support a power law form of the relationship between overconsolidated and normally consolidated undrained shear strength:

$$(s_u/\sigma'_{vc})_{OC} = S(OCR)^m$$
$$S = (s_u/\sigma'_{vc})_{NC}$$

(2.7)

Data by Ladd (1991) indicate that this relationship applies for various modes of shearing (TC, TE, and DSS), although both S and m vary for the different modes. Equation 2.7 also appears to apply irrespective of the mechanism of preconsolidation. For example, Ladd presents data demonstrating the validity of Equation 2.7 for a mechanically overconsolidated clay (Atlantic Generating Station, Koutsoftas and Ladd, 1983) and a cemented, sensitive clay (James Bay, Lefebvre et al., 1983). Critical state soil mechanics suggests that m should equal $1 - C_s/C_c$, where C_s and C_c are the swell and compression indices measured in a one-dimensional consolidation test. Actual data show some departure from critical state theory. Equation 2.7 is an integral component of the SHANSEP method developed by Ladd and Foott (1974), which is described in detail later in this chapter.

For preliminary estimates of average (of TC, TE, and DSS) strength as well as for evaluating the reliability of site-specific strength determinations of the S and m parameters in Equation 2.7, Ladd (1991) provides the ranges $S = 0.22 \pm 0.03$ and $m = 0.8 \pm 0.1$ for most soil types. He also provides more specific guidance on mean and standard deviation (SD) values for specific soil groups:

- Sensitive marine clays (PI < 30, liquidity index LI > 1):
  $S = 0.20 \pm$ nominal SD $= 0.015$
  $m = 1.0$
- Homogeneous lean clay (this is actually a Unified Soil Classification System soil group symbol) (CL) and fat clay (this is actually a Unified Soil Classification System soil group symbol) (CH) sedimentary clays of low-moderate sensitivity (PI = 20 – 80):
  $S = 0.20 + 0.05$ PI, or simply 0.22
  $m = 0.88(1 - C_s/C_c) \pm 0.06$ SD or simply 0.8
- Sedimentary silts and organic soils (excluding peat) and clays with shells:
  $S = 0.25 \pm$ nominal SD $= 0.05$
  $m = 0.88(1 - C_s/C_c) \pm 0.06$ SD or simply 0.8
- Northeastern US varved clays:
  $S = 0.16$ (assumes DSS failure mode predominates)
  $m = 0.75$

2.1.8 General conditions of consolidation

The preponderance of undrained strength tests are conducted under the K_0-consolidation conditions shown in Figure 2.1. This assumption frequently provides a good approximation to actual conditions, but it should not be uncritically accepted without justification. Deviation from the K_0 stress state can occur owing to a sloping seabed, nearby fill placement or proximity to a foundation. With six independent components of stress, the possible consolidation stress states are limitless. However, some insight into the degree to which deviation from the K_0-consolidated can affect undrained stress–strain-strength behavior by conducting special

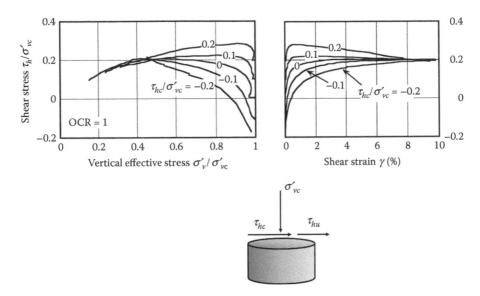

Figure 2.10 Effect of consolidation shear stress on DSS behavior. (After DeGroot DJ et al. 1996, *J Geotech Eng*, ASCE, 122(2), 91–98, with permission from ASCE.)

DSS tests in which shear stress τ_{hc} is applied during consolidation, as shown in Figure 2.10. After consolidation, shearing τ_{hu} continues under undrained conditions. Consolidation shear stresses τ_{hc} applied in the same direction as τ_{hu} are denoted as positive, and vice versa. Tests by DeGroot et al. (1996) on resedimented Boston Blue Clay show that, at sufficiently high levels, consolidation shear stress τ_{hc} can significantly impact stress–strain strength behavior. For example, for a consolidation shear stress $\tau_{hc}/\sigma'_{vc} = 0.2$, the undrained shear strength is about 27% greater than that measured in a CK_0U-DSS test. Additionally, positive consolidation shear stresses produce increasingly brittle stress–strain behavior in the soil.

The DeGroot et al. (1996) data provide a glimpse of the effects of deviation of the consolidation stress conditions from the commonly assumed condition of K_0-consolidation. However, an infinite number of consolidated stress states are possible. As noted earlier in connection with undrained shear behavior, generalized constitutive soil models (e.g., Whittle and Kavvadas, 1993; Pestana and Whittle, 1999) can provide insight into the effects of general conditions of consolidation which could otherwise only be obtained through specialized testing.

2.1.9 Strain rate effects

The dependence of undrained shear strength on applied rate of strain has long been recognized (Casagrande and Wilson, 1951). Early studies characterized this dependence in terms of a semi-logarithmic equation relating shear strength s_u measured at strain rate $\dot{\gamma}$ to a reference strength s_{u_ref} at strain rate $\dot{\gamma}_{ref}$ as follows:

$$s_u = s_{u\_ref}[1 + \mu_{ref} \log_{10}(\dot{\gamma}/\dot{\gamma}_{ref})] \tag{2.8}$$

To facilitate comparisons to load duration, rate effects sometimes expressed in terms of the time t_f to failure:

$$s_u = s_{u\_ref}[1 + \mu_{ref} \log_{10}(t_{f\_ref}/t_f)] \tag{2.9}$$

The parameter μ_{ref} is an experimentally determined strain rate multiplier. Since its magnitude depends on the selected reference strain rate, the strain rate to which μ_{ref} applies should always be specified. For example, $\mu_{0.5}$ denotes the strain rate multiplier corresponding to a reference strain rate of 0.5%/hour. A review of 13 studies of rate effects using consolidated undrained TC tests by Sheahan et al. (1996) showed $\mu_{0.5}$ varying from 3% to 17%, with an average value of about 10%. Other studies indicate a similar range of strain rate multipliers (Dayal and Allen, 1975; Graham et al., 1983; Biscontin and Pestana, 1999). Data by Sheahan et al. (1996) suggest the existence of a threshold strain rate $\dot{\gamma}_0$ below which strain rate effects are negligible. In resedimented Boston Blue Clay, reported threshold strain rates were 0.5%/hour and 5%/hour for OCR of 4 and 8, respectively. For lower OCR, thresholds were not encountered for strain rates down to 0.05%/hour. Hence, for normally and lightly overconsolidated soils, the lower limit of strain rate, or threshold strain rate for which Equation 2.8 is valid, is a matter of some uncertainty. The data by Sheahan et al. (1996) also show μ_{ref} to increase with increasing strain rate, indicating that Equation 2.8 is only approximately valid over a restricted range of strain rates. Lunne and Andersen (2007) observed similar trends for Norwegian clays.

It is instructive to compare strain rates in common laboratory tests to typical wave loading conditions. Lunne and Andersen (2007) report typical strain rates of 2%–3% per hour in consolidated-undrained triaxial tests and 4.5% per hour in DSS tests. Therefore, if failure occurs at 10% strain in a DSS test, the time to failure is 8000 seconds, in contrast to a typical period of wave loading of 10–20 seconds. Thus, 2–3 orders of magnitude difference in loading rates can exist between laboratory and field conditions, with the corresponding difference in undrained shear strength owing to rate effects being on the order of 20%–30%.

Laboratory shear tests (CK_0UC, CK_0UE, and CK_0U-DSS) have a relatively restricted range of strain rates at which the specimen can be loaded. With specially lubricated end platens, Sheahan et al. (1996) were able to test for strain rate effects over three orders of magnitude. The vane test (Section 2.1.3) has capability for testing over nine orders of magnitude (Biscontin and Pestana, 1999; Peuchen and Mayne, 2007). When considering wider ranges of strain rate, an alternative power law Equation 2.10 provides a better description of strain rate dependence. Reported values of the power law exponent β are in the range 0.05–0.1.

$$s_u = s_{u\_ref}(\dot{\gamma}_{ref} / \dot{\gamma}_{ref})^\beta \qquad (2.10)$$

More recent research directed toward applications involving high velocity, such as those associated with dynamic pile penetration (normalized velocity $v/D \approx 25$ s^{-1}), is leaning more toward a fluid mechanics framework to characterize rate-dependent shearing resistance. For such applications various forms of the Herschel–Bulkley law shown in Equation 2.11 have been adopted (Raie and Tassoulas, 2009; Zhu and Randolph, 2011; Kim et al., 2015).

$$s_u = \left[1 + \eta\left(\frac{\dot{\gamma}}{\dot{\gamma}_{ref}}\right)^\beta\right]\frac{s_{u,ref}}{1+\eta} \qquad (2.11)$$

Kim et al. (2015) report values of $\eta = 0.1$–2.0 and $\beta = 0.01$–0.15.

Equations 2.8 through 2.11 express strength increase in terms of elemental strain rate. Many studies investigating rate effects are actually based on penetration tests (Section 2.1.13) involving a well-defined penetration velocity, with high spatial variability in strain rates. Similarly, many anchor penetration problems are similarly posed in terms of a well-defined velocity, with no obvious representative strain rate. In these cases, the rate equations are frequently expressed in terms of velocity. For example, a semi-logarithmic rate law can expressed as follows:

$$s_u = s_{u\_ref}[1 + \lambda_{ref} \log_{10}(v/v_{ref})] \tag{2.12}$$

This equation replaces velocity with strain rate in Equation 2.8 and a macroscopic strain rate multiplier λ_{ref} with the elemental multiplier μ_{ref}. Finite element studies of penetrating horizontal cylinders employing a rate-dependent plasticity model (Aubeny and Shi, 2007) indicate that μ_{ref} provides a reasonable approximation to λ_{ref}. A series of subsequent studies summarized by Kim et al. (2015) indicates that this approximation is acceptable for bearing type problems, but underestimates rate dependence for frictional resistance. Their suggested equation for strain rate dependence introduces an additional parameter n as follows:

$$s_u = s_{u\_ref}[1 + \lambda_{ref} \log_{10}(nv/v_{ref})] \tag{2.13}$$

When bearing resistance dominates n can be considered as unity. However, n is an order of magnitude higher for frictional resistance, or the strain rate multiplier effectively doubles. Equation 2.13 is an adaptation of the semi-logarithmic rate formulation. Similar adaptations can be applied to the power law Equation 2.8 and Herschel–Bulkley Equation 2.11 formulations.

2.1.10 Remolded strength

The remolded strength s_{ur} is a critical parameter for predicting soil resistance to pile or anchor penetration during installation. This parameter is either reported directly or expressed in terms of sensitivity S_t defined as follows:

$$S_t = \frac{s_u}{s_{ur}} \tag{2.14}$$

where s_u is intact undrained shear strength. Remolded shear strength can be measured in the laboratory using the miniature vane shear test (ASTM, 2003), the fall cone test (Hansbo, 1957) or unconsolidated undrained (UU) triaxial shear tests. Additionally, the flow-around penetrometers and field vane (FV) tests discussed in Section 2.1.14 can provide *in situ* measurements of remolded strength. Remolded strength can be heavily dependent on method of testing. Lunne and Andersen (2007) find that s_{ur} measured in the UU test can be much lower than fall cone and FV values, which they attribute to differences in the rate of straining amongst these tests. Sensitivity S_t is less dependent on test method; since, both s_u and s_{ur} are measured by the same method, the strain rate effect tends to cancel. Since sample disturbance (Section 2.1.12) can lead to significant underestimates of s_u, laboratory tests tend to underestimate sensitivity. Details on recommended practice and differences in measured remolded strength amongst the various test methods is provided by DeGroot et al. (2012).

2.1.11 Thixotropy

As defined by Mitchell (1960), thixotropy is an isothermal, reversible, time-dependent process where a material stiffens while at rest softens or liquefies upon remolding. Thixotropy refers to time-dependent strength gain in soil at constant water content after remolding. This process is in contrast to the strength gain in soil associated with consolidation, which occurs under conditions of changing water content. Following remolding of the soil during anchor penetration, both thixotropy and consolidation contribute to setup, the time-dependent recovery of strength. Consolidation is generally the much more dominant process. Nevertheless, during the early stages of setup, the effects of thixotropy can be significant.

The thixotropy strength ratio is defined as the undrained shear strength at a certain time divided by the strength immediately following remolding. A compilation of thixotropic strength gain data for Gulf of Mexico clays by Jeanjean (2006) shows most thixotropic strength gain occurring within 30–60 days, with thixotropy strength ratios ranging from approximately 1.5 to 2.2. Lunne and Andersen (2007) present thixotropy strength ratios for various clay minerals. Thixotropic strength gain is virtually zero for kaolin, moderate (~40%) in illite, and large (>100%) in bentonite. Lunne and Andersen (2007) present data showing a general trend of increasing thixotropy strength ratio with activity [PI/clay fraction (CF)], but considerable scatter exists in this correlation.

2.1.12 Sample disturbance

High quality soil samples are essential to obtaining meaningful laboratory test measurements of soil stress–strain-strength properties. While samples obtained using mud rotary drilling methods and thin-wall samples are normally considered "undisturbed," careful consideration of the sampling process (Figure 2.11a) shows that a soil sample can undergo severe distortions and changes in effective stress on the path from its *in situ* stress state to its placement in a laboratory test device. Detailed discussion of recommended practice for soil sampling is beyond the scope of this discussion, although interested readers may refer to Ladd and DeGroot (2003). However, in Section 2.1.13, choice of the framework for characterizing stress–strain-strength behavior is influenced by an assessment of sample quality. Therefore, selection of strength parameters for anchor design requires a basic understanding of the processes inducing sample disturbance as well as methods for assessing sample quality.

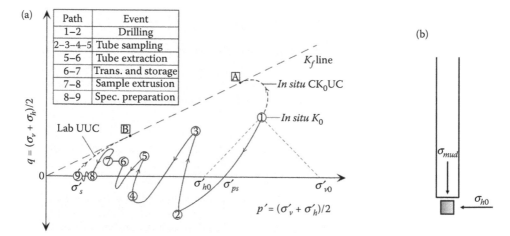

Figure 2.11 Stress path during soil sampling and preparation. (a) Stress path (Adapted from Ladd CC and DeGroot DJ, 2003, Recommended practice for soft ground site characterization: Arthur Casagrande Lecture, *Proceedings 12th Panamerican Conference on Soil Mechanics and Geotechnical Engineering*, Massachusetts Institute of Technology Cambridge, MA, USA, pp 3–57.) and (b) stress in Borehole.

In early studies on sample disturbance, Ladd and Lambe (1963) postulated a "perfect sampling" state, where the soil is unloaded from the *in situ* K_0 stress state to an isotropic stress state σ'_{ps}. Since the total stress acting on an extracted sample is zero, the effective stress σ'_{ps} is due to negative pore pressure, or suction, in the pore fluid. The perfect sampling stress state is an idealized condition that is essentially impossible to achieve, but it provides a useful benchmark for assessing sample quality. The effective stress σ'_s in a soil sample is generally well below σ'_{ps}, owing to a number of mechanisms of disturbance described by Baligh et al. (1987) and Ladd and DeGroot (2003):

1. *Boring and stress relief.* If the hydrostatic stress in the drilling mud is less than the *in situ* horizontal stress, the sample becomes subjected to shearing and potential failure in a TE mode, as depicted by negative q values in Figure 2.11a.
2. *Tube sampling.* Analytical simulations of tube penetration using the strain path method (Section 3.4.2) by Baligh et al. (1987) show that an element of soil on the centerline of the sampler experiences a load cycle of TC as the tube approaches the soil element, followed by a cycle of TE as the soil element enters the tube. The severity of the shear distortion depends on aspect ratio (D/t_w, where D is outer diameter and t_w is wall thickness) of the sampler as well as its internal clearance ratio (ICR). The internal diameter of a sampling tube at its cutting edge D_e is typically slightly less than the diameter D_i in the interior of the tube. The ICR is defined as $(D_i - D_e)/D_e$. Disturbance increases with increasing ICR, so the ICR should be kept to a minimum. The Baligh et al. study included predictions for aspect ratios $D/t_w = 40$–50 and ICR = 0.79–0.98, ranges which are generally representative of a standard 76.2 mm (3-in.) Shelby tube. Predicted strains in both TC and TE approach 1%. Path 2–5 in Figure 2.11a depicts this loading sequence.
3. *Tube extraction.* Following tube insertion, the soil as the bottom of the tube is still intact and can resist pullout through reverse end bearing. Sample recovery is achieved (a) by allowing time for recovery of adhesion between the soil and the sampling tube and (b) by rotating the tube to shear the intact soil at the base. Disturbance to the soil sample during this process is difficult to quantify, so path 5–6 in Figure 2.11a is largely speculative. An additional, potentially very severe source of sample disturbance can occur when sampling in deep water sediments, when depressurization causes dissolved gases in the pore water to come out of solution.
4. *Transport and storage.* Assuming proper handling and storage, the primary process occurring during this stage (Path 6–7) is migration of disturbance-induced pore pressures from the sides to the center of the specimen. Although this process reduces the effective stress in the sample, is does not involve shear distortion and damage to the soil structure, so it is less detrimental to sample quality.
5. *Sample extrusion.* Bonding at the soil–tube interface should be overcome during sample extrusion, which can severely disturb the sample.
6. *Specimen preparation.* Further, swelling owing to stress relief and suction of water from wet pore stones induce further reductions in effective stress.

Ladd and DeGroot (2003) estimate that the net result of all sources of disturbance is a sample suction σ'_s/σ'_{ps} on the order of 0.25–0.5 for shallow samples in moderately over-consolidated clay and 0.05–0.25 for deeper samples in lightly overconsolidated clay. Since the disturbed sample is effectively overconsolidated, a dilative response and strength gain occurs during shearing. Even then, the strength measured in an unconfined compression (UUC) test falls well short of the *in situ* TC strength. Several factors tend to mitigate this disparity between a strength measured in UUC test and the *in situ* strength. Firstly, as discussed in Section 2.1.6, the DSS strength s_{uDSS} is typically on the order of two-thirds of the TC strength s_{uTC}. Since $s_{uDSS} \approx s_{u-avg}$ is frequently considered as a suitable "average" strength for use in design, the UUC strength may coincidentally provide a realistic measure of average strength. Secondly, the standard UU test procedure shears the specimen at a relatively high strain rate (1%/minute). The increase in measured strength associated with increased strain rate tends to further narrow the gap between disturbed and *in situ* strength. Thus, use of the UUC strength requires an implicit assumption that anisotropy and strain rate effects offset the strength loss owing to sampling disturbance, an assumption that is questionable in view of the erratic and unpredictable nature of disturbance. For this reason, a more prudent course is to reconsolidate the soil to a state more representative of the *in situ* stress state prior to undrained shearing.

Specimen quality designation (SQD) is a widely used approach for assessing sample disturbance (Terzaghi et al., 1996). This approach is based the vertical strain ε_{v0} measured when loading a soil specimen to its *in situ* vertical effective stress σ'_{v0} in a one-dimensional consolidation test, with increased sample disturbance resulting in increased ε_{v0} (Andresen and Kolstad, 1979). Sample quality SQD of B or better is considered necessary for reliable laboratory results. Along the same lines, Lunne et al. (1997a) relate sample quality to the change in void ratio $\Delta e/e_0$ associated with one-dimensional consolidation to σ'_{v0}, with OCR also being considered in the sample evaluation. Table 2.3 summarizes both methods. Other procedures and tests for assessing sample quality include radiography (Ladd and DeGroot, 2003), strength index tests such as the Torvane and UUC test, measurement of sample suction (DeGroot et al., 2010), and shear wave velocity (Landon et al., 2007). Radiography (Figure 2.12) is a particularly useful tool for both assessing sample disturbance and identifying the optimal locations within the sample for performing tests.

Large-diameter long piston cores (Jumbo Piston Corer, JPC) can provide an alternative to mud rotary drilling in soft deep water sediments. The JPC comprises a long steel barrel with a 0.102 m PVC inner diameter liner. The barrel length varies according to specific design with 25–30 m reported by Silva et al. (1999) and 20–25 m reported by Young et al. (2000) and Wong et al. (2008). The piston is maintained stationary, while the barrel free falls to its final penetration depth. The JPC is a very cost effective means of sampling in deep water, although the sampling depth is insufficient for certain anchor systems, such as piles for TLPs. Studies of JPC performance showed that sample quality (assessed in terms of $\Delta e/e_0$) in the fair to excellent category is possible (Borel et al., 2002). Comparative studies in clays from offshore West Africa (Wong et al., 2008) showed favorable comparisons of JPC samples to those obtained by rotary drilling. Considering UU strength as a measure of sample quality, JPC sampling appeared to produce a slightly higher sample quality.

Table 2.3 Evaluation of sample quality

Terzaghi et al. (1996)		Lunne et al. (1997a)		
ε_{v0} (%)	SQD	$\Delta e/e_0$ OCR = 1–2	$\Delta e/e_0$ OCR = 2–4	Rating
<1	A	<0.04	<0.03	Very Good to excellent
1–2	B	0.04–0.07	0.03–0.05	Good to fair
2–4	C	0.07–0.14	0.05–0.10	Poor
4–8	D	>0.14	>0.10	Very poor
>8	E			

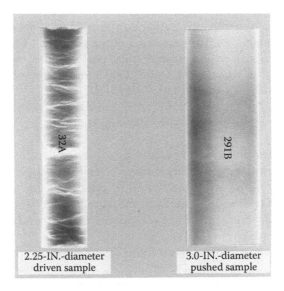

Figure 2.12 Radiography for assessing sample condition. (Courtesy Furgo-McClelland Marine Geosciences, Inc.)

2.1.13 Sample reconsolidation

As is evident from the preceding discussion, undrained testing of a soil specimen at its postsampling stress state σ'_s seriously misrepresents the actual undrained shear behavior. Bjerrum (1973) therefore proposed laboratory reconsolidation of the soil to a stress σ'_{vc} equal the *in situ* stress σ'_{v0}. This procedure is depicted in the conceptual sketch (Ladd et al., 1977; Ladd, 1991) of a one-dimensional compression curve shown in Figure 2.13. Recompression to σ'_{v0} inevitably densifies the soil, but Berre and Bjerrum (1973) maintain

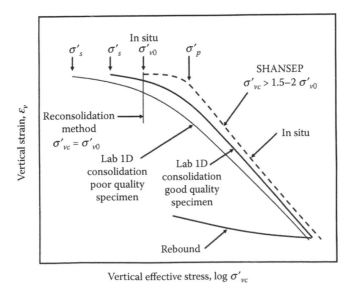

Figure 2.13 Sample reconsolidation prior to undrained shearing.

that, if the volumetric strains are less than 1.5%–4%, the reduction in shearing resistance owing to destructuring of the soil during sampling more than offsets the increase in shearing resistance owing to reconsolidation. Figure 2.13 shows the densification of the soil specimen to be relatively benign for overconsolidated soils, but becomes increasingly severe as the OCR approaches unity. For this reason, Ladd (1991) does not recommend the reconsolidation method for truly normally consolidated soils.

An alternative method is the SHANSEP approach developed by Ladd and Foott (1974) and Ladd (1991). The method employs the following steps:

1. Establish the preconsolidation stress profile, σ'_p versus depth.
2. Perform CK_0U tests at a consolidation stress σ'_{vc} well (Figure 2.13) beyond the preconsolidation stress σ'_p. Strength tests should be conducted for a normally consolidated stress state and on rebounded specimens measured overconsolidated behavior.
3. Use the strength data from Step 2 to back-calculate the parameters S and m in Equation 2.7.
4. Use the results of Steps 1 and 3 to establish the strength profile s_u versus depth.

The SHANSEP method is predicated on normalized soil behavior, in particular, that s_u/σ'_{vc} is constant for a given OCR. Therefore, the method is not well suited to highly structured, sensitive soils and cemented soils, for which normalized behavior cannot be assumed. Additionally, since the SHANSEP description of overconsolidated behavior is based on mechanical overconsolidation of laboratory specimens, the method is questionable for field situations where overconsolidation is due to other mechanisms (see Section 2.1.2). Nevertheless, the SHANSEP method has some strong advantages. Firstly, it is reliable for truly normally consolidated soils, since strength estimates are not affected by densification that occurs during recompression. Secondly, poorer sample quality is tolerable, an important consideration in deep water projects where obtaining high quality samples can be difficult. Thirdly, a single test series (i.e., strength evaluation at different OCRs) is applicable to a range of consolidation stress σ'_{vc}, which allows a single set of SHANSEP parameters (S, m) to be used not only to predict the *in situ* strength profile versus depth, but to predict future changes in strength that may occur as consolidation occurs in response to applied loads.

Since recompression well past the preconsolidation stress σ'_p destroys cementation and interparticle bonds, SHANSEP tends to provide low estimates of undrained shear strength. Thus, SHANSEP is sometimes said to provide "conservative" estimates of strength. This is generally true in regard to anchor load capacity calculations. However, the other side of anchor design is installation, for which underestimates of strength can lead to serious underestimates of resistance to anchor penetration—more suction caissons have failed during installation than during operational loading. Thus, the dual requirements in anchor design to both install the anchor and provide adequate pullout capacity during operational loading impose a requirement to evaluate the strength profile as accurately as possible.

2.1.14 *In situ* tests

In spite of the difficulties associated with sampling disturbance, soil characterization by means of sampling and laboratory testing offers the advantages of well understood boundary conditions, relatively uniform (or at least well understood) fields of stress and strain within the specimen, and well controlled drainage conditions. Since soil samples can be viewed through direct visual observation or radiography, little uncertainty exists in regard to soil type or macrofabric. Accordingly, sampling/lab testing arguably provides the most reliable measurements of stress–strain-strength behavior in fine-grained soils. However, the high cost and time-consuming nature of sampling and testing typically limits laboratory strength determinations to a limited number of discrete locations along the depth of a given boring. Additionally, drilling a large number of borings to evaluate horizontal variability in soil properties can be prohibitively expensive.

In situ testing is an important complementary approach to the sampling and lab testing approach described in previous sections. With the exception of the field vane (FV), continuous vertical profiling is possible. They can generally be conducted relatively rapidly and inexpensively, so greater horizontal coverage is economically feasible. Some *in situ* strength tests, the FV and flow-around penetrometers, have capabilities for measure both peak and remolded strength to determine soil sensitivity S_t. *In situ* tests are not without their limitations. The conditions of drainage are frequently uncertain, the stress fields associated with a penetrating probe are complex, and *in situ* measurements are influenced not only by undrained shear strength, but by soil properties that can include sensitivity, stain rate effects, and elastic behavior. With multiple soil properties affecting a single measurement, unique correlations to strength are not strictly possible and uncertainty is inherent to *in situ* test interpretation. These limitations can be mitigated by correlating *in situ* test measurements to nearby laboratory test measurements to develop site-specific relationships of shear strength to the *in situ* test measurement under consideration. Given that lab testing and *in situ* testing both have their own advantages and disadvantages, the two approaches are arguably best used as complementary tools for characterizing cohesive soil profiles, with lab testing providing high-quality stress–strain-strength measurements at relatively widely spaced locations and *in situ* measurements providing information on spatial variability of strength.

Piezocone. The standard piezocone (Cone Penetration Test [CPT] with pore pressure [U] measurements [CPTU]) has a cross-sectional area of 10 cm² with a tip angle of 60°. A somewhat larger version with a cross-sectional area of 15 cm² is commonly used in offshore applications. The probe is jacked in the ground at a penetration rate of 2 cm/second. Continuous records of three measurements are collected: the tip resistance q_c, the sleeve friction f_s and pore pressure at the shoulder of the tip u_2. A recessed area exists at the top of the cone tip to accommodate the friction sleeve shown in Figure 2.14a. Therefore, cone tip resistance should be corrected for water pressure acting at the top of the cone tip:

$$q_t = q_c + u_2(1 - a) \tag{2.15}$$

The net area ratio a is nominally equal to the ratio of inner to outer area $(D_i/D)^2$, but it should be measured in a pressure vessel. The CPTU test is used for soil classification (Robertson, 1990; Lunne et al., 1997b), estimating undrained shear strength in cohesive soils, and estimating friction angle in sands. For various test interpretation purposes, the pore pressure measurement is frequently expressed in terms of a pore pressure parameter B_q defined as follows:

$$B_q = \frac{u_2 - u_0}{q_t - \sigma_{v0}} \tag{2.16}$$

where u_0 is hydrostatic pressure and σ_{v0} is total vertical stress. Figure 2.14b shows the use of the B_q and q_t to group the soil into the zones defined in Table 2.4.

Undrained shear strength of cohesive soils is typically estimated from the following equation:

$$s_{uavg} = \frac{q_t - \sigma_{v0}}{N_{kt}} \tag{2.17}$$

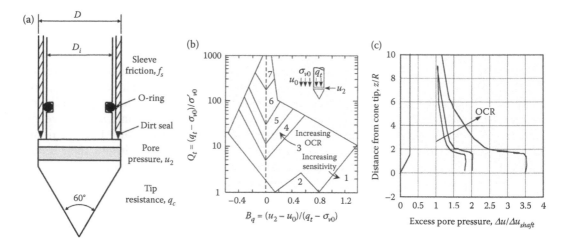

Figure 2.14 Piezocone test (CPTU). (a) Piezocone, CPTU, (b) CPTU soil classification zones. (After Lunne T, Robertson PK, and Powell JJM, 1997b, *Cone Penetration Testing in Geotechnical Practice*, Chapman & Hall, London.) and (c) pore pressure distributionalong shaft of CPT.

It is important to specify which shearing mode to which cone tip resistance is being correlated. The most common correlations are to TC s_{uTC} and the average of compression, extension, and DSS modes, s_{uavg}. The latter strength is typically considered as representative of the DSS strength, $s_{uDSS} \approx s_{uavg}$. Unless otherwise stated in the discussion that follows, N_{kt} refers to average peak undrained shear strength. Reported values of the cone factor generally in the range $N_{kt} = 10$–20 (Aas et al., 1986). A more recent study by Low et al. (2010) more or less confirms this range ($N_{kt} = 10.56$–17.39), although from their statistical study they recommend using $N_{kt} = 13.5 \pm 2.0$. Uncertainty in CPTU interpretation can be reduced considerably through site-specific correlations to available laboratory measurements. However, as pointed out by Ladd and DeGroot (2003), variability in site-specific determinations of N_{kt} can occur between different devices and operators, and at different locations within the same deposit. Insight into the variability in N_{kt} is aided by noting the following sources of uncertainty in its determination:

1. The CPT involves a mode of penetration (discussed in Section 3.4) for which both soil shearing resistance s_u and elastic properties of the soil influence penetration resistance. Elastic behavior can roughly be characterized in terms of a rigidity index $I_r = G/s_u$, where G is shear modulus. Taking a range $I_r = 100$–500 as broadly representative of

Table 2.4 Soil classification from CPTU test

Soil zone	Description
1	Sensitive, fine grained
2	Organic soils—peats
3	Clays: clay to silty clay
4	Silt mixtures: clayey silt to silty clay
5	Sand mixtures: silty sand to sand silty
6	Sands: clean sands to silty sands
7	Gravelly sand to sand

clays in conjunction with the analysis of probe penetration by Baligh (1985) shows that elastic effects can cause about 30% variation in the bearing factor N_{kt}. More refined analyses that introduce nonlinear stress–strain behavior into the analysis, if anything, increase the potential variability in the cone tip resistance factor. Since rigidity index varies with OCR, nonuniqueness of N_{kt} can be expected even a homogeneous soil profile.

2. Use of Equation 2.17 requires accurate determination of total vertical stress σ_{v0} which, in turn, requires accurate measurement of soil unit weight in the entire soil column. Reliable unit weight data typically comes from a borehole, which the *in situ* test is intended to replace. Therefore, determination of σ_{v0} typically involves some degree of guesswork or extrapolation of data from the nearest borehole.

3. Tip resistance measurements should be corrected using Equation 2.15; thus, determination of q_t relies on accurate pore pressure measurements. Possible malfunctions (e.g., desaturation of the filter) may simply produce incorrect measurements of u_2. Additionally, the pore pressure measurement u_2 is positioned in the vicinity of a singularity in the stress field at the shoulder of the cone. The high gradients of pore pressure in this region (Figure 2.14c) introduce considerable uncertainty on the validity of the correction.

The first two sources of variability in N_{kt} are inherent to the mechanics of cone penetration, making it unlikely that improved hardware or test interpretation can improve the accuracy of the strength prediction. By contrast, the need for a pore pressure correction can be avoided by modifying the cone design to omit the friction sleeve. While this would arguably detract from the capabilities of the CPTU for soil classification, an alternative means of soil classification is still possible based on pore pressure measurements, as shown in Figure 2.14b.

Flow-around penetrometers. The T-bar was proposed as an *in situ* strength measurement tool by Randolph et al. (1998). The ball penetrometer (Randolph et al., 2000) follows a similar principle, but substitutes the cylindrical probe tip of the T-bar for a spherical tip. The T-bar and ball penetrometers shown in Figure 2.15 penetrate the soil in a flow-around mode, where resistance to penetration is independent of both the elastic properties of the soil and total vertical stress σ_{v0}. Thus, these types of penetrometers avoid much of the uncertainty inherent in the cone penetration test in regard to correlating penetration resistance to undrained shear strength. Additionally, these penetrometers may be cyclically loaded to obtain estimates of remolded undrained shear strength, a capability not offered by cone penetration testing. Overall, these advantages will generally lead to flow-around penetrometers

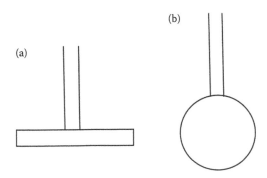

Figure 2.15 Flow-around penetrometers. (a) T-bar and (b) ball.

being the *in situ* tool of choice in soft clay profiles (Lunne et al., 2011). In regards to the limitations of this type of test, flow-around penetrometers lack the versatility of the piezo-cone in that they are not used for soil classification or for characterizing sands. Therefore, in heterogeneous deposits, the piezocone may still be a competitive alternative, in spite of its limitations for undrained shear strength determination.

A typical T-bar (Figure 2.15a) is a 4-cm diameter by 25-cm long cylinder. The ball pen-etrometer (Figure 2.15b) is spherical with a diameter of 11.3 cm. Both penetrometers are advanced at the same rate as the piezocone, 2 cm/second, which should ensure undrained penetration. The recommended area of the connecting shaft should be less than 15% of the projected T-bar or ball area. For a cylinder of length L and diameter D, to average und-rained shear strength s_u relates to total measured load P_{ult} as follows:

$$s_{uavg} = \frac{P_{ult}}{LDN_{\text{Tbar}}} \tag{2.18}$$

As stated earlier, penetration measurements can be correlated to either to either TC or average strength, but the current discussion focuses on average strength. Randolph and Houlsby (1984) provide plastic limit solution for the bearing factor N_{Tbar} varying between 9.14 and 11.94 as a function of surface roughness. However, owing to strain softening and rate effects, the actual T-bar penetration mechanism is considerably more complex than that assumed in the rigid-plastic Randolph-Houlsby model. From careful correlations to high-quality laboratory and vane shear tests, Low et al. (2010) recommend a mean N_{Tbar} value of 12 with a range of 10–14 for interpreting T-bar measurements.

Estimation of undrained shear strength from a ball penetrometer (Figure 2.15b) of radius R parallels that for the T-bar. The relevant bearing factor now is defined as N_b:

$$s_{uavg} = \frac{P_{ult}}{\pi R^2 N_b} \tag{2.19}$$

Plastic limit analyses by Randolph et al. (2000) indicate that ball penetrometer resistance should exceed that of the T-bar by 17%–26%, depending on the yield criterion used in the analysis. However, laboratory measurements indicate only small differences in average bearing pressure between the spherical and cylindrical geometries; reasons for the discrep-ancy are unclear. Current guidance recommends use of the same bearing factor for the ball and T-bar penetrometers, $N_b \approx N_{Tbar}$.

Flow-around penetrometers are well suited to cyclic loading to measure remolded strength. The usual practice uses 10 load cycles although, strictly speaking, the soil is actually still softening at a low rate at this stage of cycling. Remolded strength computation is along the same lines as that for intact strength:

$$s_{ur} = \frac{P_{ult}}{LDN_{Tbar-r}} \tag{2.20}$$

For theoretical reasons discussed by Randolph and Andersen (2006), the bearing fac-tor N_{Tbar-r} during remolding differs than that for penetration into intact soil. Additionally, N_{Tbar-r} depends on the method of measuring remolded strength; the possibilities include UU testing, the fall cone, and the vane. Low et al. (2010) provide $N_{Tbar-r} = 14.0$ for correlation to the vane shear strength. Details on correlation to other testing modes are provided by Low et al. (2010) and Lunne et al. (2011).

Field vane. The FV test uses a vane with diameter $d = 40–65$ mm and height-to-diameter ratio $h/d = 2$. To minimize disturbance, the blade thickness e should produce a perimeter ratio $4e/\pi d$ less than 3% (Lunne et al., 2011). To obtain intact undrained shear strength s_u, the vane is rotated at a rate of 0.1–0.2°/second until a peak torque is measured. Remolded undrained shear strength s_{ur} should be measured by additional rotation at a rapid rate, 24°/second, for at least 10 rotations. Lunne et al. (2011) report that offshore equipment for vane testing does not have the capability for such a rapid rotation rate, so it is impractical to do 10 rotations. Consequently, offshore FVT measurements tend to overestimate remolded undrained shear strength. Both intact and remolded undrained shear strength are computed from measured torque from the equation

$$s_{uFV} = \frac{T}{\pi \left(\dfrac{d^2 h}{2} + \dfrac{d^3}{6} \right)} \tag{2.21}$$

Vane shear involves complex shearing modes, but it is typically used as a basis for estimating average undrained shear strength. In land-based applications, the FV strength is typically adjusted using a correction by Bjerrum (1973) based on calibration of s_{uFV} to observed slope failures. This correction is not normally applied in offshore practice (DeGroot et al., 2010; Lunne et al., 2011), possibly because increased disturbance during vane insertion in offshore site investigation already leads to reduced estimates of strength.

In contrast to the piezocone and flow-around penetrometers, the FV test does not provide continuous strength profiling. Interpreted FV strength profiles also show more scatter, since they are affected by disturbance during vane insertion. Accordingly, Lunne et al. (2011) recommend the use of flow-around penetrometers as the primary tool for *in situ* undrained strength determination in soft clays, with the FV test being considered as a supplementary tool to increase the reliability in the strength estimate.

Free-fall CPT. The "CPT Stinger" (Young et al., 2011) is a free falling cone penetrometer that is deployed in soft clay profiles using the same equipment as the JPC described in Section 2.1.12. The test uses the same hardware described earlier for the CPTU test and records tip resistance q_c, resistance f_s, and pore pressure u_2. The test has two stages: (1) free fall into the seabed at impact velocities approaching 10 m/second and (2) additional quasi-static penetration at the standard CPTU insertion rate of 2 cm/second. The depth to which penetration resistance data can be obtained depends on the barrel length of the device. For the longest barrel reported by Young et al. (2011) dynamic insertion extends to a depth of 18.3–19.2 m, which an additional 15.2 m of penetration obtained by quasi-static penetration. This provides a total penetration of 33.5–34.4 m. Multiple tests at adjacent locations can be performed using long and short barrels to create overlapping depths of static and dynamic penetration, from which dynamic penetration resistance can be correlated to static resistance. Since the apparatus uses JPC equipment, it frequently proves convenient to collect JPC samples for laboratory testing, which provides an additional source of data for site-specific correlations of penetration resistance to soil strength. The dynamic measurements involve variable penetration velocity as the probe decelerates, requiring careful attention to the strain rate correction. Buhler and Audibert (2012) investigated three strain rate correction methods, the logarithmic and power law equations described in Section 2.1.9, as well as a square root power law method proposed by Abelev and Valent (2009). The latter two methods were most promising, although none proved capable of accurately correcting strengths over the full range of penetration velocities. In spite of the need for improving the rate correction, penetration measurements from the free-fall CPT compare favorably to those from conventional quasi-static penetration measurements (Jeanjean et al., 2012).

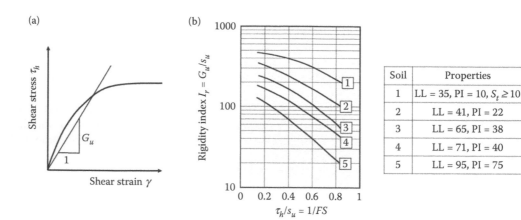

Figure 2.16 Undrained secant stiffness of normally consolidated soil. (a) Secant shear modulus and (b) typical variation in I_r with stress level. (After Foott R and Ladd CC, 1981, *J Geotech Eng Div*, ASCE, 107(GT8), 1079–1094, with permission from ASCE.)

2.1.15 Undrained stiffness

Prediction of immediate deformation under applied loading frequently proceeds in terms of a simplified linearly elastic analysis using undrained elastic parameters, the Young's modulus E_u or shear modulus G_u. Since undrained deformation implies pure distortion with zero volume change, Poisson's is considered as $\mu = 0.5$ or, depending on analysis method, the system may be modeled as nearly incompressible with $\mu = 0.49$. Additionally, various aspects of anchor performance are frequently characterized in terms the shear modulus normalized by undrained shear strength; that is, the rigidity index $I_r = G_u/s_u$. Selection of a representative shear modulus is complicated by the nonlinear stress–strain relationship of soils; therefore, elastic moduli are commonly expressed in terms of a secant stiffness associated with a given level of strain. Shear strain is seldom known in advance of the analysis. However, shear strain times rigidity index is simply the inverse of the local factor of safety for the soil element in question, $\gamma I_r = \tau_h/s_u = 1/FS$. Thus, the inverse of the factor of safety provides a useful basis for estimating an appropriate representative strain level for selection of elastic parameters. The computed global factor of safety is used for this purpose. Foott and Ladd (1981) present curves for stiffness degradation versus the inverse of safety factor for various soil types; Figure 2.16 shows some selected curves. Elastic stiffness is also dependent on OCR. Figure 2.17 shows selected curves from Foott and Ladd (1981). It should also be noted that the determination of stiffness using the SHANSEP method (Section 2.1.13) tends to underestimate stiffness in comparison to the recompression method (Ladd and DeGroot, 2003).

2.2 CONSOLIDATION

Drained behavior of clays is relevant to several aspects of anchor performance:

- *Long-term deformation under sustained loading.* Although anchor movements in the seabed are small relative to mooring line motions, the design basis for anchor systems frequently places limits on anchor displacements. Under sustained loading, drainage

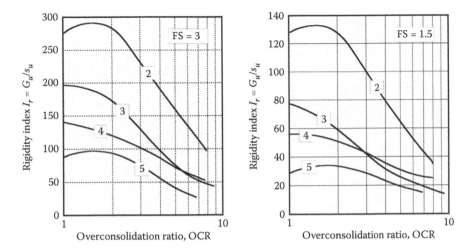

Figure 2.17 Undrained secant stiffness of normally overonsolidated soil. (From Foott R and Ladd CC, 1981, *J Geotech Eng Div*, ASCE, 107(GT8), 1079–1094, with permission from ASCE.)

will produce displacements that will likely far exceed undrained displacements associated with short-term loading.

- *Dissipation of excess pore pressures.* Regain of soil strength following anchor installation, termed "setup," is largely controlled by the decay of excess pore pressures associated with the process of consolidation. In addition, redistribution of negative excess pore water pressures under sustained loading can lead to loss of "suction"; that is, swelling of the soil with a concomitant reduction in soil strength, which adversely affects anchor uplift capacity.
- *Interrelationship to undrained behavior.* A noted throughout the previous section, undrained stress–strain-strength behavior is driven by the conditions of consolidation. Most notable are the preconsolidation stress σ'_p, which is measured in the one-dimensional consolidation test discussed in this section, and the consolidation stress σ'_{vc}, which evolves as consolidation occurs.

2.2.1 Primary consolidation deformations

During primary consolidation under one-dimensional loading, vertical effective stress σ'_v relates to vertical strain ε_v by the following well-known semi-logarithmic equation:

$$\varepsilon_v = \frac{C_r}{1+e_0}\log_{10}\left(\frac{\sigma'_p}{\sigma'_{v0}}\right) + \frac{C_c}{1+e_0}\log_{10}\left(\frac{\sigma'_{vf}}{\sigma'_p}\right) \qquad (2.22)$$

where C_r is the recompression index, C_c is the compression index, e_0 is initial void ratio and σ'_p is preconsolidation stress. The subscripts "0" and "f" refer to effective stress levels at the beginning and end of the load episode under consideration. Deformation or settlement is evaluated by integrating the strain ε_v along any path of interest.

The parameters C_r, C_c, and σ'_p are traditionally measured in the incrementally loaded (IL) one-dimensional consolidation, or oedometer, test. In this test a 24-hour consolidation time interval permitted after the application of each load increment. The increase in stress

$\Delta\sigma'_v$ in a given increment of applied loading is expressed in terms of a load increment ratio $LIR = \Delta\sigma'_v/\sigma'_v$. Conventional testing typically uses $LIR = 1$, although this may be reduced (say to $LIR = \frac{1}{2}$) to get better resolution in the consolidation curve, particularly in the vicinity of the preconsolidation stress. Disturbance effects preclude accurate determination of C_r based on measurements from the initial stage of recompression, so the test procedure typically employs an unload–reload cycle at an elevated stress level to obtain this parameter. This procedure implicitly assumes that C_r is independent of stress level. Hysteresis and plastic strain accumulation occurring during the unload–reload cycle are typically neglected in monotonic loading applications; however, these effects can be significant in situations involving cyclic loading as discussed in Section 3.3.3. The parameters C_c and C_r are determined from direct measurement of the slopes of the consolidation curve. A number of methods have been proposed to determine the preconsolidation stress σ'_p. In spite of the subjective judgment required in its implementation, the Casagrande construction appearing in most elementary texts on soil mechanics (e.g., Holtz and Kovacs, 1981) continues to be the most widely used method.

The conventional IL consolidation test has limitations for obtaining a high resolution measurement of the consolidation curve, particularly in regard to obtaining an accurate measurement of σ'_p. For this reason, the constant rate of strain (CRS, Wissa et al., 1971) consolidation test, which provides continuous measurement of stress, is arguably a preferable approach (DeGroot et al., 2010). The CRS test has one-way drainage conditions. Guidance on the CRS test varies, but the excess pore pressure u_e at the base of the soil specimen maintained at about 15% of σ_v. Typical strain rates in normally consolidated clays are 0.5%–1%, which is sufficiently rapid to avoid secondary creep, but slow enough to avoid large variations of stress and void ratio in the specimen (Ladd and DeGroot, 2003).

Consolidation deformation are sometimes expressed in terms of a recompression ratio, RR, and a compression ratio, CR, defined as follows:

$$RR = \frac{C_r}{1 + e_0} \tag{2.23}$$

$$CR = \frac{C_c}{1 + e_0} \tag{2.24}$$

Characterizing primary consolidation deformations in terms compression ratios is frequently useful for judging the validity of laboratory test results or for preliminary calcultions, since they are relatively well bounded for inorganic, noncalcareous soils. Compression ratios are generally in the range $CR \approx 0.25 \pm 0.10$ for lean clays and for 0.35 ± 0.10 for fat clays. The recompression index is typically on the order of 1/10–1/5 of the compression index, $C_r/C_c = RR/CR \approx 0.1$–0.2.

Equation 2.22 is strictly applicable to one-dimensional loading under K_0-consolidation conditions. This type of loading may be considered as a rough approximation to relatively simple cases, such as consolidation above and below a vertically loaded horizontal plate anchor. Actual consolidation problems involving anchors typically involve more complex loading and geometries that are not amenable to analysis in a one-dimensional framework (Figure 2.18). In these cases, generalized constitutive models of soil behavior (Sections 3.3.2 and 3.3.3) are employed, with volumetric response characterized in a manner that parallels that described by Equation 2.22. However, the relationship is now cast in terms of the relationship between mean effective stress σ' (plotted on a natural logarithmic scale) and

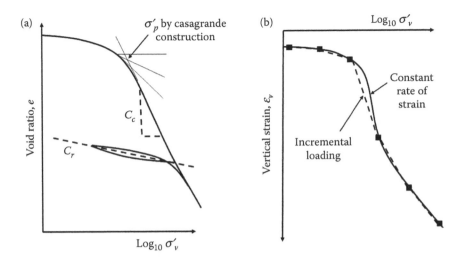

Figure 2.18 One-dimensional consolidation behavior. (a) Basic test interpretation and (b) CRS vs IL tests.

volumetric strain ε_{vol} (Figure 2.19). Experimental evidence (Schofield and Wroth, 1968) demonstrates that the slope of the VCL is independent of the consolidation stress ratio $K = \sigma'_h / \sigma'_v$. Thus, the slope of VCL measured under K_0 consolidation conditions in the one-dimensional consolidation test accurately represents the slope of the VCL under isotropic loading. Conversion from a common to a natural logarithmic basis (by a factor 0.434) is shown in Figure 2.19.

2.2.2 Secondary compression

The compression described in the previous section occurs while excess pore pressures dissipate (Figure 2.20a). The continued compression that occurs after excess pore pressures have substantially dissipated is termed secondary compression, or drained creep. The change in

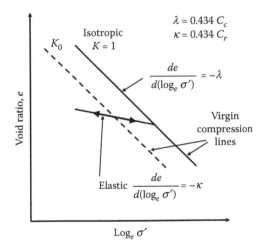

Figure 2.19 Description of volumetric behavior of fine-grained soils.

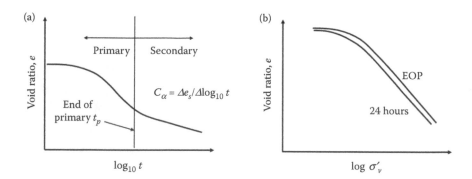

Figure 2.20 Secondary consolidation. (a) Settlement versus log time and (b) consolidation curves at EOP and 24 hours.

void ratio Δe_s secondary consolidation versus time (t) relationship is well described by a semi-logarithmic relationship:

$$\Delta e_s = C_\alpha \Delta \log_{10}(t) \tag{2.25}$$

where C_α is defined as the coefficient of secondary consolidation. Some controversy exists as to when secondary consolidation initiates (Ladd et al., 1977). One school of thought (Hypothesis A) holds that secondary compression initiates following the completion of primary consolidation, while the second school (Hypothesis B) maintains that secondary compression occurs throughout primary consolidation. The assumption as to when secondary compression initiates has little impact on interpretation of consolidation tests on thin specimens, but can significantly affect field scale settlement predictions (Ladd and DeGroot, 2003). An implication of Hypothesis A is that the end-of-primary (EOP) consolidation curve is unique and independent of clay layer thickness. By contrast, Hypothesis B implies that the end of primary consolidation (EOP) curve depends on layer thickness. Discussions of arguments supporting the two hypotheses are contained in various sources (e.g., Mesri, 2001).

The IL consolidation test generates the curves illustrated in Figure 2.20a. Secondary consolidation behavior is generally well defined; thus, the IL test is well suited to determining C_α. Having said that, in a majority of anchor design applications the preconsolidation stress σ'_p is typically more critical than C_α. Accordingly, unless drained creep is identified as a critical concern, the advantages of the CRS test will typically outweigh the advantage of being able to measure C_α in the IL test. If the IL test is selected, drained creep effects should be considered in the interpretation of σ'_p. For example, a conventional IL oedometer test with a 24-hour duration of load increments will typically involve one to two log cycles of secondary compression; thus, the 24-hour consolidation curve will be offset from the EOP curve as depicted in Figure 2.20b. The estimated preconsolidation stress will therefore depend on which curve is used; Ladd et al. (1977) report that σ'_p determined from the 24-hour curve will be about 10%–20% lower than that obtained from the EOP curve. Ladd and DeGroot (2003) provide guidance on determining the EOP curve.

The coefficient of secondary consolidation C_α has long been recognized to be proportional to the compression index C_c. Summarizing an extensive body of research on this topic for various geotechnical materials, Mesri (2001) reports $C_\alpha/C_c = 0.04 \pm 0.01$ for inorganic clays and silts, and $C_\alpha/C_c = 0.05 \pm 0.01$ for organic clays and silts. The averages reported here apply to conditions of K_0-consolidation. The rate of drained creep actually varies according to the level of shear or deviator stress (q/q_f, where q_f is the failure stress) in the soil. Ladd et al. (1977) report data indicating that the drained creep rate under isotropic loading is about one-tenth that under K_0-loading, and increases linearly with increasing shear stress up to about $q/q_f = 0.7$. Drained creep becomes more rapid as the failure stress q_f is approached.

2.2.3 Coefficient of consolidation and permeability

As discussed in more detail in Chapter 3, consolidation is actually a coupled process involving unsteady flow of the pore water and the mechanical response of the soil skeleton. In the case of one-dimensional vertical consolidation, the problem decouples to a diffusion equation of the following form:

$$c_v \frac{\partial^2 u_e}{\partial^2 z} = \frac{\partial u_e}{\partial t} \tag{2.26}$$

where u_e is excess pore pressure, z is vertical coordinate, t is time, and c_v is the coefficient of consolidation (units of length squared over time). While the decoupled form expressed by Equation 2.26 is strictly valid for one-dimensional consolidation and several other special cases, the formulation is commonly extended to two and three dimensions. While such analyses can provide meaningful solutions, the limitations of such analyses (Section 3.5.3) should be kept in mind.

The coefficient of consolidation c_v is actually not a fundamental soil parameter, but the product of permeability k_v and the coefficient of volume change m_v as defined in Equation 2.27:

$$c_v = \frac{k_v}{m_v \gamma_w} \tag{2.27}$$

The coefficient of volume change m_v is the slope of the strain–stress curve plotted on a natural scale. It is related to the compression index C_c (or recompression index C_r) illustrated in Figure 2.21a and defined by Equation 2.28:

$$m_{vNC} = \frac{\Delta \varepsilon_v}{\Delta \sigma_v'} = 0.434 \frac{C_c}{(1+e_0)\sigma_v'} \quad \text{Normally consolidated}$$

$$m_{vOC} = \frac{\Delta \varepsilon_v}{\Delta \sigma_v'} = 0.434 \frac{C_r}{(1+e_0)\sigma_v'} \quad \text{Overconsolidated} \tag{2.28}$$

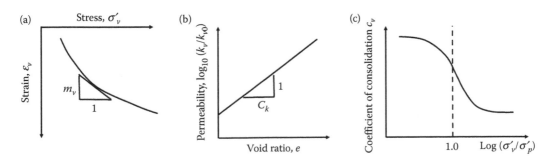

Figure 2.21 Coefficient of consolidation. (a) Compressibility, (b) permeability, and (c) coefficient of consolidation.

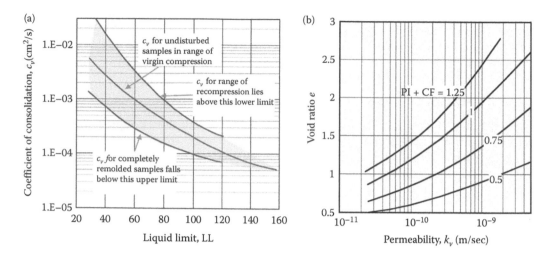

Figure 2.22 Correlation of flow properties to index properties. (a) Coefficient of consolidation versus liquid limit (Adapted from Navfac DM-7.I, 1986, *Soil Mechanics, Facilities Engineering Command*, U.S. Dept. of the Navy, Alexandria, VA, 364p.) and (b) permeability versus liquid limit and clay fraction.

Clay permeability k_v strongly correlates to void ratio. Tavenas et al. (1983a,b) and Terzaghi et al. (1996) describe the relationship in terms of the following exponential equation:

$$k_v = k_{v0}10^{(e-e_0)/C_k} \tag{2.29}$$

where k_{v0} is permeability at the *in situ* void ratio e_0, and the empirical coefficient for soft clays is $C_k \approx (0.45 \pm 0.1)e_0$. The net effect of the dependencies in Equations 2.28 and 2.29 is a relatively modest dependence of c_v on effective stress σ'_v, except in the transition from an overconsolidated to a normally consolidated state, as illustrated in Figure 2.21c. In addition to significant differences in c_v according to whether the soil is normally or overconsolidated, c_v also correlates well to the soil liquid limit LL. Navfac DM 7.1 (Figure 2.22a) shows empirical correlations of c_v to LL for normally and overconsolidated soils.

Laboratory determinations of c_v from IL oedometer tests employ the well-known Casagrande log-time and Taylor root-time graphical constructions presented in introductory soil mechanics textbooks (e.g., Holtz and Kovacs, 1981). The former bases the c_v estimate on the time to 50% consolidation t_{50}, while the latter uses t_{90}. Estimates of c_v by the Taylor method systematically exceed those based on the Casagrande method by a factor 2.0 ± 0.5. As noted in the previous section, a LIR = 1 with a 24-hour consolidation time involves 1–2 log cycles of secondary compression; thus, the soil will behave as overconsolidated during the early stages of consolidation in the following load increment. Ladd and DeGroot (2003) attribute the differences in the Taylor and Casagrande estimates of c_v to the effects of this secondary compression. As the CRS consolidation test avoids secondary compression, it arguably provides a preferable means of measuring c_v, albeit at the expense of the loss of ability to measure C_α.

While laboratory consolidation tests measure c_v, representative of vertical flow, the horizontal coefficient of consolidation c_h is more relevant to many actual field situations; for example, dissipation of penetration pore pressures surrounding a driven pile. In certain soils, such as varved clays, the ratio c_h/c_v can be as high as 5–10, but in marine clays it is essentially one (Ladd and DeGroot, 2003). If required, site-specific determinations of c_h are possible by performing consolidation tests on vertical specimens.

In coupled consolidation analyses (Section 3.5.2), hydraulic behavior of soil is expressed directly in terms of soil permeability k. Indirect determination of k is possible using Equation 2.27, but Tavenas et al. (1983a) report this approach to be questionable for structured clays, irrespective of the type of consolidation test (IL or CRS) or the method of interpretation (Taylor versus Casagrande). They report satisfactory measurements from both the constant and the falling head permeability tests. A somewhat contrary view is provided by Ladd and DeGroot (2003), who suggest that a properly conducted CRS test can provide satisfactory estimates of permeability. For estimating permeability from basic soil index properties, Tavenas et al. (1983b) correlate vertical permeability k_v to PI and the CF (CF = percentage of particles finer than 2 μm) as shown in Figure 2.22b.

2.2.4 *In situ* tests

Piezocone penetration induces high excess pore pressures in the region of soil disturbance surrounding the probe. When penetration is halted, these pore pressures dissipate at a rate controlled by the coefficient of consolidation of the soil. Since the pore water flows primarily in the horizontal direction, the horizontal coefficient of consolidation c_h is the dominant parameter influencing the rate of pore pressure decay. It should be noted that the cone penetration process in normally and lightly overconsolidated soils reduces the effective stress level in the soil in the vicinity of the cone, leading to an artificially overconsolidated stress state. Thus, the early stages pore pressure dissipation involve reloading of the soil, such that the rate of pore pressure decay is controlled by the overconsolidated coefficient of consolidation, c_{hOC} (Levadoux and Baligh, 1986). Thus, the coefficient of consolidation derived from a piezocone "holding test" largely represents c_{hOC}, irrespective of whether the soil is normally or overconsolidated.

Interpretation of piezocone holding test data requires an analytical framework that (1) provides realistic predictions of the initial distribution of excess pore pressures induces by cone penetration and (2) prediction of the rate of dissipation of pore pressure after penetration is interrupted. To predict initial pore pressures, Levadoux and Baligh (1986) employed the strain path method (Section 3.4.2), in conjunction with a total stress model (MIT-T1) capable of modeling the complex strain reversals occurring during penetration as well as stress-induced anisotropy and strain softening. To model the pore pressure decay process, they employed both coupled and uncoupled analyses of consolidation. For the case of the conventional 60° piezocone, they found that an uncoupled formulation (Section 3.5.2) is adequate for interpretation of holding tests. Figure 2.23 shows their solutions for pore pressure decay for measurement locations at the tip, face, shoulder and shaft of the cone. Baligh and Levadoux (1986) point out that measurements on the shaft actually produce the least scatter in interpreted c_{hOC} values, but the consolidation times (approximately 10 times slower than consolidation at the tip) require test durations that are unacceptable in most practical situations.

As discussed in Section 2.1.1, the time required for full dissipation of excess pore pressures surrounding the conventional piezocone can exceed 24 hours, which is impractical for most offshore site investigations. Fortunately, sufficiently accurate measurement of c_{hOC} is possible from incomplete dissipation measurements. Baligh and Levadoux (1986) report that interpretation dissipation measurements approximately 50% consolidation provide acceptable estimates of c_{hOC}. In the case of the Boston Blue Clay (c_{hOC} = .02–.04 cm²/second) test site where they conducted their tests, this corresponds to a holding times of approximately 10–15 minutes.

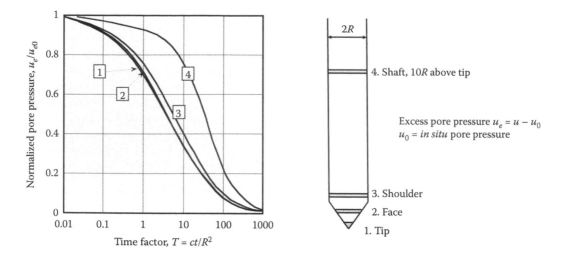

Figure 2.23 Piezocone holding test.

Subsequent studies Lacasse and Lunne (1982) compared laboratory coefficients of consolidation (essentially c_{hOC}) to those obtained from holding tests using the Baligh–Levadoux method of interpretation at two test sites, Onsoy and Drammen. They found excellent agreement between the holding test and laboratory measurements.

2.3 SAND BEHAVIOR

This section presents fundamental aspects of sand behavior that can be relevant to anchor design. Noting that undrained conditions may possibly prevail under certain conditions of seismic loading or rapidly applied loads on large anchors, the discussion will nonetheless focus on the more common case of fully drained loading. Sections 2.1.1 through 2.1.4 cover mechanical behavior of sands, while Section 2.1.5 addresses hydraulic behavior.

2.3.1 Drained stress–strain-strength behavior

Figure 2.24a illustrates basic aspect of sand stress–strain strength behavior. Shearing resistance q in dense sands rises to an early peak at low strain and softens to its ultimate value under increased strain. By contrast, in loose sands shearing resistance increases monotonically to an ultimate value in loose sands. As shown in the lower plots of volume change and void ratio versus axial strain, stress–strain response is closely linked to the volumetric response of the soil during shearing. Dense sands tend to dilate when subjected to shear, with the maximum rate of dilation ($-d\varepsilon_{vol}/d\varepsilon_a$) corresponding to peak shearing resistance, with an associated friction angle ϕ'_p. Under continued straining the rate of dilation tends to zero and the soil shears at constant volume. In a loose sand the volumetric behavior is completely contractive. A given sand placed at different initial densities will, under continuous shearing, approach a unique void ratio, with an associated unique constant volume friction angle ϕ'_{cv}.

Repeated tests at different consolidated stresses σ'_c permits construction of the strength envelope shown in Figure 2.24b. Equivalent representations of the strength envelope are commonly made in either $\sigma - \tau$ space or $p' - q$ space, with the latter being somewhat more

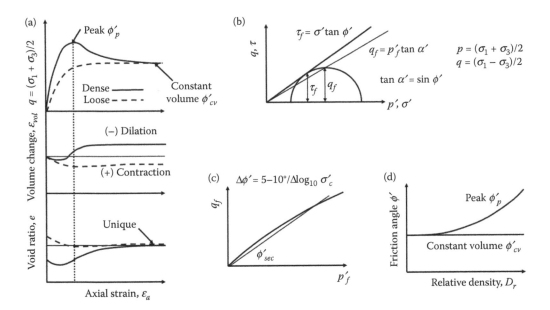

Figure 2.24 Drained stress–strain-strength behavior of sands. (a) Stress-strain behavior, (b) represetantion of strength envelope, (c) stress level dependency, and (d) effect of relative density.

convenient for data reduction. While the strength envelope is typically idealized as being linear, tests conducted over a wide range of confining stress σ'_c will actually show some curvature, which is attributed to particle crushing and suppression of dilation at elevated stress levels. For dense sands and gravels, the reduction in friction angle is typically on the order of 5–10° per log-cycle increase in stress, with σ'_c having little effect on the friction angle of loose sands (Ladd et al., 1977). Strength is normally reported in terms of secant friction (Figure 2.24c).

Cohesionless soil behavior is effectively described in terms of relative density D_r described below:

$$D_r = \frac{e_{max} - e}{e_{max} - e_{min}} \tag{2.30}$$

where e is void ratio, and e_{max} and e_{min} are reference void ratios for the soil in loose and dense states, respectively. Test procedures for determination of e_{max} and e_{min} vary amongst organizations, but ASTM specifies standard D 4253-16 (ASTM, 2016a) for maximum density and 4254-91 (ASTM, 2016b) for minimum density. As depicted in Figure 2.24d, peak friction angle increases with relative density, while (at a given consolidation stress σ'_c) the constant friction angle is independent of density. The constant volume friction angle ϕ'_{cv} derives from sliding friction plus interference; that is, particles moving around one another. These mechanisms plus dilation contribute the peak friction ϕ'_p. As discussed by Bolton (1986), ϕ'_{cv} is a function of mineralogy, while the dilatant component of strength, $\phi'_p - \phi'_{cv}$, is a function of relative density.

Most measurements of friction angle ϕ' are based on TC tests, where the intermediate principal stress equals the minor principal stress, $\sigma_2 = \sigma_3$. The relative magnitude of the intermediate principal stress σ_2 actually has some influence on friction angle ϕ'. Ladd et al. (1977) report that the plane strain friction angle exceeds that for TC by 4–9° in dense

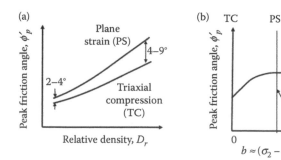

Figure 2.25 Effect of intermediate principal stress on friction angle. (a) Plane strain friction angle ϕ'_{ps} and (b) intermediate principle stress σ_2.

sands and by 2–4° in loose sands (Figure 2.25a). Defining an intermediate stress parameter $b = (\sigma_2 - \sigma_3)/(\sigma_1 - \sigma_3)$, under plane strain conditions at the failure state b is on the order of 0.25–0.4. Reported data consistently show ϕ' increasing between $b = 0$ and the plane strain condition. Ladd et al. (1977) report divergent trends are reported as b increases toward a TE state; however, Kulhawy and Mayne (1990) report that the friction angle in TE exceeds that in TC or, on average, $\phi'_{te} = 1.12\ \phi'_{tc}$.

2.3.2 Correlations to density and stress level

To predict the dilative component of frictional resistance, Bolton (1986) introduces a relative dilatancy index I_{RD}, which is a function of relative density D_r and stress level at failure p_f. In nondimensional form (Kulhawy and Mayne, 1990) the expression takes the following form:

$$I_{RD} = D_r[Q - \log_e(100p'_f/p'_a)] - R \tag{2.31}$$

The fitting parameter R equals unity, and Q is a function of mineralogy that equals 10 for quartz and feldspar, 8 for limestone, 7 for anthracite, and 5.5 for chalk. Bolton relates I_{RD} to the dilatancy component of friction through the following correlations:

$$\phi'_p - \phi'_{cv} = 5I_{RD} \quad \text{Plane strain compression} \tag{2.32}$$

$$\phi'_p - \phi'_{cv} = 3I_{RD} \quad \text{Triaxial compression} \tag{2.33}$$

For quartz sands, Bolton suggests $\phi'_{cv} = 33°$.

Andersen and Schjetne (2013) analyzed an extensive database of sands, which included over 500 TC tests on 54 sands from 38 different sites, approximately 40%–50% of which were from offshore sites. Most of the sands classified as poorly graded (SP in the Unified Soil Classification System), with 80% of the samples containing no fines and the average uniformity coefficient being $C_u = 1.95$. The D_{50} particle size for their database was 0.23 mm ± SD 0.11 (fine sand). Calcareous soils are not included in their study. Figure 2.26 shows the trend lines of peak and constant volume friction angles derived from their study. The relative density of the samples in their study was bracketed in a range $D_r = 20–110$, and the confining stress level ranged from $\sigma'_c < 15$ kPa to > 500 kPa. Andersen and Schjetne showed Bolton's equation for peak friction angle in TC (Equation 2.33) to compare favorably to the trends shown in Figure 2.26a.

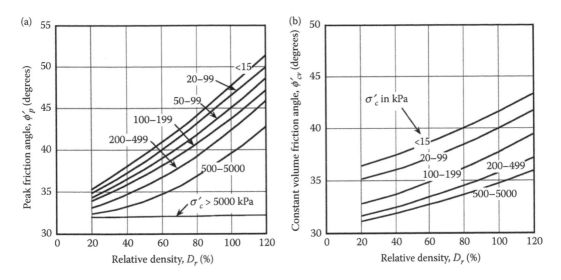

Figure 2.26 Friction angle as function of density and stress level. (a) Peak friction and (b) constant volume friction. (After Andersen KH and Schjetne K, 2013, *ASCE J Geotech Eng*, ASCE, 139(7), 1140–1155.)

2.3.3 Dilatancy

In addition to friction angle, some analyses of anchor capacity require input of a dilatancy parameter. A widely used measure of volume change induced by shear is the dilatancy angle ψ_{max} (Bolton, 1986; Figure 2.27a), defined by the ratio of incremental volumetric strain $d\varepsilon_{vol}$ to incremental shear strain $d\gamma_{13}$ by

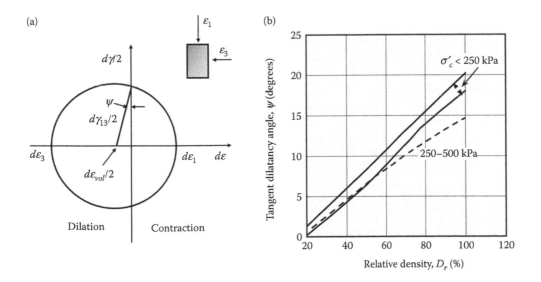

Figure 2.27 Dilatancy angle. (a) Dilation angle for plane strain compression and (b) relation to D_r. (Adapted from Andersen KH and Schjetne K, 2013, *ASCE J Geotech Eng*, ASCE, 139(7), 1140–1155.)

$$\sin \psi_{max} = \left(-\frac{d\varepsilon_{vol}}{d\gamma_{13}} \right)_{max} = -\left(\frac{d\varepsilon_1 + d\varepsilon_3}{d\varepsilon_1 - d\varepsilon_3} \right)_{max} = -\left(\frac{d\varepsilon_{vol}}{2d\varepsilon_1 - d\varepsilon_{vol}} \right)_{max} \qquad (2.34)$$

It should be noted that this measure of dilatancy is strictly applicable only to plane strain testing conditions. A measure applicable to both plane strain and triaxial shear tests is the ratio $d\varepsilon_{vol}/d\varepsilon_1$. Bolton (1986) shows that for both triaxial and plane strain compression tests, this measure of dilatancy relates to the relative dilatancy index I_{RD} (Equation 2.31) by

$$\left(-\frac{d\varepsilon_{vol}}{d\varepsilon_1} \right) = 0.3\, I_{RD} \qquad (2.35)$$

Andersen and Schjetne (2013) combine Equation 2.34 with Equation 2.35 to obtain the following relationship between dilatancy angle and I_{RD}:

$$\sin \psi_{max} = \frac{0.3 I_{RD}}{2 + 0.3 I_{RD}} \qquad (2.36)$$

Equation 2.36 provides a useful empirical estimate of dilation angle ψ_{max} from relative density D_r and confining stress σ'_c.

Andersen and Schjetne (2013) applied Equation 2.34 to their database of TC tests (127 data points) to establish independent correlations between ψ_{max}, relative density D_r and confining stress σ'_c. Figure 2.27b shows the trend lines derived from their study. The empirical curves show ψ_{max} to be relatively insensitive to confining stress. Andersen and Schjetne also found that Equation 2.36 slightly overestimates the curves from the Norwegian Geotechnical Institute (NGI) database.

2.3.4 Correlations to CPTU tip resistance

Penetration resistance measurements are normally the primary means for *in situ* determination of *in situ* density, strength and stiffness of seafloor sands. The piezocone (CPTU) test has already been described in Section 2.1.14. In contrast to penetration in clays, fully drained conditions should prevail in sands, but this should be confirmed by monitoring the water pressure measurements. The most widely used correlations between penetration resistance and sand properties are based on calibration chamber tests, where penetration tests are conducted under controlled conditions of relative density and vertical stress. In some cases, the effects of overconsolidation are also considered. Soil compressibility significantly affects penetration resistance, which some methods explicitly consider. Useful summaries of the various correlations are provided by Lunne et al. (1997b) and Kulhawy and Mayne (1990).

Commonly used correlations of cone tip resistance q_c to relative density D_r are summarized in Table 2.5. The correlations are largely based on data for fine to medium, clean silica sands (Lunne et al., 1997b). The Baldi et al. (1986) correlations were for a Ticino sand. The Kulhawy and Mayne (1990) correlation accounts for compressibility and aging effects. Silica sands with rounded grains may be considered as having low compressibility, silica sands with some feldspar and/or some fines may be considered to have medium compressibility, and sands with a high mica or fines content may be considered as having a high

Table 2.5 CPTU correlations to relative density

Reference	Equation	Fitting parameters
Baldi et al. (1986)	$$D_r = \frac{1}{C_2}\log_e\left[\frac{q_c}{C_0\left(\sigma'\right)^{C_1}}\right]$$ σ' in kPa	Normally consolidated (use $\sigma' = \sigma'_{v0}$): $C_0 = 157, C_1 = 0.55$ and $C_3 = 2.41$ Overconsolidated (use $\sigma' = \sigma'_m$): $C_0 = 181, C_1 = 0.55$ and $C_3 = 2.61$
Kulhawy and Mayne (1990)	$$D_r^2 = \frac{1}{305\,Q_c\,OCR^{0.18}Q_A}\frac{q_c/p_a}{\left(\sigma'_{v0}/p_a\right)^{0.5}}$$ $Q_A = 1.2 + \log_{10}(t/100)$	$Q_c = 0.91$ low compressibility $= 1.00$ medium compressibility $= 1.09$ high compressibility $p_a =$ atmospheric pressure $(\sim 100$ kPa$)$

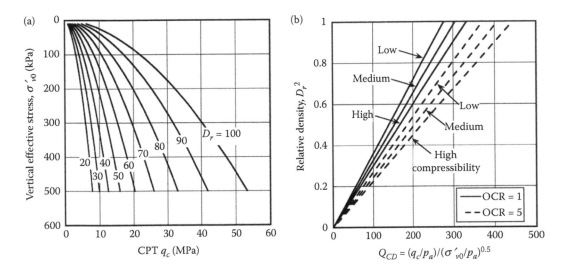

Figure 2.28 Correlations of CPTU tip resistance to relative density. (Adapted from Baldi G et al., 1986, Interpretation of CPTs and CPTUs; 2nd part: Drained penetration of sands, *Proceedings of the 4th International Geotechnical Seminar*, Singapore, pp 143–156; Kulhawy F and Mayne PW, 1990, *Manual on Estimating Soil Properties for Foundation Design*, Electric Power Research Institute, EPRI EL-6800, Palo Alto, California.)

compressibility. The Kulhawy and Mayne (1990) correlation characterizes the effect of compressibility through the Q_c parameter shown in Table 2.5, and OCR is directly incorporated into the correlation equation. The Kulhawy–Mayne formula also includes the effect of aging through the parameter Q_A. Figure 2.28 shows example curves derived from the equations in Table 2.5.

Cone tip resistance q_c can also directly correlated to friction angle ϕ'. Figure 2.29a shows a graphical correlation by Robertson and Campanella (1983). Kulhawy and Mayne (1990) provide the following approximation to this correlation:

$$\phi'_{tc} = \tan^{-1}[0.1 + 0.38\log_{10}(q_c/\sigma'_{v0})] \tag{2.37}$$

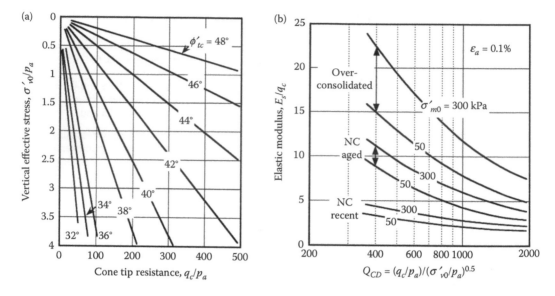

Figure 2.29 Correlation of CPTU tip resistance to sand strength and elastic modulus. (a) Friction angle (Adapted from Robertson PK and Campanella RG, 1983, *Canad Geotech J*, 20(4), 718–733.) and (b) elastic modulus (Adapted from Bellotti R, et al. 1989, Shear strength of sand from CPT, *Proceedings of the 12th International Conference on Soil Mechanics and Foundation Engineering*, Rio de Janeiro, vol. I, pp 179–184.)

Elastic modulus of sand E_s has also been correlated to cone tip resistance. Bellotti et al. (1989) provide the correlation of q_c to E_s shown in Figure 2.29b for overconsolidated sands, normally consolidated aged ($t > 1000$ years) sands and normally consolidated young sands. The correlations cover a range of mean *in situ* stress level σ'_{m0} 50–300 kPa. The elastic modulus E_s shown represents a strain level of about $\varepsilon_a = 0.1\%$.

2.3.5 Permeability

Darcy's law relates bulk velocity v to hydraulic gradient i (Equation 2.38), where k is coefficient of permeability. The bulk velocity v is defined as flow rate q divided by gross area A, which includes both void area and areas occupied by solid particles. The linear relationship described by Darcy's law is generally valid so long as flow turbulent flow does not initiate, which is generally a safe assumption for silts through medium sands (USACE, 1993).

$$v = -ki \tag{2.38}$$

Permeability of sands can be measured directly in the constant head permeability test. Additionally, correlations exist relating permeability to the gradation and soil density. An early, widely used empirical relationship between permeability k and the D_{10} particle size is the Hazen (1911) equation provided by Equation 2.39. For units of k in cm/second and D_{10} in cm, C_1 is between 90 and 120 cm^{-1} second^{-1} ($C_1 = 100$ is commonly assumed) for clean, loose sands.

$$k = C_1 D_{10}^2 \tag{2.39}$$

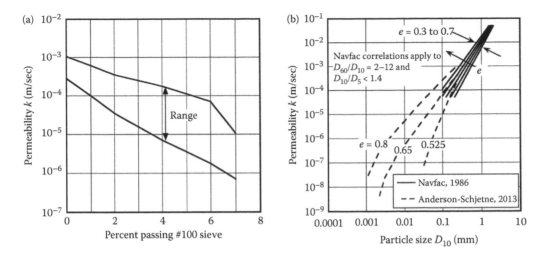

Figure 2.30 Permeability of sands. (a) Effect of finer fraction and (b) correlations D_{10} to and e.

The permeability of clean and borderline coarse-grained soil groups (e.g., SP, SP-SM) is highly sensitive to minor variations in the fine-grained fraction of particle sizes. For example, Cedergren (1967) reports that the permeability of sand–gravel mixtures (maximum particle size 25 mm) reduces by two orders of magnitude as the percent passing the #100 sieve increases from 0% to 7% (Figure 2.30). Such a high sensitivity to small variations in gradation highlights the difficulty in estimating permeability in these soil groups.

Density also has a marked influence on permeability of coarse-grained soils. An unpublished equation by Casagrande cited by Terzaghi and Peck (1967) provides an empirical, parabolic relationship Equation 2.40 between void ratio and permeablity, where k_1 is permeability at a void ratio $e = 0.85$. Figure 2.30b shows other empirically based relationships, including correlations extracted from Navfac (1986) and Andersen and Schjetne (2013).

$$k = 1.4\,k_1 e^2 \tag{2.40}$$

Fundamental studies of the nature of flow through porous media can provide useful insights to complement to the empirical relationships discussed above. Such studies are frequently framed in terms of tubes considered to be representative of the pore spaces. By this model, resistance to flow arises primarily from the side resistance at the solid–fluid interfaces. However, Bakhmeteff and Feodoroff (1937) argue that, owing to the highly irregular geometry of the flow channels, resistance to flow is primarily from form resistance (arising from differential pressures on front and aft of the soil particles), rather than tangential resistance. Thus, resistance to flow is more effectively characterized in terms of a drag coefficient C_d, defined as follows:

$$Di = C_d \frac{v^2}{2g} \tag{2.41}$$

where D is effective particle diameter, i is hydraulic gradient, and v is bulk velocity. To account for density effects, the Bakhmeteff–Feodoroff analysis employs scaling laws for velocity and particle diameter in term of porosity n, resulting in the following equations:

$$v_p = v / n^{2/3} \quad D_p = Dn^{1/3} \tag{2.42}$$

A generalized drag coefficient C_{dp} can now be defined as follows:

$$C_{dp} = \frac{2giD_p}{v_p^2} \tag{2.43}$$

Following the approach of classical fluid mechanics for this category of problem, the drag coefficient is expressed as a function of Reynolds number R_p, defined as follows:

$$R_p = \frac{v_p D_p}{v_k} \tag{2.44}$$

where v_k is kinematic viscosity. Figure 2.31a shows the $R_p - C_{dp}$ relationship derived from tests by Bakhmeteff and Feodoroff at Columbia University on assemblages of lead shot of various diameters and porosity. Figure 2.31b expresses the same relationship, except that the vertical axis is the product of R_p and C_{dp}. The experimental data show that at low Reynolds numbers, that is, in the range of laminar flow, the product $C_p = R_p C_{dp}$ is constant, implying:

$$v_p = \left(\frac{2g D_p^2}{v_k} \frac{1}{C_p} \right) i \quad R_p < 10 \tag{2.45}$$

In terms of unscaled variables, the relationship between bulk velocity and hydraulic gradient becomes:

$$v = \left(\frac{2g}{v_k C_p} D^2 n^{4/3} \right) i \quad R_p < 10 \tag{2.46}$$

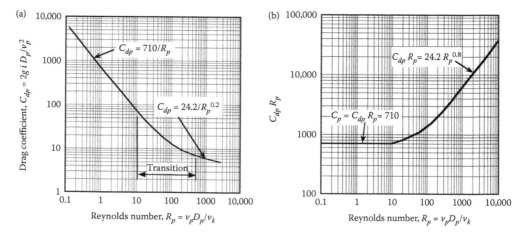

Figure 2.31 Drag resistance in porous media. (a) Reynolds number versus drag coefficient and (b) Reynolds number versus $C_{dp}R_p$

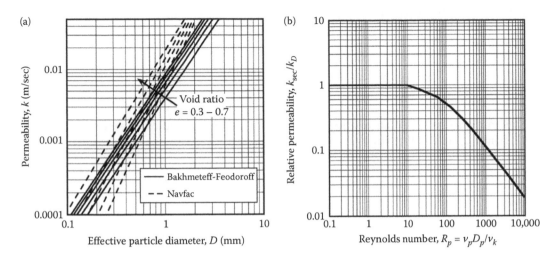

Figure 2.32 Hydraulic characteristics of porous media. (a) Effective porosity and particle size and (b) effect of Reynolds number.

Equation 2.46 shows the drag resistance formulation conforming to Darcy's law at low Reynolds numbers, $R_p < 10$. The model further implies that permeability varies with the square of particle diameter, which is consistent with the Hazen equation. Substituting the experimentally determined $C_p = 710$ and water viscosity $v_k = 0.01$ cm²/second (at temperature 20°C) into Equation 2.46 for porosities in the range $n = 0.35$–0.5, predicts a range of Hazen coefficients $C_1 = 70$–110. This is consistent with the expected range for this coefficient. In regard to the effect of soil density on permeability, the drag resistance model implies a simple power law relationship between k and porosity n, with an exponent of 4/3. Comparison to the Navfac (1986) correlations (Figure 2.32a) indicates that the Bakhmeteff–Feodoroff model tends to match the sensitivity of permeability to variations in density for large particle diameters, but underestimates the effect of density for finer particle sizes. The differences between the model and the empirical curves may be due to the uniform particle sizes used in the Bakhmeteff–Feodoroff tests, in contrast to the variable particle size distributions in real soils.

Figure 2.31 shows laminar flow conditions prevail up to a Reynolds number $R_{pL} = 10$. Imposing this constraint together with Equations 2.42 and 2.45 produces the following equation for estimating the maximum particle dimension D_L for which laminar flow and Darcy's law may be assumed to unequivocally apply:

$$D_L^3 ni = \frac{R_{pL} C_p v_k^2}{2g} \tag{2.47}$$

The onset of the transition between laminar and turbulent flow thus depends on particle diameter, porosity, and hydraulic gradient. As noted earlier, the Bakhmeteff–Feodoroff tests provide $C_p = 710$, and water viscosity at 20°C is $v_k = 1$ mm²/second. For the case of porosity $n = 0.4$ and hydraulic gradient $i = 1.0$, Equation 2.47 indicates the maximum particle diameter for which Darcy's law may be considered to strictly apply is on the order of $D_L = 1$ mm, which lies within the particle size range of a medium sand. This result is generally consistent with the common assumption that Darcy's law is applicable to soil types of medium sands or finer.

Turning now to the region beyond the transition zone where full turbulent flow develops, Figure 2.31 indicates that the limiting drag factor C_{dT} under turbulent flow conditions is described by

$$C_{dT} = \frac{24.2}{R_p^{0.2}} \quad R_p > 500 \tag{2.48}$$

This implies the following relationship between hydraulic gradient under turbulent flow conditions i_{turb} and bulk velocity v:

$$i_T = \frac{24.2}{2g} \frac{v_p^2}{D_p} \frac{1}{(v_p D_p / \nu_k)^{0.2}} \quad R_p > 500 \tag{2.49}$$

Thus, at Reynolds numbers $R_p > 500$ hydraulic gradient is expected to vary in proportion to $v^{1.8}$; that is, slightly less than the square of velocity. To establish the particle dimension at which fully developed turbulent flow can occur, Equation 2.48 together with Equations 2.42 and 2.43 lead to the following expression:

$$D_T^3 ni = \frac{24.2 R_{pT}^{1.8} \nu_k^2}{2g} \tag{2.50}$$

where $R_{pT} = 500$ and D_T is the particle dimension at which full turbulence develops. To gain a sense of what particle size range can be associated with turbulent flow, one can consider a hydraulic gradient $i = 1$ and porosity $n = 0.4$. In this case, Equation 2.50 provides $D_T = 6$ mm, which lies in the range of a fine gravel. Overall, Equations 2.47 and 2.50 provide a general picture of laminar flow prevailing for medium sands and finer, potential transitional behavior occurring in coarse sands, and potential full turbulent flow in fine gravels and coarser.

The above discussion applies to a single-gravity environment. At elevated g-levels occurring in centrifuge tests, the threshold particle diameters marking the transition from laminar to turbulent flow decrease in inverse proportion to the cube root of centrifugal acceleration.

Insight into the influence of increased energy loss at high Reynolds numbers is possible by considering a secant permeability $k_{sec} = v/i$. Equations 2.41 through 2.46 can be invoked to show that the ratio of permeability k_{sec} to Darcian permeability K_D (corresponding to laminar flow at low Reynolds numbers) is $k_{sec}/k_D = 710/C_{dp} R_p$. Based on the relationship between $C_{dp} R_p$ and R_p shown in Figure 2.31b, the decrease in effective permeability at high Reynolds numbers can be established, as shown in Figure 2.32b.

2.4 CYCLIC BEHAVIOR

Loading of anchors from wind and waves induces a cyclic component of loading in addition to any sustained load components, requiring an assessment of the effects of cycling on soil stress–strain-strength behavior. As much of anchor design revolves around ultimate capacity considerations, this discussion focuses on the cyclic strength aspect of the problem. The term "cyclic" denotes repeated patterns with some degree of regularity in amplitude and period. While actual wind and wave loading on anchors typically involve random amplitudes, current best practice in offshore engineering (Andersen et al., 2013; Andersen, 2015) typically does not explicitly model random load sequences. More commonly, "rain flow" procedures are employed, which group a random time series of loading into a series of packets of load cycles, with all load cycles within a given packet having a uniform magnitude. Accordingly, laboratory studies to characterize cyclic behavior almost invariably subject soil specimens to repeated load cycles (typically stress control) having uniform magnitude. The period of loading relates directly to the rate of loading; thus, strain rate effects (Section 2.1.9) can be very significant.

Laboratory testing programs to characterize cyclic load behavior of soils typically parallel the general approach of those used to characterize monotonic undrained behavior described in Sections 2.1.1 through 2.1.4. Following this approach, cyclic loading proceeds from a preselected stress state, typically K_0-consolidation at a selected OCR. The initial stress τ_0 denotes the drained shear stress in the plane of applied cyclic stresss τ_{cy}. For triaxial tests the consolidation shear stress is therefore $\tau_0 = 0.5(1 - K_0)\sigma'_{vc}$. For the conventional K_0-consolidated DSS testing conditions $\tau_0 = 0$. Following consolidation, the soil is subjected to cyclic stresses having a single amplitude τ_{cy} (Figure 2.33a).

A description of soil behavior under cyclic loading requires a "memory parameter" to track the degradation in soil stiffness and strength. Both excess pore pressure and strain have been applied to this purpose. Relevant measurements during cyclic loading include the cyclic, average, and permanent pore pressures (u_{cy}, u_a, and u_p, respectively) illustrated by

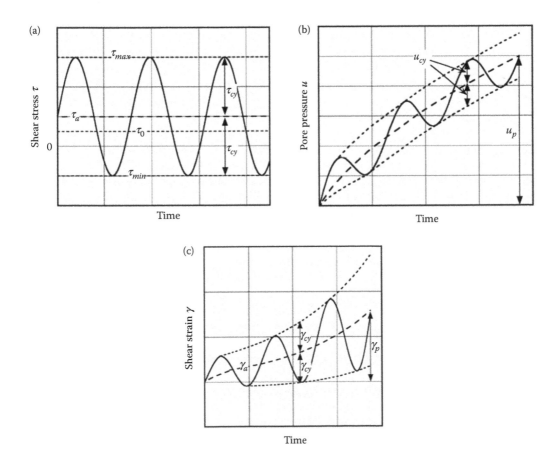

Figure 2.33 Definition of terms for cyclic loading. (a) Stress, (b) pore water pressure, and (c) strain.

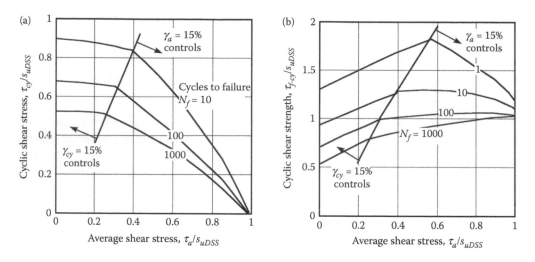

Figure 2.34 Cyclic DSS strength. (a) Average-cyclic stress interaction and (b) cyclic strength.

Figure 2.33b. The corresponding shear strains are denoted by γ_{cy}, γ_a, and γ_p (Figure 2.33c). The cyclic pore pressure or strain is half the trough-to-peak amplitude within a load cycle, while average pore pressure or strain is the average of the trough and peak magnitudes. Permanent pore pressure or strain is the value occurring at the end of a given load cycle. When drainage effects are potentially significant, often the case in sands, characterizing cyclic behavior in terms of pore pressure is advantageous. In clays, where pore pressures are difficult to measure reliably in the laboratory and where pore pressure dissipation is typically insignificant, cumulative strain typically provides an adequate basis for characterizing cyclic loading effects.

Cyclic shear strength can be represented as a function of number of load cycles, average stress and cyclic stress, as illustrated in Figure 2.34a (Andersen, 2015). When strain is considered as the criterion, failure is typically considered to occur when either the average or cyclic strain, γ_a or γ_{cy}, is 15%. Figure 2.34a shows example diagrams for DSS tests on clays; Andersen (2015) shows examples of similar plots derived from triaxial test data. Normalized average stress τ_a/s_u is plotted on the horizontal axis with normalized cyclic stress τ_{cy}/s_u on the vertical axis. The contours show the loci of points corresponding to equal numbers of cycles to failure. Stresses are normalized $s_u = s_{uDSS}$. An average strength equal to the undrained strength corresponds to monotonic loading, with a single load cycle to failure. Cyclic shear strength $\tau_{f\text{-}cy}$ is the sum of the average and cyclic stresses, illustrated in Figure 2.34b. Since cyclic loading τ_{cy} is applied at a much greater rate than monotonic loads, the normalized cyclic strength $\tau_{f\text{-}cy}/s_u$ exceeds unity for a wide range of loading combinations. The most severe reductions in strength owing to cycling occur at low values of τ_a/s_u; that is, when load reversals occur.

The irregular nature of most load histories require a scheme for determining an equivalent number of uniform load cycles N_{eq} for use in diagrams such as those illustrated in Figure 2.34. The approach developed by NGI for determining N_{eq} first groups irregular load histories into parcels of loads having equal magnitude. The rain flow method commonly used in fatigue analyses (ASTM, 2011) has traditionally been used for this purpose. However, Andersen (2015) reports that this approach can overestimate load amplitudes, and a more recent approach proposed by Noren-Cosgriff et al. (2015) indicates that details of the method of load counting can significantly affect cyclic strength estimates. The NGI

(a)

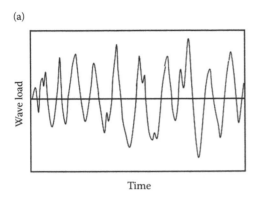

Time

(b)

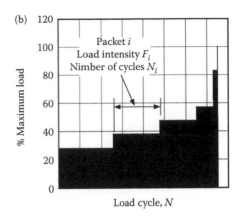

Load cycle, N

Figure 2.35 Grouping of random loads into parcels. (a) Random wave loading and (b) packets of equivalent uniform loads.

assumption of load parcels occurring in ascending order provides conservative estimates of cyclic strength, but Andersen (2015) cautions that it is not necessarily conservative for estimating soil stiffness for dynamic analyses.

Estimating cyclic strength for variable load amplitudes requires a procedure for tracking the damage induced by episodes of loading at variable stress levels. As noted earlier, accumulated pore pressure or strain may be used for this purpose, where pore pressure may be more suited to sands and strains more suited to clays (Andersen, 2015). Both approaches assume the load history from a rain flow procedure can be grouped into parcels having N_i repetitions of load intensity F_i, where i is the parcel number (Figure 2.35). The analyses further assume that cyclic stress intensity τ_{cy_i} is proportional to load intensity F_i, such that

$$\frac{\tau_{cy\_i}}{\sigma'_{vc}} = s_f \frac{F_i}{F_{max}} \tag{2.51}$$

F_{max} is the maximum force in the time history of loading, and s_f is an arbitrary scaling factor. In the procedure that will be described the scaling factor s_f is successively adjusted by trial-and-error until failure ($\gamma_{cy} = 15\%$) is reached. In the pore pressure accumulation method, the analysis starts by plotting experimentally determined contours of permanent pore pressure as a function of load ecycle N and cyclic stress intensity τ_{cy}/σ'_{vc}. Figure 2.36 illustrates the construction for the case of DSS loading with zero average cyclic stress, $\tau_a = 0$. Proceeding in ascending order of load intensity and starting with an arbitrarily selected scale factor s_f, the first parcel of N_1 load cycles of intensity τ_{cy_1}/σ'_{vc} follows path AB in Figure 2.36a, with u_{p1} being the permanent pore pressure at the end of this load parcel. Load parcel 2 initiates with a memory of u_{p1}, as depicted by path BC. N_2 load cycles at intensity τ_{cy_2}/σ'_{vc} increases the permanent pore pressure along path CD to u_{p2}. These steps are repeated for the entire suite of load parcels. For the selected scale factor the process leads to an end point E. The entire process is now successively repeated with different scale factors to generate the locus of end points illustrated in Figure 2.36a. By superimposing the locus of end points onto the strain contour plot in Figure 2.36b, one can determine the intersection with the failure strain contour, $\gamma_{cy} = 15\%$, which defines the equivalent number of uniform load cycles to failure N_{eq}.

To gain a sense of the potential significance of symmetric ($\tau_a = 0$) cyclic loading effects for normally consolidated clays, Figure 2.37a (after Andersen, 2015) shows the trend line for

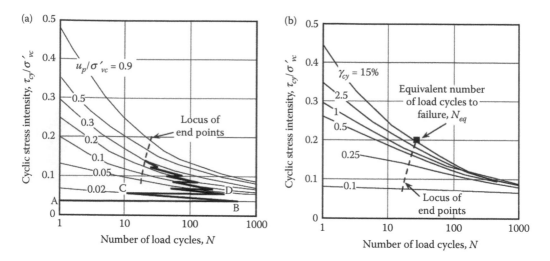

Figure 2.36 Equivalent number of load cycles by pore pressure accumulation procedure. (a) Analysis of pore accumulation and (b) determination of N_{eq} at failure.

the reduction of undrained DSS simple shear strength after 10 cycles of loading. Andersen (2015) provides Equation 2.52 as a best fit curve for this line. Experimental data indicated that cyclic strength degradation effects are most prominent at low PI. Figure 2.36a shows strength degradation for the worst case condition of symmetric loading. The influence of asymmetric loading ($\tau_a > 0$) was illustrated earlier in Figure 2.34. Figure 2.37b from Andersen (2015) shows strength degradation effects for load cycles other than 10 for the case of Drammen clay.

$$\frac{\tau_{f,cy}}{s_{uDSS}} = 0.41\, PI^{0.224} \tag{2.52}$$

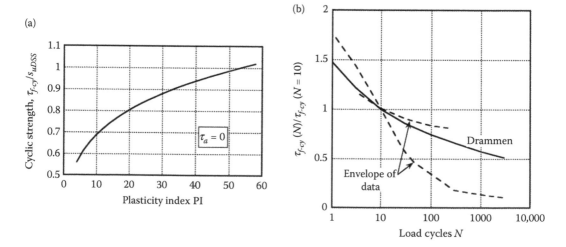

Figure 2.37 Typical cyclic strength degradation in clays. (a) Symmetric loading with 10 load cycles and (b) effect of number of load cycles.

REFERENCES

Aas G, Lacasse S, Lunne T, and Hoeg K, 1986, Use of *in situ* tests for foundation design on clays, *Proceedings of the ASCE Specialty Conference*, In Situ '86, ASCE GSP 6, pp 1–30.

Abelev A and Valent P, 2009, *Strain-rate dependency of strength of soft marine deposits of the Gulf of Mexico*, MTS/IEEE Biloxi—Marine Technology for Our Future: Global and Local Challenges, OCEANS 2009.

Andresen A and Kolstad P, 1979, The NGI 54-mm sampler for undisturbed sampling of clays and representative sampling of coarser materials, *Proceedings of the International Conference on Soil Sampling*, Singapore, pp 1–9.

Andersen KH, 2015, Cyclic soil parameters for offshore foundation design, *Proceedings of the International Symposium on Frontiers in Offshore Geotechnics ISFOG15*, Oslo, Taylor & Francis Group, London, pp 5–82.

Andersen KH, Puech A, and Jardine R, 2013, Cyclic resistant geotechnical design and parameter selection for offshore engineering and other applications, *Proceedings of XVIIIth ICSMGE, Design for Cyclic Loading: Piles and Other Foundations*, Paris, pp 9–44.

Andersen KH and Schjetne K, 2013, Database of friction angles of sand and consolidation characteristics of sand, silt, and clay, *ASCE J Geotech Eng*, ASCE, 139(7), 1140–1155.

Arthur JRF, Chua KS, and Dunstan T, 1977, Induced anisotropy in sand, *Geotechnique*, 7(1),13–36.

ASTM, 2003, Standard test method for laboratory miniature vane shear test for saturated fine-grained clayey soil, *ASTM D 4648-00*, West Conshohoken, PA.

ASTM, 2011, Standard practices for cycle counting in fatigue analysis, *ASTM E1049-85*, West Conshohocken, PA.

ASTM, 2016a, Standard test method for maximum index density and unit weight of soils using a vibratory table, *ASTM D 4253-16*, West Conshohoken, PA.

ASTM, 2016b, Standard test method for minimum index density and unit weight of soils and calculation of relative density, *ASTM D 4254-16*, West Conshohoken, PA.

Aubeny CP and Shi H, 2007, Penetration of cylinders in cohesive soils: Strain rate effects, *IEEE J Oceanic Eng*, 32(1), 49–56.

Bakhmeteff BA and Feodoroff NV, 1937, Flow through granular media, *Joint Meeting of the Applied Mechanics and Hydraulic Division*, ASCE, pp A97–A104.

Baldi G, Bellotti R, Ghionna V, Jamiolkowski M, and Pasqualini E, 1986, Interpretation of CPTs and CPTUs; 2nd part: Drained penetration of sands, *Proceedings of the 4th International Geotechnical Seminar*, Singapore, pp 143–156.

Baligh MM, 1985, *Fundamentals of deep penetration, I: Shear stresses*, MIT Research Report, R85-9, Order No. 776, 64p.

Baligh MM, Azzouz AS, and Chin CT, 1987, Disturbances due to ideal tube sampling, *J Geotech Eng*, 113(7), 739–757.

Baligh MM and Levadoux J-N, 1986, Consolidation after undrained piezocone penetration: II interpretation, *ASCE J Geotech Eng*, ASCE, 112(7), 727–745.

Biscontin G and Pestana J, 1999, *Influence of Peripheral Velocity on Undrained Shear Strength and Deformability Characteristics of a Bentonite-Kaolinite Mixture*, Geotechnical Engineering Report No. UCB/GT/99-16, University of California, Berkeley.

Bishop AW and Henkel DJ, 1976, *The Measurement of Soil Properties in the Triaxial Shear Test*, Edward Arnold (Publishers) Ltd., London, 2nd Edition, 227p.

Berre T and Bjerrum L, 1973, Shear strength of normally consolidated clays, *Proceedings of the 8th International Conference on Soil Mechanics and Foundation Engineering*, Moscow, vol. 1, pp 39–49.

Bellotti R, Ghionna V, Jamiolkowski M, and Robertson PK, 1989, Shear strength of sand from CPT, *Proceedings of the 12th International Conference on Soil Mechanics and Foundation Engineering*, Rio de Janeiro, vol. 1, pp 179–184.

Bjerrum L, 1973, Problems of soil mechanics and construction on soft clays, *Proceedings of the 8th International Conference Soil Mechanics and Foundation Engineering*, Moscow, vol. 3, pp 111–159.

Bjerrum L and Landva A, 1966, Direct simple shear tests on a Norwegian quick clay, *Geotechnique*, 16(1), pp 1–20.

Bolton MD, 1986, The shear strength and dilatancy of sands, *Geotechnique*, 36(1), pp 65–78.

Borel D, Puech A, Dendani H, and de Ruijter M, 2002, High quality sampling for deepwater engineering: The STACOR® experience, *Ultra Deep Engineering and Technology (UDET)*, Brest, France.

Brooker EW and Ireland HO, 1965, Earth pressures at rest related to stress history, *Canad Geotech J*, 2(1), 1–15.

Buhler RL and Audibert JME, 2012, Rate-effect correction methods for free-fall CPT data in deepwater Gulf of Mexico—An operator's perspective, *Society of Underwater Technology, Offshore Site Investigations and Geotechnics, Offshore Site Investigations and Geotechnics: Integrated Technologies – Present and Future*, 12–14 September, London, UK.

Campanella and Vaid, 1972, A simple K_0 triaxial cell, *Canad Geotechn J*, 9(3), pp 249–260.

Casagrande A and Wilson SD, 1951, Effect of rate of loading on the strength of clays and shales at constant water content, *Geotechnique*, London, U.K., 2(3), 251–263.

Cedergren HR, 1967, *Seepage, Drainage, and Flow Nets*, John Wiley and Sons, New York, 489p.

Dayal U and Allen JH, 1975, The effect of penetration rate on the strength of remolded clay and sand samples, *Canad Geotech J*, 336, 336–348.

DeGroot DJ, Ladd CC, and Germaine JT, 1996, Undrained multidirectional direct simple shear behavior of cohesive soil, *J Geotech Eng*, ASCE, 122(2), pp 91–98.

DeGroot DJ, Lunne T, Andersen KH, and Boscardin AG, 2012, Laboratory measurement of the remoulded shear strength of clays with application to design of offshore infrastructure, *Proceedings of the 7th International Conference on Offshore Site Investigations and Geotechics*, London, UK, pp 355–366.

DeGroot DJ, Lunne T, and Tjelta TI, 2010, Recommended best practice for geotechnical site characterisation of cohesive offshore sediments, *Proceedings of the 2nd International Symposium on Offshore Geotechnics*, Perth, Australia, pp 33–57.

Duncan JM and Seed HB, 1966, Anisotropy and stress reorientation in clay, *J Soil Mech Found Div*, ASCE, 92(SM5), 21–50.

Dyvik R, Berre R, Lacasse A, and Raadim B, 1987, Comparison of truly undrained and constant volume direct simple shear tests, *Geotechnique*, 37(1), pp 3–10.

Ehlers CJ, Chen J, Robert HH, and Lee YC, 2005, The origin of near-seafloor "crust zones" in deepwater, *Frontiers in Offshore Geotechnics: ISFOG 2005*, Gourvernec and Cassidy, (eds.), Taylor & Francis Group, London, pp 927–933.

Foott R and Ladd CC, 1981, Undrained settlement of plastic and organic clays, *J Geotech Eng Div*, ASCE, 107(GT8), 1079–1094.

Graham J, Crooks JHA, and Bell AL, 1983, Time effects on the stress–strain behaviour of soft natural clays, *Geotechnique*, 33(3), 327–340.

Hansbo S, 1957, *A New Approach to the Determination of the Shear Strength of Clay by the Fall Cone Test*, Swedish Geotechnical Institute, Publication No. 14, Stockholm.

Hazen A, 1911, Discussion of dams on sand foundations, by A.C. Koenig, *Transactions*, ASCE, 73, 199–203.

Henkel DJ and Wade NH, 1966, Plane strain tests on a saturated remolded clay, *J Soil Mech Found Div* ASCE, 92(SM6), 67–80.

Holtz RD and Kovacs WD, 1981, *An Introduction to Geotechnical Engineering*, Prentice-Hall, Inc., Englewood Cliffs, New Jersey, 733p.

Jaky J, 1944, The coefficient of earth pressure at rest, *Journal of Society of Hungarian Architects and Engineers*, Budapest, Hungary, pp 355–358.

Jeanjean, P, 2006, Setup characteristics for suction anchors in Gulf of Mexico Clays: Experience from field installation and retrieval, OTC-18005-MS, *Offshore Technology Conference*, Houston, TX, pp 1–9 (electronic format).

Jeanjean P, Spikula D, and Young A (2012) Technical vetting of free-fall cone penetrometer, *Society of Underwater Technology, Offshore Site Investigations and Geotechnics, Offshore Site Investigations and Geotechnics: Integrated Technologies – Present and Future*, 12–14 September, London, UK, OSIG-12-16.

Kim Y, Hossain MS, Wang D, and Randolph MF, 2015, Numerical investigation of dynamic installation of torpedo anchors in clay, *Ocean Eng*, 108, 820–832.

Koutsoftas DC and Ladd CC, 1985, Design strengths for an offshore clay, *J Geotech Eng*, ASCE, 111(3), 337–355.

Kulhawy F and Mayne PW, 1990, *Manual on Estimating Soil Properties for Foundation Design*, Electric Power Research Institute, EPRI EL-6800, Palo Alto, California.

Kuo MY-H and Bolton MD, 2009, Soil characterization of deep sea West African Clays: Is biology a source of mechanical strength? *Proceedings of the Nineteenth International Offshore and Polar Engineering Conference*, Osaka, Japan, pp 488–494.

Lacasse S and Lunne T, 1982, Penetration tests in two Norwegian clays, *Proceedings of the Second European Symposium on Penetration Testing*, ESOPT-II, Amsterdam, pp 661–669.

Ladd CC, 1991, Stability evaluation during staged construction, *J Geotech Eng*, ASCE, 117(4), 540–615.

Ladd, CC, Bovee R, Edgers L, and Rixner JJ, 1971, *Consolidated Undrained Plane Strain Tests on Boston Blue Clay*, MIT Research Report R71-13.

Ladd CC and DeGroot DJ, 2003, Recommended practice for soft ground site characterization: Arthur Casagrande Lecture, *Proceedings 12th Panamerican Conference on Soil Mechanics and Geotechnical Engineering*, Massachusetts Institute of Technology Cambridge, MA, USA, pp 3–57.

Ladd, CC and Foott R, 1974, New design procedure for stability of soft clays, *J Geotech Eng*, ASCE, 100(7), 736–786.

Ladd CC, Foott R, Ishahara K, Schlosser F, and Poulos HG, 1977, Stress-deformation and strength characteristics, *Proceedings of the 9th International Conference on Soil Mechanics and Foundation Engineering*, Tokyo, Japan, Vol. 2, pp 421–494.

Ladd, CC and Lambe TW, 1963, The strength of 'undisturbed' clay determined from undrained tests, *Symposium on Laboratory Shear Testing of Soils*, ASTM, STP 361, pp 342–371.

Landon MM, DeGroot DJ, and Sheahan TC, 2007, Nondestructive sample quality assessment of a soft clay using shear wave velocity, *ASCE J Geotech Geoenviron Eng*, ASCE, 133(4), 424–432.

Lefebvre G, Ladd CC, Mesri G, and Tavenas F, 1983, *Report of the Testing Subcommittee, Committee of Specialists on Sensitive Clays on the NBR Complex*, SEBJ, Montreal, Canada, Annexe I.

Levadoux J-N and Baligh MM, 1986, Consolidation after undrained piezocone penetration: I prediction, *ASCE J Geotechl Eng*, ASCE, 112(7), 707–726.

Low HE, Lunne T, Andersen KH, Sjursen MA, Li X, and Randolph, MF, 2010, Estimation of intact and remoulded undrained shear strengths from penetration tests in soft clays, *Geotechnique*, 60(11), 843–859.

Lunne T and Andersen KH, 2007, Soft clay shear strength parameters for deepwater geotechnical design, *Proceedings.of the 6th OSIG*, SUT, London, UK, pp 151–176.

Lunne T, Andersen KH, Low HE, and Randolph MF, 2011, Guidelines of offshore *in situ* testing and interpretation in deepwater soft clays, *Canad Geotech J*, 48, 543–556.

Lunne T, Berre T, and Strandvik S, 1997a, Sample disturbance effects in soft low plasticity Norwegian clay, *Proceedings of the Conference on Recent Developments in Soil and Pavement Mechanics*, Rio de Janeiro, pp 81–102.

Lunne T, Robertson PK, and Powell JJM, 1997b, *Cone Penetration Testing in Geotechnical Practice*, Chapman & Hall, London.

Mayne PW and Kulhawy FH, 1982, K0-OCR relationships in soil, *J Geotech Eng Div*, ASCE, 108(GT6), 851–872.

Mesri G, 2001, Primary compression and secondary compression, *Soil Behavior and Soft Ground Construction*, Germaine JT, Sheahan TS and Whitman RV, (eds.), ASCE Special Geotechnical Publication No. 119, pp 122–166.

Mitchell JK, 1960, 1982, Fundamental aspects of thixotropy in soils, *J Soil Mech Found Div*, ASCE, 86(SM3), 19–52.

Navfac DM-7.1, 1986, *Soil Mechanics, Facilities Engineering Command*, U.S. Dept. of the Navy, Alexandria, VA, 364p.

Noren-Cosgriff K, Jostad HP, and Madshus C, 2015, Idealized load composition for determination of cyclic undrained degradation of soils, *Proceedings of the International Symposium on Frontiers in Offshore Geotechnics*, ISFOG, Oslo, Norway.

Pestana JM and Whittle AJ, 1999, Formulation of a unified constitutive model for clays and sands, *Int J Numer Anal Meth Geomech*, 23, 1215–1243.

Peuchen J and Mayne PW, 2007, Rate effects in vane shear testing, *Proceedings of the Sixth International Conference on Offshore Site Investigation and Geotechnics Conference: Confronting New Challenges and Sharing Knowledge*, London, UK, pp 187–194.

Raie M and Tassoulas J, 2009, Installation of torpedo anchors: Numerical modeling, *J Geotech Geoenviron Eng*, ASCE, 135(12), 1805–1813.

Randolph MF, Hefer PA, Geise JM, and Watson PG, 1998, Improved seabed strength profiling using T-bar penetrometer, *Proceedings of the International Conference on Offshore Site Investigation and Foundation Behaviour - "New Frontiers"*, London, September 1998, Society for Underwater Technology, London, pp 221–235.

Randolph MF and Andersen KH, 2006, Numerical analysis of Tbar penetration in soft clay, *Int J Geomech*, ASCE, 6(6), 411–420.

Randolph MF and Houlsby GT, 1984, The limiting pressure on a circular pile loaded laterally in cohesive soil, *Geotechnique*, 34(4), 613–623.

Randolph MF, Martin CM, and Hu Y, 2000, Limiting resistance of a spherical penetrometer in cohesive material, *Geotechnique*, 50(5), 573–582.

Robertson PK, 1990, Soil classification using the cone penetration test, *Canad Geotech J*, 27(1), 151–158.

Robertson PK and Campanella RG, 1983, Interpretation of cone penetration tests, Part I: Sand, *Canad Geotech J*, 20(4), 718–733, © Canadian Science Publishers or its licensors.

Roscoe KH and Burland JB, 1968, On the generalized stress-strain behavior of 'wet' clay, *Engineering Plasticity*, Ed. J Heyman, (ed.), Cambridge University Press, Cambridge, pp 535–609.

Schmidt B, 1966, Earth pressures at rest related to stress history, *Canad Geotech J*, 3(4), 239–242.

Schofield AN and Wroth CP, 1968, *Critical State Soil Mechanics*, McGraw Hill, London, 310p.

Seah TH, 1990, Anisotropy of resedimented Boston Blue Clay, PhD Thesis, Massachusetts Institute of Technology, Cambridge, Massachusetts.

Sheahan TC, Ladd CC, and Germaine JT, 1996, Rate-dependent undrained shear strength behavior of saturated clay, *J Geotech Eng*, ASCE, 122(2), 99–108.

Silva AJ, Bryant WR, Young AG, Schultheiss P, Dunlap WA, Sykora G, Bean D, and Honganen C, 1999, Long coring in deep water for seabed research, geohazard studies and geotechnical investigations, *1999 Offshore Technology Conference*, OTC 10923, Houston, Texas, USA, pp 1–17 (electronic format).

Strout JM and Tjelta TI, 2007, Excess pore pressure measurements and monitoring for offshore instability problems, *Proceeding Offshore Technology Conference*, Houston, TX, Paper18706, pp 1–10 (electronic format).

Tavenas R, Jean P, Leblond P, and Leroueil S, 1983a, The permeability of natural soft clays, Part I: Laboratory measurement, *Canad Geotech J*, 20(4), 629–644.

Tavenas R, Jean P, Leblond P, and Leroueil S, 1983b, The permeability of natural soft clays, Part II: Permeability characteristics, *Canad Geotech J*, 20(4), 645–660.

Terzaghi K and Peck RB, 1967, *Soil Mechanics in Engineering Practice*, John Wiley and Sons, NY.

Terzaghi K, Peck RB, and Mesri G, 1996, *Soil Mechanics in Engineering Practice*, 3rd Edition, John Wiley and Sons, NY.

USACE, 1993, *Seepage Analysis and Control for Dams (CH1)*, U.S. Army Corps of Engineers EM 1110-2-1901.

Vaid YP and Campanella RG, 1974, Anisotropy and stress reorientation in clay, *J Geotech Eng Div*, ASCE, 100(GT5), 207–224.

Whittle AJ and Aubeny CP, 1992, The effects of installation disturbance on interpretation of in-situ tests in clays, *Predictive Soil Mechanics, Proceedings of the Wroth Memorial Symposium*, Oxford, England, Thomas Telford Ltd., London, pp 742–767.

Whittle AJ, DeGroot DJ, Ladd CC, and Seah T-H, 1994, Model prediction of anisotropic behavior of Boston Blue Clay, *J Geotech Eng*, ASCE, 120(1), 199–224.

Whittle AJ and Kavvadas MJ, 1993, Formulation of MIT-E3 constitutive model for overconsolidated clays, *J Geotech Eng*, ASCE, 120(1), 173–198.

Whittle AJ, Sutabutr T, Germaine JT, and Varney A, 2001, A prediction and interpretation of pore pressure dissipation for a tapered piezoprobe, *Geotechnique*, 51(7), 601–617.

Wissa AEZ, Christian JT, and Davis EH, 1971, Consolidation at constant rate of strain, *Journal Soil Mechanics and Foundations Division*, ASCE, 97(SM10), pp 1393–1413.

Wong PC, Taylor BB, and Audibert JME, 2008, Differences in shear strength between Jumbo Piston Core and conventional rotary core samples, *2008 Offshore Technology Conference*, OTC 19683, Houston, Texas, USA, pp 1–17 (electronic format).

Wood DM, 1981, True triaxial tests for Boston Blue Clay, *Xth International Conference on Soil Mechanics and Foundation Engineering*, Stockholm, pp 825–830.

Young AG, Bernard BB, Remmes BD, Babb LV, Detail Design, and Brooks JM, 2011, CPT Stinger—An innovative method to obtain CPT data for integrated geoscience studies, *2011 Offshore Technology Conference*, OTC 21569, Houston, Texas, USA, pp 1–10 (electronic format).

Young AG, Honganen CD, Silva AJ, and Bryant WR, 2000, Comparison of geotechnical properties from large-diameter long cores and borings in deep water Gulf of Mexico, *2000 Offshore Technology Conference*, OTC 12089, Houston, Texas, USA, pp 1–12 (electronic format).

Zhu H and Randolph MF, 2011, Numerical analysis of a cylinder moving through rate dependent undrained soil, *Ocean Eng*, 38, 943–953.

Chapter 3

Analytical methods

Analytical tools for characterizing anchor behavior include plastic limit analysis (PLA), cavity expansion methods, and finite element (FE) techniques. The PLA methods provide estimates of ultimate load capacity, which is central to many anchor problems; hence, they are prominent in research and design practice related to anchors. However, they are limited in that they cannot characterize processes that are inherently affected by elastic behavior, a prominent example being pile penetration. Cavity expansion methods—including their original one-dimensional form and their extension to two and three dimensions using the strain path method—are one approach for characterizing such types of processes. A second limitation of PLA methods is that while they can predict the effects of installation disturbance—for example, by specifying a surface roughness condition on an anchor—they do not predict the disturbance itself. Cavity expansion methods again offer substantial capabilities in this regard when they are used in conjunction with a realistic description of soil constitutive (stress–stain strength) behavior. Finally, FE techniques can be applied to a multiplicity of problems relevant to anchors including (1) poro-elastic problems such as postinstallation setup and sustained loading, (2) collapse load calculations, particularly for combined loading and complex geometries and material conditions, and (3) large deformation problems. In principle, FE analyses can be applied toward any of the problems discussed above in association with PLA and cavity expansion methods. However, FE analysis has its own limitations. Displacement-based FE analyses tend to overestimate collapse loads, sometimes severely; consequently, FE solutions frequently should be calibrated to benchmark PLA solutions.

This chapter provides an overview of the following analytical methods that have been applied to characterizing anchor performance:

- Plasticity theory, which is central to the various methods of the analysis.
- PLA, including upper bound virtual work analysis and lower bound analysis by method of characteristics.
- Soil elasto-plastic constitutive models, which comprise a key component of both cavity expansion and FE simulations.
- Cavity expansion analysis for characterizing processes that are affected by elastic behavior and for simulating installation disturbance.
- Finite element analysis for problems involving consolidation and drainage; collapse loads for complex conditions of loading, geometry, or material heterogeneity; and large deformation problems. The FE formulations can include displacement-based analyses or more recently developed optimization methods.

3.1 PLASTICITY THEORY

Plastic equilibrium deals with stresses in the soil mass at the yield state. As many anchor problems involve determinations of ultimate load capacity, plasticity theory frequently provides a useful framework for characterizing their behavior. The basic components of a plasticity model include (1) a yield criterion describing the stress combinations at which yield occurs, (2) a flow rule describing the direction of plastic flow in stress space, and (3) hardening laws describing the evolution in the size and shape of the yield surface as the material undergoes plastic straining. In a general elasto-plastic formulation, the elastic properties of the material within the yield surface should also be described. The following discussion focuses on the yield criterion and flow rule, which comprise a key component of PLAs based on the bounding theorems of classical plasticity theory. Discussion of hardening behavior and elastic effects is addressed later in this chapter, in the section covering soil constitutive models.

3.1.1 The yield criterion

Laboratory tests typically provide data on the yield stress for a single mode of loading such as triaxial or simple shear. As actual physical systems involve combinations of the various loading modes, a fundamental issue in plasticity involves defining the load combinations that induce yield. For undrained loading, a commonly applied yield criterion is a maximum shear stress criterion, also known as the Tresca criterion:

$$\tau_{max} = \frac{\sigma_1 - \sigma_3}{2} = k \tag{3.1}$$

The term k is a strength measure selected for the analysis. In principle, the undrained shear strength measured in any shearing mode is valid, but typically the direct simple shear mode is selected, $k = s_{uDSS}$, since this shearing mode frequently represents a reasonable rough "average" of the actual anisotropic strength characteristics of the soil. An alternative option is the von Mises yield criterion, which considers the effects of all shearing modes, and can be expressed in the form of the second invariant of the deviatoric stress tensor:

$$\sqrt{J_{2s}} = k$$
$$J_{2s} = \frac{1}{6}\left[(\sigma_{xx} - \sigma_{yy})^2 + (\sigma_{yy} - \sigma_{zz})^2 + (\sigma_{zz} - \sigma_{xx})^2\right] + \sigma_{xy}^2 + \sigma_{yz}^2 + \sigma_{zx}^2 \tag{3.2}$$

In this case, more care is required in the selection of k. For example, if one chooses to match the yield locus to the soil strength measured in a direct simple shear mode (the usual choice), the examination of Equation 3.2 shows $k = s_{uDSS}$ to be the correct choice. However, if triaxial shear strength is the selected measure, Equation 3.2 dictates $k = \left(2/\sqrt{3}\right)s_{uTX}$. Differences in the results from analyses using Tresca and von Mises yield criteria are relatively small. From the standpoint of how the two criteria compare to strength measurements, the comparisons tend to be obscured by the fact that both criteria ignore the anisotropy of actual soils; therefore, neither criterion matches the measurements particularly well. Moreover, the selection of a criterion is frequently on the basis of computational convenience. Fortunately, the simple shear strength frequently approximates the average of the anisotropic strength

Table 3.1 Transformed stress and strain measures

Mode	Deviatoric or shear stress	Deviatoric or shear strain
Triaxial	$S_1 = (2/\sqrt{6})[\sigma_{yy} - (\sigma_{xx} + \sigma_{zz})/2]$	$E_1 = (2/\sqrt{6})[\varepsilon_{yy} - (\varepsilon_{xx} + \varepsilon_{zz})/2]$
Cavity expansion	$S_2 = (1/\sqrt{2})(\sigma_{zz} - \sigma_{xx})$	$E_2 = (1/\sqrt{2})(\varepsilon_{zz} - \varepsilon_{xx})$
Simple shear, xy	$S_3 = \sqrt{2}\,\sigma_{xy}$	$E_3 = \sqrt{2}\,\varepsilon_{xy}$
Simple shear, yz	$S_4 = \sqrt{2}\,\sigma_{yz}$	$E_4 = \sqrt{2}\,\varepsilon_{yz}$
Simple shear, zx	$S_5 = \sqrt{2}\,\sigma_{zx}$	$E_5 = \sqrt{2}\,\varepsilon_{zx}$

variations; therefore, matching either yield criterion to s_{uDSS} is the approach adopted in a majority of cases.

Despite the widespread utility of an isotropic analysis, cases can certainly be envisioned where a simple average of the shear strength in different shearing modes is not representative of the problem under consideration. For example, if gapping occurs on the back side of a short caisson, the dominant shearing mode will be triaxial extension, with a strength s_{uTE} lower than s_{uDSS} in most normally consolidated clays. In these cases, a more rigorous treatment of soil strength anisotropy may be warranted. Characterization of soil strength anisotropy is complex, but the analysis can be considerably simplified by transforming the deviatoric stress and strain tensors into the measures shown in Table 3.1 and Figure 3.1. The coefficients applied to the basic deviator or shear components in Table 3.1 ensure energy conjugate stress–strain pairs in a manner that parallels the form that applies to conventional stress–strain measures:

Distortional strain energy:

$$U = s_{ij}e_{ij} = S_i E_i \tag{3.3}$$

In this equation the conventional deviatoric stress tensor is $s_{ij} = \sigma_{ij} - \delta_{ij}\,\sigma$, and the deviatoric strain tensor is $e_{ij} = \varepsilon_{ij} - \delta_{ij}\,\varepsilon$.

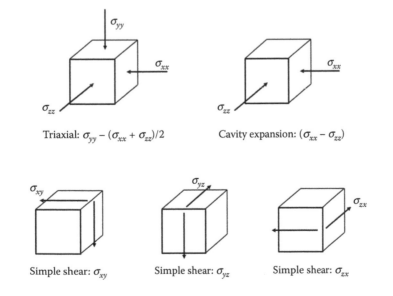

Triaxial: $\sigma_{yy} - (\sigma_{xx} + \sigma_{zz})/2$ Cavity expansion: $(\sigma_{xx} - \sigma_{zz})$

Simple shear: σ_{xy} Simple shear: σ_{yz} Simple shear: σ_{zx}

Figure 3.1 Deviator and shear stress components.

Similarly, the second invariants of the deviatoric stress and strain tensors follow the same form as the conventional equations:

Second invariant of deviator stress:

$$J_{2s} = \frac{1}{2} s_{ij} s_{ij} = \frac{1}{2} S_i S_i \qquad (3.4)$$

Second invariant of deviator strain:

$$J_{2e} = \frac{1}{2} e_{ij} e_{ij} = \frac{1}{2} E_i E_i \qquad (3.5)$$

When required, the mean stress σ and volumetric strain ε provide the sixth independent components of stress and strain to complete the description of these tensors. For a general discussion of tensorial description of stress and strain, the reader may consult a text on continuum mechanics (e.g., Malvern, 1969).

A description of anisotropy in terms of transformed stress measures now becomes fairly straightforward. For example, if an elliptical form of the yield locus is selected, a yield function f can be constructed as follows:

$$f = \frac{1}{2} \sum_{i=1}^{5} \left(\frac{S_i - b_i}{a_i} \right)^2 - k^2 = 0 \qquad (3.6)$$

The strength measure can normally be matched to the soil strength under pure simple shear loading, $k = s_{uDSS}$. The coefficients a_i control anisotropy amongst the various shearing modes (e.g., triaxial vs. direct simple shear), while b_i control anisotropy within a given shearing mode according to the sense of loading (e.g., triaxial compression versus extension). Equation 3.6 may be considered a general form of special cases developed by various authors. Setting all coefficients a_i to unity and all b_i to zero produces the von Mises yield surface. The Hill (1950) model permits independently specified soil strengths according to shearing mode ($a_i \neq 1$) but centers the yield ellipse at the origin of the stress space (all $b_i = 0$). The Hill model, therefore, cannot independently specify different strengths in triaxial compression and extension with a single yield locus. This limitation can be overcome by adopting a composite yield locus comprising two ellipses, as shown in Figure 3.2. The Davis and Christian (1971) model fits an ellipse to the undrained shear strengths in triaxial compression, triaxial extension, and simple shear. Figure 3.2 compares the three yield functions discussed above for a typical condition of anisotropy in a K_0-normally consolidated clay: $s_{uTC}/s_{uDSS} = 4/3$ and $s_{uTE}/s_{uDSS} = 2/3$. Differences and the Hill and Davis-Christian loci are relatively minor. The von Mises yield surface, when matched to the simple shear strength s_{uDSS}, is observed to substantially underestimate the triaxial compression strength and overestimate the strength in triaxial extension. The example shown depicts the $S_1 - S_3$ anisotropic condition typically observed in K_0-normally consolidated clays. This is a common consolidation condition, but not necessarily the only one that should be considered. For example, consolidation on a sloping seabed can generate anisotropy in the $S_3 - S_4$ plane in stress space.

3.1.2 The flow rule and internal energy dissipation

In a perfectly plastic material, yield and flow can occur at a constant stress state, and there is no unique plastic strain corresponding to a given stress state. However, a unique relationship

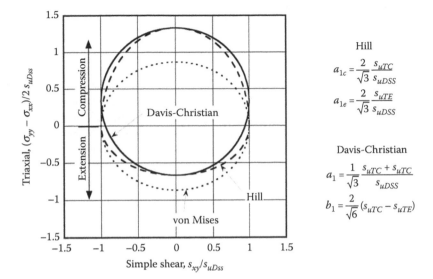

Figure 3.2 Undrained yield loci.

does exist between the direction (in stress space) of plastic flow and stress state, which can be described in terms of a scalar plastic potential function $g(\sigma_{ij})$ in conventional stress space. When incompressible materials are considered, consideration of the mean stress σ may be omitted and the plastic potential function can be expressed purely in terms of deviatoric stresses, or $g(S_i)$ in transformed stress space. Taking the gradient of g produces a plastic strain increment normal to the plastic potential surface, the components of which are described as follows:

$$\dot{E}_i^p = \lambda \frac{\partial g}{\partial S_i} \tag{3.7}$$

The scalar λ is indeterminate; however, Equation 3.7 provides useful information with regard to determining the direction of plastic flow, as shown in Figure 3.3. The yield

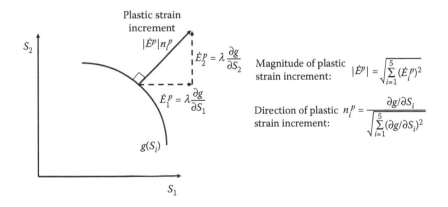

Figure 3.3 Flow rule.

function f is frequently selected as the plastic potential function, which is termed "associated flow."

The flow rule has several important applications. In generalized elasto-plastic constitutive models of soil behavior, it is used to separate strain increments into elastic and plastic components. In classical plasticity theory, it is a critical component of virtual work upper bound estimates of collapse loads. The upper bound analysis will be discussed later in this chapter; however, the key aspect from the standpoint of this discussion is that strain rates are determined in advance of the energy dissipation calculation. The flow rule enables a determination of the kinematic information necessary to evaluate the rate of energy dissipation occurring in a plastically deforming region, which in an incompressible material will be as follows:

$$\dot{D} = \sigma_{ij}\dot{\varepsilon}_{ij} = s_{ij}\dot{e}_{ij} = S_i\dot{E}_i \tag{3.8}$$

The superscript designating plastic deformation is omitted here, because in this context all deformations are plastic. Moreover, it is reiterated that repeated indices imply summation. For a von Mises material with an associated flow rule ($f = g$), the rate of energy dissipation per unit volume can now be determined from a simple series of steps:

$$\dot{D} = S_i\dot{E}_i = S_i\frac{\partial f/\partial S_i}{\sqrt{(\partial f/\partial S_i)^2}}\sqrt{\dot{E}_i^2} = S_i\frac{S_i}{\sqrt{S_i^2}}\sqrt{\dot{E}_i^2} = k\sqrt{2\dot{E}_i^2} \tag{3.9}$$

This result can be expressed in terms of conventional stress–strain measures to produce the well-known form of the equation $\dot{D} = k\sqrt{2\dot{\varepsilon}_{ij}\dot{\varepsilon}_{ij}}$. An alternative assumption of a Tresca yield criterion produces a second well-known expression for internal energy dissipation:

$$\dot{D} = 2k\,|\dot{\varepsilon}_{max}| = k\,|\dot{\gamma}_{max}| \tag{3.10}$$

If the material is anisotropic, a choice of the Hill form of an anisotropic yield surface requires a relatively simple modification to the derivation for a von Mises material. In this case, it is advantageous to redefine the stress and strain measures into alternative energy conjugate pairs: $S_{ai} = S_i\,a_i$, and $E_{ai} = E_i/a_i$. The sequence followed in Equation 3.9 can now be retraced to produce the following energy dissipation function for a Hill anisotropic material:

$$\dot{D} = k\sqrt{2\dot{E}_{ai}^2} = k\sqrt{2\sum_{i=1}^{5}(\dot{E}_ia_i)^2} \tag{3.11}$$

The advantage of transformed stress–strain measures for anisotropic materials now becomes very apparent: up to five degrees of anisotropy can be treated with a relatively simple equation. As noted earlier, the Hill description of anisotropy requires different coefficients according to the sense of loading (e.g., compression vs. extension). This is a minor issue in virtual work analysis calculations where strain fields are determined in advance of dissipation calculations.

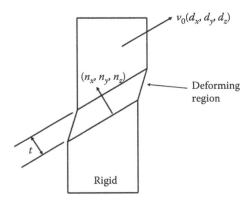

$v_0(d_x, d_y, d_z)$

(n_x, n_y, n_z)

Deforming region

t

Rigid

Figure 3.4 Velocity jump across a slip surface.

The energy dissipation rate equations described above can be applied to any continuously deforming region. However, plasticity theory also permits discontinuous velocity fields across a slip boundary provided that the normal components of velocity are continuous across the boundary. This can be analyzed in terms of the sketch provided in Figure 3.4. For an isotropic soil, the dissipation rate is easily analyzed by recognizing that the maximum strain is simply the velocity jump divided by the thickness of the zone of deformation v_0/t. The energy dissipation is then computed as the limit of the total energy dissipation in the deforming region as the thickness of the region approaches zero:

$$\dot{D} = \lim_{t \to 0} \int_t (kv_0/t)\, dt = kv_0 \tag{3.12}$$

The same result applies for both the Tresca and von Mises yield criteria. If the material is anisotropic, the strain rates within the deforming region must be decomposed into the constituent components according to the axes of anisotropy. In three dimensions the plane of slippage can be described in terms of a unit vector normal to the plane $\mathbf{n} = (n_x, n_y, n_z)$. Similarly, the velocity of the upper block in Figure 3.4 can be described in terms of the magnitude of velocity v_0 and a unit direction vector (d_x, d_y, d_z). Along the lines of Murff (1980), strain rates in the deforming region using the soil mechanics sign convention can now be determined as follows:

$$\dot{\varepsilon}_{ij} = -\frac{1}{2}\left(\frac{\partial v_j}{\partial x_i} + \frac{\partial v_i}{\partial x_j}\right) = -\frac{v_0}{2t}(d_j n_i + d_i n_j) \tag{3.13}$$

Transforming the strain rates in Equation 3.13 and invoking the requirement that the velocity acts parallel to the slip plane produces the strain rate measures in Table 3.2. With strain rates thus defined, the rate of energy dissipation for an anisotropic soil can be evaluated from Equation 3.11. In the limit for an infinitely thin deforming region, t vanishes as per Equation 3.12. The relatively simple expressions in Table 3.2 and Equation 3.11 demonstrate the effectiveness in treating an anisotropic medium in terms of transformed stress–strain measures together with the Hill description of anisotropy. This formulation is developed further in the analysis of caisson capacity in an anisotropic soil in Chapter 7.

Table 3.2 Strain rates across a velocity jump

Component	Strain rate
Triaxial	$\dot{E}_1 = \dfrac{v_0}{t}\sqrt{\dfrac{3}{2}}(d_x n_x + d_z n_z)$
Cavity expansion	$\dot{E}_2 = -\dfrac{v_0}{t}\dfrac{1}{\sqrt{2}}(d_z n_z - d_x n_x)$
Simple shear, xy	$\dot{E}_3 = -\dfrac{v_0}{t}\dfrac{1}{\sqrt{2}}(d_x n_y + d_y n_x)$
Simple shear, yz	$\dot{E}_4 = -\dfrac{v_0}{t}\dfrac{1}{\sqrt{2}}(d_y n_z + d_z n_y)$
Simple shear, zx	$\dot{E}_5 = -\dfrac{v_0}{t}\dfrac{1}{\sqrt{2}}(d_z n_x + d_x n_z)$

3.1.3 Generalized stresses and strains

The discussion above pertains to stress and strain for an infinitesimal element of material. The analysis is frequently conducted at the level of an entire structure or some element of a structure, in which case it becomes advantageous to employ the following generalized principles of plasticity (Prager, 1959):

1. Forces and moments are considered as generalized stresses.
2. Displacements and rotations are considered as generalized strains.
3. Force–moment interaction diagrams are considered as generalized yield surfaces.
4. Displacements and rotations relate to forces and moments by a flow rule.

An example is given by the eccentrically loaded plate anchor shown in Figure 3.5. In this case, the plate is subjected to combined force–moment loading. The ultimate capacity under combined loading is governed by an interaction diagram that corresponds to the yield locus for stress combinations. Plastic deformation is governed by a flow rule relating generalized stresses to generalized deformation increments, in this case translation and rotation. The generalized analysis provides a convenient basic for analysis at the "macroscopic" level and forms the basis for a large number of practical design methods for anchors.

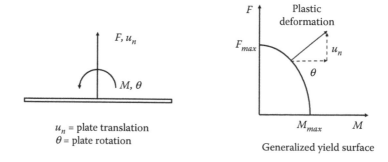

Figure 3.5 Generalized plasticity example.

3.2 PLASTIC LIMIT ANALYSIS

A rigorous solution to a plastic equilibrium problem should at every point in the soil mass be statically admissible, kinematically admissible, and not violate the yield condition. A statically admissible solution satisfies the partial differential equations of equilibrium at all points. Kinematic admissibility requires that the displacement field satisfy all displacement boundary conditions and that no gaps open up in the interior. Internal slip surfaces having discontinuous tangential components of displacement are admissible provided that the normal components of displacement at the slip surface are continuous. Aside from the permitted tangential discontinuities across slip surfaces, the displacement fields should have continuous first partial derivatives.

3.2.1 Lower and upper bound theorems

The requirements outlined above are commonly so onerous that rigorous solutions are frequently not possible for anything beyond a few simple special cases. Fortunately, useful solutions are possible when either the static admissibility or the kinematic admissibility requirements are relaxed. For the latter instance, the lower bound theorem states that a statically admissible stress field that balances external loads and does not violate the yield condition will produce a lower bound estimate of the collapse load. Regarding the former case, the upper bound theorem states that, for a kinematically admissible collapse mechanism, the collapse load obtained by equating the internal rate of energy dissipation to the rate of work from external loads leads to an upper bound estimate of load capacity.

While lower bound estimates can be obtained in some instances by simply guessing a stress field, a general lower bound solution normally requires a more powerful framework. The method of characteristics, which operates on the partial differential equations of equilibrium together with the yield function of the soil, provides such a framework. As will be observed in the example presented in this chapter, the application of the method of characteristics for the solution of boundary value problems generally requires insight and experience; therefore, it is not commonly used as a routine analysis tool. Nevertheless, lower bound solutions have been proved extremely valuable in providing benchmark solutions for evaluating solutions derived by other methods.

An upper bound analysis involves postulating a kinematically admissible collapse mechanism, from which a collapse load is computed by equating the internal rate of energy dissipation to the rate of work performed by external loads; that is, a virtual work analysis. The method has the advantage that the collapse mechanism can be systematically varied to seek an optimal mechanism producing a least upper bound. While the "least upper bound" for a presumed collapse mechanism is still not necessarily equal to the exact solution, the optimization process can nonetheless provide increased confidence in the result. As the calculations are often relatively straightforward, upper bound methods are readily adaptable to routine design calculations. Occasionally, the upper and lower bound collapse load estimates yield identical results, in which case the solution may be considered exact. Otherwise, we need to content ourselves with a bounded estimate of load capacity, which is nevertheless frequently useful.

3.2.2 Virtual work analysis

An example upper bound analysis is now presented for the case of a deeply embedded strip plate anchor of length L and finite thickness t_f. Figure 3.6 shows a postulated slip mechanism comprising two rigid wedges (ABB′ and FEE′), four continuously deforming circular shear

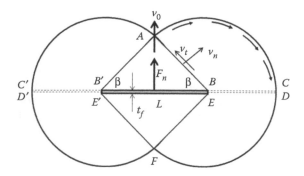

Figure 3.6 Upper bound analysis of strip anchor.

fans (ABC, AB′C′, DEF and D′E′F) and two rigid blocks (BCDE and B′C′D′E′). Bransby and O'Neill (1999) analyzed this for the case of full adhesion at the soil–anchor interface, $\alpha = 1$. The foregoing discussion expands the analysis slightly for general adhesion conditions. The shape of the triangular wedge is defined by the angle β. Taking advantage of the problem symmetry, only five unique dissipation functions in the upper right quadrant need to be defined: along slip surfaces, AB, BE, CD, and AC; within the continuously deforming region, ABC.

If the wedge ABB′ translates upward at velocity v_0, then the normal and tangential components of velocity for surface AB become $v_n = v_0 \cos \beta$ and $v_t = v_0 \sin \beta$, respectively. The rotation of arc AC, as shown in Figure 3.6, requires a velocity discontinuity equal to v_t across the slip surface AB. Thus, the rate of energy dissipation at this surface becomes

$$\dot{D}_{AB} = s_u v_0 \sin \beta \frac{L/2}{\cos \beta} = s_u v_0 L \left\{ \frac{1}{2} \tan \beta \right\} \tag{3.14}$$

As the magnitude of the velocity along arc AC should equal v_n, the corresponding rate of energy dissipation becomes

$$\dot{D}_{AC} = s_u v_0 \cos \beta \left\{ \frac{L/2}{\cos \alpha} \times (\pi - \beta) \right\} = s_u v_0 L \left\{ \frac{1}{2} (\pi - \beta) \right\} \tag{3.15}$$

With the fluke moving upward at velocity v_0 and block BCDE moving downward at velocity v_n, the relative velocity between the fluke and surface BE is $v_0 + v_n$, while the relative velocity between block surface CD and the adjacent rigid soil is v_0. If α is the adhesion factor at the soil–fluke interface, the rate of energy dissipation along surfaces BE and CD are as follows:

$$\dot{D}_{BE} = \alpha s_u v_0 (1 + \cos \beta) \frac{t_f}{2} \tag{3.16}$$

$$\dot{D}_{CD} = s_u v_0 \cos \beta \frac{t_f}{2} \tag{3.17}$$

Considering a polar coordinate $(r - \theta)$ system in the shear fan ABC, the radial component of velocity v_r must be zero everywhere, while the circumferential component v_θ equals

$v_0 \cos \beta$ at all points. The single nonzero strain rate component (Malvern, 1969, Section II.4) is, therefore, as follows:

$$\dot{\gamma}_{r\theta} = 2\dot{\varepsilon}_{r\theta} = 2\frac{1}{2}\left(\frac{1}{r}\frac{\partial v_r}{\partial \theta} + \frac{\partial v_\theta}{\partial r} - \frac{v_\theta}{r}\right) = -\frac{v_0 \cos \beta}{r} \tag{3.18}$$

A choice of the Tresca yield criterion will lead to the following expression for the total rate of energy dissipation within the shear fan:

$$\dot{D}_{ABC} = \int_V s_u \mid \dot{\gamma} \mid_{max} dV = \int_\beta^\pi \int_0^{L/2\cos\beta} s_u \frac{v_0 \cos \beta}{r} r\,dr\,d\theta = s_u v_0 L\left\{\frac{1}{2}(\pi - \beta)\right\} \tag{3.19}$$

Equating the external rate of work $F_n v_0$ to the sum of the dissipation terms ultimately leads to the following expression for ultimate uplift capacity under pure normal loading:

$$\frac{F_n}{s_u L} = N_{nmax} = 4\left\{\left[(\pi - \beta) + \frac{\tan \beta}{2}\right] + \frac{t_f}{2L}[\alpha + (1+\alpha)\cos \beta]\right\} \tag{3.20}$$

Minimizing F_n with respect to β produces an optimal $\beta_{opt} = 45°$ for a fluke of zero thickness and increases to about $48°$ for a thicker fluke $t_f/L = 1/7$. The impact on F_n for this range of β turns out to be negligible. Therefore, the use of a single value $\beta = 45°$ for all cases should be considered to yield sufficient accuracy for the typical range of fluke thickness ratios $L/t_f > 7$. Equation 3.20 indicates a general range $N_{nmax} = 11.5–12.1$ depending on fluke thickness ratio and adhesion factor (Figure 3.7). The bearing factor N_{nmax} is denoted as being a maximum because no eccentricity exists.

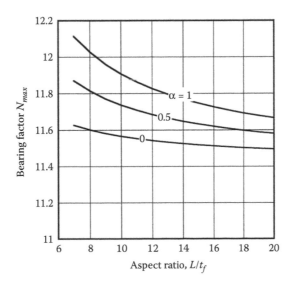

Figure 3.7 Upper bound bearing factors for strip anchor.

3.2.3 Method of characteristics

The method of characteristics is a technique for solving hyperbolic partial differential equations by transforming them into a family of ordinary differential equations that can be solved by integration along so-called characteristic surfaces (Rao, 2002). In bearing capacity problems, the equations of equilibrium coupled with the yield criterion generate a system of hyperbolic differential equations amenable to solution by this method. As this solution approach enforces static equilibrium everywhere, but does not ensure strain compatibility, computed collapse loads should be considered as lower bounds. The example presented below is for a plane strain problem, but solutions to axisymmetric problems are also possible.

For a frictionless material with shear strength s_u, the relevant plane strain equilibrium equations and yield condition are shown below, where f_x and f_y are body forces (e.g., soil unit weight).

$$\frac{\partial \sigma_{xx}}{\partial x} + \frac{\partial \sigma_{xy}}{\partial y} = f_x$$

$$\frac{\partial \sigma_{xy}}{\partial x} + \frac{\partial \sigma_{yy}}{\partial y} = f_y \qquad (3.21)$$

$$\left(\frac{\sigma_{xx} - \sigma_{yy}}{2}\right)^2 + \sigma_{xy}^2 = s_u^2$$

Defining $\sigma_m = (\sigma_{xx} + \sigma_{yy})/2$ and θ as the angle between the orientation of the major principal stress and the horizontal axis leads to

$$\sigma_{xx} = \sigma_m + s_u \cos 2\theta$$

$$\sigma_{yy} = \sigma_m - s_u \cos 2\theta \qquad (3.22)$$

$$\sigma_{xy} = s_u \sin 2\theta$$

The equilibrium equations can now be restated as follows:

$$\frac{\partial \sigma_m}{\partial x} - 2s_u \sin 2\theta \frac{\partial \theta}{\partial x} + 2s_u \cos 2\theta \frac{\partial \theta}{\partial y} = f_x - \cos 2\theta \frac{\partial s_u}{\partial x} - \sin 2\theta \frac{\partial s_u}{\partial y}$$

$$\frac{\partial \sigma_m}{\partial y} + 2s_u \sin 2\theta \frac{\partial \theta}{\partial y} + 2s_u \cos 2\theta \frac{\partial \theta}{\partial x} = f_y + \cos 2\theta \frac{\partial s_u}{\partial y} - \sin 2\theta \frac{\partial s_u}{\partial x} \qquad (3.23)$$

Two additional identities are now introduced:

$$d\sigma_m = \frac{\partial \sigma_m}{\partial x} dx + \frac{\partial \sigma_m}{\partial y} dy$$

$$d\theta = \frac{\partial \theta}{\partial x} dx + \frac{\partial \theta}{\partial y} dy \qquad (3.24)$$

Equations 3.23 and 3.24 can be expressed in the following matrix form:

$$\begin{bmatrix} 1 & 0 & -2s_u\sin 2\theta & 2s_u\cos 2\theta \\ 0 & 1 & 2s_u\cos 2\theta & 2s_u\sin 2\theta \\ dx & dy & 0 & 0 \\ dx & dy & 0 & 0 \end{bmatrix}\begin{bmatrix} \partial\sigma_m/\partial x \\ \partial\sigma_m/\partial y \\ \partial\theta/\partial x \\ \partial\theta/\partial y \end{bmatrix} = \begin{bmatrix} f_1 \\ f_2 \\ d\sigma_m \\ d\theta \end{bmatrix} \tag{3.25}$$

where $f_1 = f_x - \cos 2\theta \, \partial s_u/\partial x - \sin 2\theta \, \partial s_u/\partial y$ and $f_2 = f_y + \cos 2\theta \, \partial s_u/\partial y - \sin 2\theta \, \partial s_u/\partial x$.

A characteristic is defined as the curve along which the determinant of the matrix in Equation 3.25 vanishes. The operation produces the following two (α and β) characteristic curves:

$$\frac{dy}{dx} = \tan(\theta \pm \pi/4) \tag{3.26}$$

The stress gradient components $\partial\sigma_m/\partial x$ and $\partial\sigma_m/\partial y$ are obtained by invoking Cramer's rule to Equation 3.25, recognizing that the determinants in both the numerators and denominator must vanish along the characteristics. This produces the governing ordinary differential equation along the respective α and β characteristics:

$$d\sigma_m \mp 2s_u d\theta = \left(f_x \mp \frac{\partial s_u}{\partial y}\right)dx + \left(f_y \pm \frac{\partial s_u}{\partial x}\right)dy \tag{3.27}$$

Solution by integration along the characteristics starts from a prescribed boundary value stress and marching into the solution domain. Equations 3.26 and 3.27 typically must be solved numerically and iteratively. As will be discussed in Chapter 4, the general approach presented here can be extended to axisymmetric problems. In the special case of a soil with uniform strength where self-weight effects are neglected, the right-hand sides of Equation 3.27 is zero where analytical solutions are sometimes possible.

Figure 3.8 shows two simple but important special cases. Figure 3.8a considers the case of linear characteristic lines in a region with minor principal stress $\sigma_3 = 0$ and $\theta = 0$ at the free boundary. As the soil is at yield, the major principal stress and mean stress are $\sigma_1 = 2s_u$ and $\sigma_m = s_u$, respectively. The orientations of the characteristic lines are constant at $\pm 45°$ throughout the region. From Equation 3.27, mean stress is constant in this region, $\Delta\sigma_m = 0$. Figure 3.8b shows the case of intersecting β characteristics, as can occur at a stress discontinuity (e.g., the edge of a plate). From Equation 3.27 mean stress varies due to the curvature of the α characteristics such that $\Delta\sigma_m = 2s_u \, \Delta\theta$.

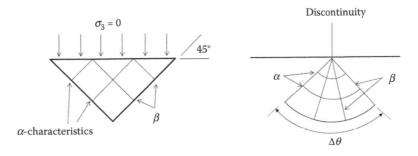

Figure 3.8 Method of characteristics special cases.

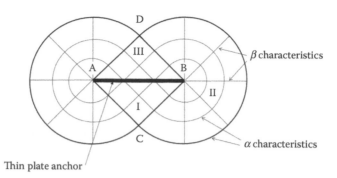

Figure 3.9 Lower bound analysis of strip anchor.

Figure 3.9 shows the application of these solutions to compute the load capacity of a deeply embedded plate, which essentially follows the solution of Meyerhof (1951). The postulated characteristic net comprises two regions with linear characteristic lines (I and III) and one region containing radial β characteristics. Arbitrarily assigning a vertical stress $\sigma_3 = 0$ at the base of the plate AB leads to a uniform mean stress $\sigma_m = s_u$ in Region I. Integrating Equation 3.27 from BC to BD in Region II leads to $\sigma_m = s_u (1 + 3\pi)$. Mean stress is constant in Region III; therefore, the mean stress along the top surface of plate AB is $\sigma_m = s_u (1 + 3\pi)$. The major principal stress on AB equals the mean stress σ_m plus the radius of Mohr's circle s_u. The method of characteristics solution for the bearing capacity of a strip anchor is then as follows:

$$\frac{q_{ult}}{s_u} = N_{n\max} = 3\pi + 2 \tag{3.28}$$

Strictly speaking, the analysis shown is actually a partial stress field contained within the zone of yield. Qualifying as a rigorous lower bound requires that a statically admissible stress field can be extended outside of the zone of yielding. Martin and Randolph (2001) actually demonstrate that such an extension is possible for this case. Therefore, the lower bound estimate is shown to be independent of surface roughness of the plate and provides an exact match to the upper bound estimate, Equation 3.20 for an infinitely thin plate.

3.3 SOIL CONSTITUTIVE BEHAVIOR

The PLAs discussed above require relatively limited information on soil behavior in the form of a yield locus and flow rule. Other analyses, such as the cavity expansion and FE methods discussed later in this chapter, require a more complete description of soil behavior that considers elastic response and—in many instances—hardening behavior. Soil constitutive (stress–strain-strength) behavior involves complexities including anisotropy, hardening and softening, nonlinearity, and permanent deformations at small strains. Entire volumes have been devoted to the topic, and full coverage is beyond the scope of this text. Accordingly, the discussion below outlines only the major points necessary to guide subsequent discussion of the analysis methods presented later in this chapter and throughout the remainder of the text. The present discussion is restricted to rate-independent elasto-plastic constitutive models. Strain rate is known to affect soil stress–strain behavior. Discussion of rate effects is omitted here, but will be introduced when relevant to anchor behavior in subsequent discussions throughout the text.

3.3.1 Linearly elastic–perfectly plastic

For analysis of anchor behavior under undrained loading conditions the simplest and most widely used model of soil behavior assumes linearly elastic–perfectly plastic behavior (LEPP) beneath either a von Mises or Tresca yield surface. At yield, an associated flow law is assumed. Figure 3.10 illustrates the LEPP model using a von Mises yield criterion. As they are vectors obeying the usual rules of vector algebra and calculus, the use of transformed stress–strain measures (Table 3.1) is helpful in visualizing the derivation. The key to the model formulation is the decomposition of the total strain increment into plastic and elastic components as shown in Figure 3.10. As discussed earlier, k is a strength parameter normally equated to the soil shear strength in a simple shearing mode s_{uDSS}. The reader may also refer to Figure 3.3 for the general rule for evaluating the direction of plastic flow. Stress increments relate directly to the elastic component of the strain increment by a factor of twice the shear modulus $2G$. Purely elastic behavior occurs when loading from a stress state beneath the yield or unloading from a point on the yield surface. Thus, the complete model is as follows:

$$\dot{S}_i = 2G\dot{E}_i \quad \text{Elastic}$$

$$\dot{S}_i = 2G\left(\dot{E}_i - \frac{\dot{E}_j S_j}{2k^2}S_i\right) \quad \text{Elasto-plastic} \tag{3.29}$$

In an undrained analysis—a common use of this model for anchors in clay—the bulk modulus is infinite; therefore, mean stress should be calculated from equilibrium considerations. In a cavity expansion analysis, the equilibrium equations from continuum mechanics are directly applied for this purpose. In FE analyses, a large but finite bulk modulus is often employed as a numerical expedient; however, mean stresses are still fundamentally controlled by equilibrium in the boundary value problem under consideration. Alternatively, a hybrid FE analysis is possible in which mean stresses are treated as nodal variables in addition to displacements.

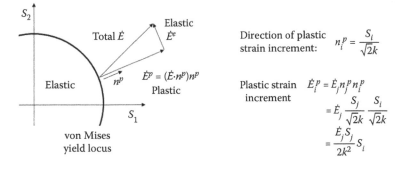

Figure 3.10 Linearly elastic–perfectly plastic model.

3.3.2 Modified Camclay

Problems involving drainage require a description of volumetric constitutive behavior in addition to shear behavior. Among the most widely used for this purpose is the Modified Camclay (MCC) model by Roscoe and Burland (1968). The MCC soil model characterizes soil behavior in a critical state framework (Schofield and Wroth, 1968; Atkinson and Bransby, 1977). The model unifies soil volumetric behavior as measured in consolidation tests to shear behavior as measured in a triaxial test. Figure 3.11 depicts the key features of the model, namely:

1. The yield locus in deviator-mean stress $(S - \sigma')$ space comprises an ellipsoid. The size of one axis of the ellipsoid is controlled by a hardening parameter α that describes the hydrostatic preconsolidation pressure. The size of the other axis is controlled by the critical state line, the slope of which (defined by parameter c) is a function of the large strain soil friction angle in triaxial compression.
2. Generalization of the critical state and yield functions to stress states other than pure triaxial shear is commonly based on a von Mises generalization as depicted in Figure 3.11. This generalization can lead to implied friction angles in triaxial extension that are either unrealistically high or physically impossible. To mitigate this problem, some analysis procedures permit an alternative generalization using a yield criterion that is also a function of the third invariant of the stress tensor (Dsimulia, 2012). In its limiting form, this yield locus approaches a triangular shape, rather than the circular shape shown in Figure 3.11.
3. Plastic strains follow an associated flow rule.
4. The elastic bulk modulus K_b is estimated from the slope of the swelling line κ as measured during the unloading stage of a consolidation test. The bulk modulus is stress level dependent, being directly proportional to the mean effective stress σ'. The shear

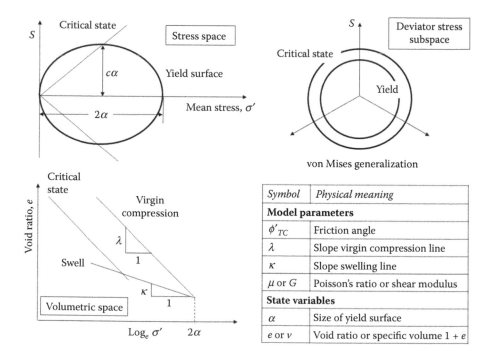

Symbol	Physical meaning
Model parameters	
ϕ'_{TC}	Friction angle
λ	Slope virgin compression line
κ	Slope swelling line
μ or G	Poisson's ratio or shear modulus
State variables	
α	Size of yield surface
e or ν	Void ratio or specific volume $1 + e$

Figure 3.11 Modified Camclay model.

Table 3.3 Modified Camclay equations

Model component	Equation
Elastic behavior	Bulk modulus: $K_b = \left(\dfrac{v_0}{\kappa}\right)\sigma'$
	Shear modulus: $G = 3\,K_b\,(1-2\mu)/(1+\mu)/2$ or $G =$ constant specified value
Yield function	$f = S_i S_i - c^2 \sigma'\left(2\alpha - \sigma'\right) = 0$
	$c = \sqrt{\dfrac{2}{3}}\,\dfrac{6\sin\phi'}{3-6\sin\phi'}$
Gradient of yield function **Q** and direction of plastic flow **P**	Mean: $Q = P = 2c^2(\sigma' - \alpha)$ Deviator: $Q_i' = P_i' = 2S_i$
Elasto-plastic modulus	$H = 4c^4\sigma'\left(\sigma' - \alpha\right)\left(\dfrac{v_0}{\lambda - \kappa}\right)\alpha$
Magnitude of plastic flow (scalar)	$\dot{\Lambda} = \dfrac{2GQ_i'E_i + K_bQ\dot{\varepsilon}}{H + 2GQ_i'\ Q_i' + K_bQ^2}$
Stress increment	Mean: $\dot{\sigma}' = K_b(\dot{\varepsilon} - \dot{\Lambda}Q)$ Deviator: $\dot{S}_i = 2G(E_i - \dot{\Lambda}Q_i')$
Hardening law	$\dot{\alpha} = \left(\dfrac{v_0}{\lambda - \kappa}\right)\dot{\Lambda}P$

modulus is commonly estimated by one of two methods. The first method assumes a constant Poisson's ratio μ such that $2G/K_b = 3(1-2\mu)/(1+\mu) = $ constant. As shear modulus is proportional to bulk modulus, it will also be dependent on stress level. The second method assumes a constant G that is independent of stress level σ'.

5. The yield surface exhibits density hardening, which is proportional to the rate of plastic volumetric strain.

Table 3.3 lists the relevant equations for numerical implementation of the MCC model. Model parameters and state variables are tabulated in Figure 3.11. For a given volumetric-deviatoric strain increment $(\dot{\varepsilon}, \dot{E})$ a mean-deviatoric stress increment $(\dot{\sigma}, \dot{S})$ is computed. The full set of equations applies to elasto-plastic load conditions. In cases of loading within or unloading from the yield surface, the plastic strain magnitude parameter $\dot{\Lambda}$ is set to zero.

The MCC model has a number of limitations including a lack of capability to model the following types of behavior: strain softening in normally consolidated soils, anisotropy, nonlinear stress–strain behavior in over-consolidated soil under undrained loading, and permanent deformations during cyclic loading. Nevertheless, in many instances such behavior does not significantly impact the process under consideration; therefore, the MCC model remains a useful analytical tool when used within its limitations.

3.3.3 Advanced elasto-plastic models

A relatively large number of soil constitutive models exist, most of which are tailored to characterize specific aspects of soil behavior. Attention here is provided to the MIT-E3 model by Whittle and Kavvadas (1993), the reason being that this model was originally developed for characterizing a common type of anchor problem, skin friction on pile anchors for tension leg platforms. In developing the model, specific focus was provided for systematically considering all stages in the life of a single pile, including pile installation with its associated soil disturbance, dissipation of excess pore pressures following anchor installation (setup), application of a monotonic mooring force to the anchor, and cyclic loading owing to storm waves. The MIT-E3 soil model introduced capabilities for modeling several critical aspects of soil behavior that cannot be modeled by the original MCC model, including softening owing to installation disturbance and excess pore pressure accumulation occurring under cyclic loading.

The MIT-E3 model makes three levels of refinement to the MCC model. The first refinement modifies the elliptical MCC yield surface along the lines of an earlier model proposed by Kavvadas (1982) as shown in Figure 3.12. The rotation of the yield surface in stress space is controlled by an anisotropy tensor $\mathbf{b}$, which in turn is controlled by the consolidation history of the soil. Thus, the MIT-E3 model features two forms of hardening: density hardening controlling the size of the yield surface through the state variable α, and rotational hardening controlling the orientation of the yield surface through the state variable $\mathbf{b}$. This modification permits two important capabilities from the standpoint of realistically describing measured behavior in normally consolidated clays: anisotropy and strain softening. The model also describes the failure state in terms of critical state cone in stress space, the axis of which is offset from the hydrostatic axis a distance O–O′ in the S_1 (triaxial) direction in stress space (Figure 3.12). This simple adjustment avoids the problem of unrealistically high friction angles in triaxial extension that were noted earlier in connection with the von Mises generalization of the failure criterion.

The second refinement introduces hysteretic behavior for unload–reload cycles beneath the yield surface (Figure 3.13). The purely hysteretic component of the model is isotropic with no coupling between shear and volumetric behavior. This feature permits a description of nonlinear stress–strain behavior at small strains. The final refinement introduces plastic strains for over-consolidated clays, that is, for unload–reload within the yield surface.

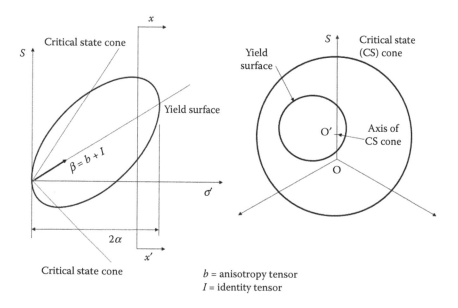

Figure 3.12 Yield and failure surface in MIT-E3 model.

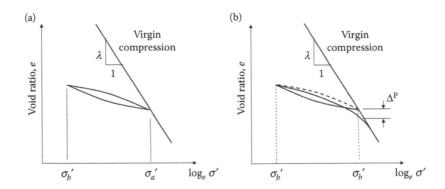

Figure 3.13 Unload–reload in the MIT-E3 model. (a) Perfect hysteresis and (b) hysteresis with bounding surface plasticity.

Plastic deformations within the yield surface provides capabilities for (1) modeling the generation of shear-induced pore pressures for undrained monotonic and cyclic shearing and (2) anisotropic stress–strain behavior at small strains. The model employs a bounding surface plasticity formulation in which the magnitude of plastic strains depend on the proximity of the current stress state to the yield surface.

The applicability of MIT-E3 soil model is restricted to normally and over-consolidated clays. A subsequent development is the MIT-S1 model (Pestana and Whittle, 1999), which extends and modifies many of the features of MIT-E3 to describe both clays and sands in a unified framework; thus, it is arguably a more powerful model.

3.4 CAVITY EXPANSION METHODS

The upper bound analysis presented earlier assumed rigid-plastic soil behavior, which implicitly assumes that elastic deformations have no influence on ultimate pullout capacity. In the case of a deeply embedded strip anchor, this assumption is justified, since a kinematically admissible collapse is made possible by assuming plastically deforming soil to flow from top to bottom around the plate as shown Figure 3.14. Shallowly embedded anchors (Figure 3.14) also lend themselves well to rigid-plastic analysis, since the free surface provides a mechanism for accommodating an uplifting volume of plastically deforming soil. The right side of Figure 3.14 shows a contrasting condition occurring during pile penetration, where a rigid-plastic analysis is inadequate. In this case, the volume ΔV displaced beneath the pile tip flow-around cannot be accommodated by flow-around owing to the rigid sidewall of the pile. With no nearby free surface, the only mechanism for accommodating ΔV under undrained loading conditions is elastic shear distortion outside of the zone of yield. Thus, computation of the collapse load must necessarily consider the elastic properties as well as the strength of the soil mass.

The simplest analytical framework for this process is to model the penetrating pile as an expanding cylinder. In spite of the obvious simplification of the actual process, this captures the elasto-plastic nature of the problem which may be adequate in many instances. However, the expanding cylinder analogy clearly neglects the two-dimensional nature of the problem. Furthermore, not all penetrating anchors are solid cylinders. They can also be plates or open-ended tubes with some portion of the displaced soil flowing inside the tube. In these instances, the expanding cylinder analogy becomes increasingly tenuous. The strain

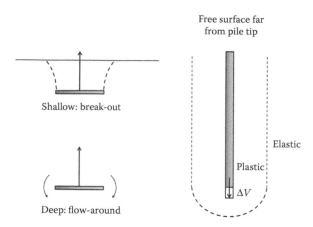

Figure 3.14 Applicability of rigid-plastic analysis.

path method, as formulated by Baligh (1985), employs an approach that parallels a simple cavity expansion analysis, but with much more versatility in simulating two- and three-dimensional penetrator geometries. Finally, some situations are too complex altogether for analyzing with any cavity expansion analogy. For example, the collapse loads for deeply embedded finite length plate anchors, in contrast to the strip anchor in Figure 3.14, show significant sensitivity to material elastic properties, notwithstanding the potential for a flow-around mechanism. This topic is taken up later in the discussion of large displacement FE analysis of collapse loads.

3.4.1 Cylindrical cavity expansion

Analysis of an expanding cylindrical cavity in an incompressible soil proceeds by considering the line source of incompressible material of intensity Q shown in Figure 3.15. If the soil behavior is considered to be rate independent, any convenient value may be selected for the source intensity (say $Q = \pi$) and time serves as a dummy variable for computing the current

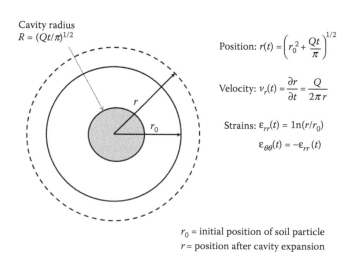

$$\text{Position: } r(t) = \left(r_0^2 + \frac{Qt}{\pi} \right)^{1/2}$$

$$\text{Velocity: } v_r(t) = \frac{\partial r}{\partial t} = \frac{Q}{2\pi r}$$

$$\text{Strains: } \varepsilon_{rr}(t) = \ln(r/r_0)$$
$$\varepsilon_{\theta\theta}(t) = -\varepsilon_{rr}(t)$$

r_0 = initial position of soil particle
r = position after cavity expansion

Figure 3.15 Cylindrical cavity expansion.

radius R of the expanding cavity. A simple mass balance computation determines the current coordinate r for any selected point in the soil mass located having an initial coordinate r_0. Once the particle coordinate is updated, the strain experienced by the soil particle under consideration can be computed by a simple logarithmic equation (Figure 3.15).

With strain histories fully defined, either effective stress or deviatoric stress is computed, depending on the selected soil constitutive model. Figure 3.16 shows an example effective stress profile using the MCC soil model to predict the stress field following installation of a pile of radius R in a normally consolidated clay. The pore water pressure p generated during the expansion is calculated by invoking the equilibrium equation (in polar coordinates) as shown by the calculation sequence in Equation 3.30. Adding pore pressure p to effective stresses produces total stress, which completes the analysis. If a total stress constitutive model is selected, a similar analysis proceeds in terms of deviatoric and mean stress in lieu of effective stress and pore pressure.

$$\frac{\partial(\sigma'_{rr}+p)}{\partial r}+\frac{\sigma'_{rr}-\sigma'_{\theta\theta}}{r}=0$$

$$g_p=\frac{\partial p}{\partial r}=-\left(\frac{\partial\sigma'_{rr}}{\partial r}+\frac{\sigma'_{rr}-\sigma'_{\theta\theta}}{r}\right)$$

$$p=\int_r g_p\,dr$$

$$\sigma_{rr}=\sigma'_{rr}+p;\quad \sigma_{\theta\theta}=\sigma'_{\theta\theta}+p$$

(3.30)

Stress fields induced by pile installation disturbance can now be prescribed as an initial condition for a consolidation analysis to simulate "setup," or strength recovery following installation. This is typically accomplished by FE or finite difference techniques, which will be discussed subsequently. If the aim of the analysis is to predict initial excess pore pressures for estimating setup time, plots such as that depicted in Figure 3.16 are not unrealistic. However, the effective stress fields shown here are far from realistic—the effective stress at the pile boundary is typically less, rather than greater, than the far-field *in situ* stress. Achieving a better match to field observations requires a more realistic characterization of installation disturbance along with more realistic modeling of soil behavior.

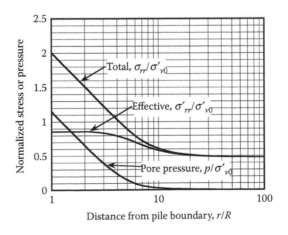

Figure 3.16 Cavity expansion using modified Camclay model.

3.4.2 The strain path method

The cylindrical cavity expansion analysis has an obvious limitation with regard to simulating pile penetration in that it models a two-dimensional problem with a one-dimensional analogy. To more closely model the two-dimensional nature of the process, Baligh (1985) developed the strain path method, the starting point of which is the expanding spherical cavity emanating from an incompressible point source as shown in Figure 3.17. The form of the equations closely parallel to that of an expanding cylinder. If the point source is now moved downward at a velocity v_0, it leaves the trail of incompressible material closely resembling a solid pile shown in Figure 3.18. The source strength required to generate the penetrating pile is $Q = \pi R^2 v_0$. The shape is termed a "simple pile" because it derives from a single point source. Velocity fields within the soil mass are generated through simple addition of v_0 to the spherical cavity expansion terms. Streamlines are recursively generated by simple numerical integration. As previously noted for cylindrical cavity expansion analyses, if the soil properties are rate-independent, any convenient selection of the time and velocity scale is acceptable, for example $v_0 = 1$.

As velocity is expressed analytically, strain rates can be computed from spatial derivatives of velocity and then integrating with respect to time along a streamline (Baligh, 1985) for determining strains. However, calculations can be greatly simplified (Aubeny and Grajales, 2015) by taking advantage of the following relationship associated with steady penetration at velocity v_0 in the z-direction:

$$\frac{\partial}{\partial z} = \frac{1}{v_0}\frac{\partial}{\partial t} \tag{3.31}$$

Strains experienced by soil elements as they traverse from their initial position (r_0, z_0) to any position (r, z) along a streamline are now as follows:

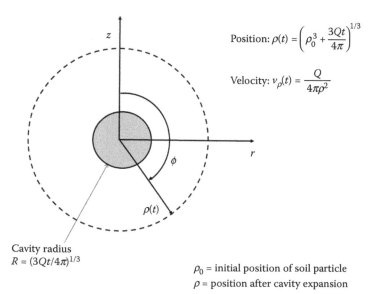

Position: $\rho(t) = \left(\rho_0^3 + \dfrac{3Qt}{4\pi}\right)^{1/3}$

Velocity: $v_\rho(t) = \dfrac{Q}{4\pi\rho^2}$

Cavity radius
$R = (3Qt/4\pi)^{1/3}$

ρ_0 = initial position of soil particle
ρ = position after cavity expansion

Figure 3.17 Spherical cavity expansion.

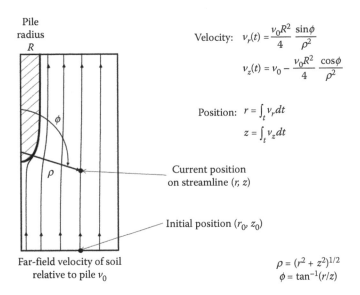

Pile radius R

Velocity: $v_r(t) = \frac{v_0 R^2}{4} \frac{\sin\phi}{\rho^2}$

$v_z(t) = v_0 - \frac{v_0 R^2}{4} \frac{\cos\phi}{\rho^2}$

Position: $r = \int_t v_r dt$

$z = \int_t v_z dt$

Current position on streamline (r, z)

Initial position (r_0, z_0)

Far-field velocity of soil relative to pile v_0

$\rho = (r^2 + z^2)^{1/2}$
$\phi = \tan^{-1}(r/z)$

Figure 3.18 Simple pile.

$$\varepsilon_{zz} = -\int^t \frac{\partial v_z}{\partial z} dt = -\int^t \frac{1}{v_0} \frac{\partial v_z}{\partial t} dt = 1 - \frac{v_z}{v_0}$$

$$\varepsilon_{\theta\theta} = -\int^t \frac{v_r}{r} dt = -\int^t \frac{1}{r} \frac{\partial r}{\partial t} dt = -\log_e(r/r_0)$$

$$\varepsilon_{rr} = -\varepsilon_{zz} - \varepsilon_{\theta\theta}$$

$$\varepsilon_{rz} = -\int^t \frac{\partial v_r}{\partial z} dt = -\int^t \frac{1}{v_0} \frac{\partial v_r}{\partial t} dt = -\frac{v_r}{v_0}$$

(3.32)

As shown in Figure 3.18, r_0 and z_0 are the coordinates of the initial position of the point under consideration, and v_0 is the velocity of pile penetration. The negative signs here arise from the soil mechanics convention of defining compressive strains as positive. The simplified form of the meridianal strain term ε_{rz} is possible by recognizing the velocity field in this case is irrotational. Strain at any location (Figure 3.19) can thus be calculated from simple expressions in terms of the current location and velocity of the soil element in question. The computational effort to determine strain at any point surrounding the penetrating pile is therefore comparable to the effort required for the simpler cylindrical cavity expansion analysis, albeit additional strain components and spatial computation points are required since the displacement field is two dimensional.

With strain histories fully defined along the streamlines, effective stresses are computed from the constitutive model selected for the analysis, and pore pressure gradients (g_{pr}, g_{pz}) are computed by invoking the conventional equilibrium equations from continuum mechanics (Malvern, 1969):

$$g_{pr} = \frac{\partial p}{\partial r} = -\left(\frac{\partial \sigma'_{rr}}{\partial r} + \frac{\sigma'_{rr} - \sigma'_{\theta\theta}}{r} + \frac{\partial \sigma'_{rz}}{\partial z} \right)$$

$$g_{pz} = \frac{\partial p}{\partial z} = -\left(\frac{\partial \sigma'_{zz}}{\partial z} + \frac{\partial \sigma'_{rz}}{\partial r} + \frac{\sigma'_{rz}}{r} \right)$$

(3.33)

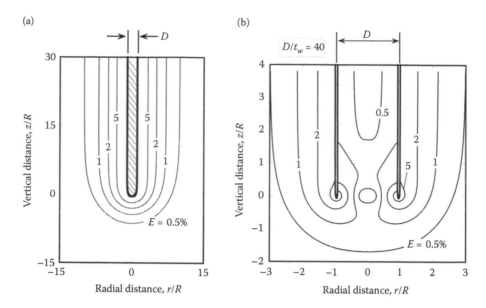

Figure 3.19 Shear strain around penetrating piles. (a) Solid or plugged and (b) open-ended.

Pore pressures may be computed by direct integration of Equation 3.33. As the velocity fields are determined independently of equilibrium considerations, pore pressures will generally be dependent on integration path; that is, they are nonunique. In essence, the strain path analysis is an upper bound analysis and shares the same advantages and limitations of that class of analysis; most notably, statically inadmissible stress fields cannot be guaranteed. A Poisson solution for pore pressure (Whittle and Aubeny, 1991) is one method of mitigating this problem. Alternatively, Teh and Houlsby (1988) use the stress field computed from a stain path analysis as the initial condition for subsequent pile penetration computed by a FE analysis. This approach effectively achieves an equilibrated state while preserving to a large extent the effects of disturbance in the soil mass induced by pile penetration.

Overall, a relatively simple modification to the cavity expansion analysis can produce a much closer approximation to not only real pile geometry but also to the actual two-dimensional strain histories experienced by soil elements close to the pile boundary. More realistic tip configurations are possible through superposition methods using multiple sources and sinks (Levadoux and Baligh, 1980). Plate and open tube penetration can be modeled in a manner similar to that shown for a solid pile by employing, respectively, a semi-finite line source (Aubeny, 1992) or a ring source (Chin, 1986). In the latter case, a ring source of intensity Q and radius R_r moves downward at velocity v_0, producing the following velocity field in the soil mass:

$$v_r = \frac{Q}{4\pi^2} \frac{1}{r\sqrt{(r+R_r)^2 + z^2}} \left[K(k)\left\{1 - \frac{2r(r-R_r)}{z^2 + (r-R_r)^2}\right\} E(k)\right]$$

$$v_z = v_0 + \frac{V}{4\pi^2} \frac{2z}{(z^2 + (r-R_r)^2)\sqrt{(r+R_r)^2 + z^2}} E(k)$$

(3.34)

where the terms v_r and v_z are the respective radial and vertical components of particle velocity. The functions K and E are elliptic integrals of the first and second kind, which can be expressed in terms of the following infinite series:

$$K = \frac{\pi}{2}\left[1 + \left(\frac{1}{2}\right)^2 k + \left(\frac{1\cdot 3}{2\cdot 4}\right)^2 k^2 + \left(\frac{1\cdot 3\cdot 5}{2\cdot 4\cdot 6}\right)^2 k^3 + \left(\frac{1\cdot 3\cdot 5\cdot 7}{2\cdot 4\cdot 6\cdot 8}\right)^2 k^4 + \cdots\right]$$

$$E = \frac{\pi}{2}\left[1 - \left(\frac{1}{2}\right)^2 \frac{k}{1} - \left(\frac{1\cdot 3}{2\cdot 4}\right)^2 \frac{k^2}{3} - \left(\frac{1\cdot 3\cdot 5}{2\cdot 4\cdot 6}\right)^2 \frac{k^3}{5} - \left(\frac{1\cdot 3\cdot 5\cdot 7}{2\cdot 4\cdot 6\cdot 8}\right)^2 \frac{k^4}{7} + \cdots\right]$$

where k is a position parameter:

$$k = \frac{4rR_r}{(r + R_r)^2 + z^2}$$

The source strength Q (units of flow rate) should match the rate at which soil is displaced by the advancing pile, the cross-sectional area of the pile times its rate of penetration:

$$Q = \pi\left[R^2 - (R - t_w)^2\right]v_0 \approx \pi D t_w v_0$$

where R is the outer radius of the pile and t_w is its wall thickness, the latter approximation being valid for thin-walled piles for which second-order terms become negligible. Post-analysis of the solution shows that the stagnation streamline occurs at an original radial coordinate r_0 located approximately three quarters inside the outer wall boundary. The target pile radius is thus achieved by setting the ring source radius at 0.75 t_w inside the outer pile wall, $R_r = R - 0.75\ t_w$.

Figure 3.19 compares the predicted field of equivalent shear strain $E = \sqrt{e_{ij}e_{ij}/2}$ for solid and open-ended ($D/t_w = 40$) piles. In a simple elasto-plastic medium, the shear strain at yield is proportional to the reciprocal of the rigidity index, $E_y = s_u/2G = 1/(2I_r)$. A typical rigidity index in a normally consolidated clay may be considered to be on the order of $I_r = 100$, which corresponds to a yield strain of about $E_y = 0.5\%$. The strain contours in Figure 3.19 can provide some sense of the expected zone of yielding during pile penetration. For solid pile penetration, the expected zone of yield is thus expected to extend laterally about $9R$ from the pile, in contrast to $2R$ for the open-ended case with $D/t_w = 40$.

The discussion to this point focuses on displacement penetration where the volume of the pile is accommodated by distortion of the soil mass during undrained penetration. Attention is now directed to jetted penetration, which has long been used for installation of conductors in the oil–gas industry, and is receiving increased attention as a potential means of pile anchor installation (Zakeri et al., 2014). A method for simulating jetted penetration using the strain path method is described by Aubeny et al. (2000) for simulation of the self-boring pressuremeter test. The approach is equally applicable to jetted pile penetration. In this approach, the ring source simulating an open-ended pile is supplemented by a point sink simulating extraction of soil from the interior of the pile by the jetting process. An extraction ratio can be defined as the ratio of soil volume extracted by jetting to the pile volume, $f = V_{extract}/V_{pile}$. For pure displacement penetration the extraction ratio becomes $f = 0$, while "perfect" jetting may be described by a state where the extracted soil exactly balances the pile volume, $f = 1$. Figure 3.20 compares shear strain contours for an

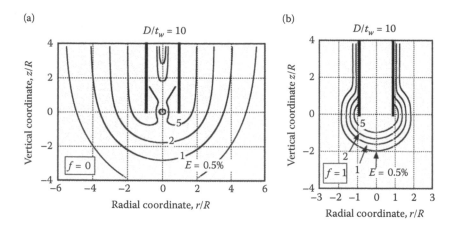

Figure 3.20 Displacement versus jetted penetration. (a) Displacement and (b) jetted.

open-ended pile with diameter-to-wall thickness ratio $D/t_w = 10$ installed by displacement versus jetted penetration. Considering a typical soft clay yield strain $E_y = 0.5\%$, yielding in the soil surrounding a pile installed by displacement penetration occurs out to about 5 radii from the pile wall. By contrast, yielding during jetted penetration occurs to about one radius from the wall in the vicinity of the pile tip. Further up the shaft, the lateral extent of shear strains induced by penetration becomes relatively small. However, examination of strains far above the pile shaft can provide a deceptive picture of disturbance in this case. As the soil near the pile tip has reached yield to about one pile radius, any soil above this region will retain a memory of this disturbance in the form of excess pore pressure and possibly strain softening.

3.4.3 Installation disturbance

Strain path analyses can be directed toward computation of collapse loads; for example, computation of tip resistance in advance of a penetration pile or caisson. However, the most fruitful use of the method has arguably been for prediction of fields of soil disturbance during installation. With strain fields defined in advance, constitutive models capable of simulating complex soil behavior—nonlinearity, evolving anisotropy, strain-softening—can be employed without the distraction and uncertainty associated with techniques requiring iterative solutions for strain fields.

Figure 3.21 shows strain path predictions of installation disturbance by Whittle (1992) for solid and open-ended pile penetration using strain path and cavity expansion analyses in conjunction with the MIT-E3 soil models. To make comparisons in terms of equal volumes of displaced soil, the coordinates for the open-ended solution are normalized by an equivalent radius $R_{eq} \approx \sqrt{Dt}$. Shown in the figure are disturbed fields of excess pore pressure Δu, mean effective stress σ', radial effective stress σ'_r, and cavity shear stress $q = (\sigma'_r - \sigma'_\theta)/2$. From the standpoint of the spatial extent of the zone of disturbance the cavity expansion solution is generally consistent with the strain path analyses. However, from the standpoint of the magnitude of the effective stresses in the vicinity of the pile shaft, substantial differences are apparent. Of particular importance is the radial effective stress σ'_r, since that strongly influences the side-shearing resistance. The cavity expansion analysis predicts a 5%–10% reduction in radial effective stress owing to installation disturbance, in contrast to an 85%–95% reduction predicted by the strain path analysis. The cavity expansion and

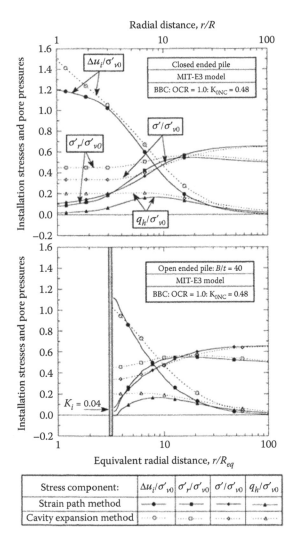

Figure 3.21 Pile installation disturbance. (Adapted from Whittle, AJ, 1992, *Proceedings of the Conference on Offshore Site Investigation and Foundation Behaviour*, London, 28, pp 607–643.)

strain path analyses actually predict exactly the same radial and tangential strains at the end of pile penetration, ε_{rr} and $\varepsilon_{\theta\theta}$ in Equation 3.32. However, the cavity expansion analysis neglects the effects of remaining two shear components, which turn out to significantly affect the field of disturbance.

3.5 FINITE ELEMENT METHODS

While PLAs and cavity expansion methods provide valuable analytical tools for predicting anchor performance, solutions to problems involving complex problem geometries and loading conditions are frequently beyond the reach of these methods. Such problems often require recourse to numerical methods, such as FE analyses. Entire texts are devoted to these topics (e.g., Bathe, 1996; Potts and Zdravkovic, 2001); therefore, no attempt is made to repeat the details here. Rather, the following discussion presents the partial differential equations that apply to a given classes of problems relevant to anchors and examples of solutions to these problems using FE techniques. An overview discussion is also included to provide intuitive insights into the approach.

3.5.1 Overview

Considering a displacement field (u_x, u_y) for plane strain conditions, the operator matrix $[B_u]$ relates displacement to engineering strain by the following:

$$\begin{bmatrix} \varepsilon_{xx} \\ \varepsilon_{yy} \\ \gamma_{xy} \end{bmatrix} = [B_u][u] = \begin{bmatrix} \partial/\partial x & 0 \\ 0 & \partial/\partial y \\ \partial/\partial y & \partial/\partial x \end{bmatrix} \begin{bmatrix} u_x \\ u_y \end{bmatrix} \tag{3.35}$$

The B-matrix essentially functions as the gradient operator (grad or ∇) from vector calculus. A useful feature of the gradient operator is that its transpose is the negative of the divergence operator (Strang, 1986):

$$\nabla \cdot = -(\nabla)^T \quad \text{or} \quad \text{div} = -(\text{grad})^T \tag{3.36}$$

As will be discussed, the divergence operator is frequently employed to express a conservation principle (momentum, mass, etc.) The statement of static equilibrium in tensor-vector format (Figure 3.22) takes the following form:

$$\nabla \cdot \boldsymbol{\sigma} + \mathbf{f}^B = 0 \tag{3.37}$$

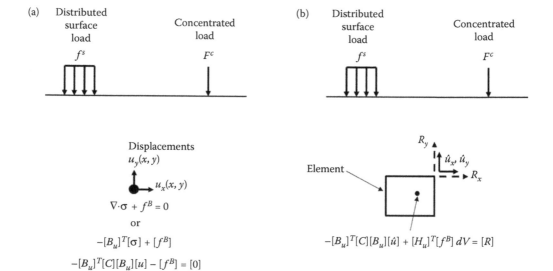

Figure 3.22 Finite element approximation to continuum mechanics problems. (a) Continuum and (b) finite element.

where σ is the stress tensor and $\mathbf{f}^B$ is a vector of body forces (force per unit volume). The equivalent statement in matrix operator form is therefore

$$-[B_u]^T[\sigma]+[f^B]=[0] \tag{3.38}$$

Recalling that the transpose of a differential operator is its own negative, $(\partial/\partial x)^T = -\partial/\partial x$, one may demonstrate that Equation 3.38 actually expresses a statement of equilibrium by writing the individual terms of the respective matrices for the plane strain case and performing the matrix multiplications as shown in Equation 3.39.

$$-\begin{bmatrix} -\partial/\partial x & 0 & -\partial/\partial y \\ 0 & -\partial/\partial y & -\partial/\partial x \end{bmatrix}\begin{bmatrix} \sigma_{xx} \\ \sigma_{yy} \\ \sigma_{xy} \end{bmatrix} + \begin{bmatrix} f_x^B \\ f_y^B \end{bmatrix} = \begin{bmatrix} 0 \\ 0 \end{bmatrix} \tag{3.39}$$

$$\begin{bmatrix} \dfrac{\partial\sigma_{xx}}{\partial x} + \dfrac{\partial\sigma_{xy}}{\partial y} \\[3mm] \dfrac{\partial\sigma_{yy}}{\partial y} + \dfrac{\partial\sigma_{xy}}{\partial x} \end{bmatrix} + \begin{bmatrix} f_x^B \\ f_y^B \end{bmatrix} = \begin{bmatrix} 0 \\ 0 \end{bmatrix} \tag{3.40}$$

Equation 3.40 is, of course, the well-known equation of equilibrium of stresses under plane strain conditions. A second illustrative example is the axisymmetric case, where strains in a system of cylindrical coordinates $(r - \theta - z)$ may be expressed as

$$\begin{bmatrix} \varepsilon_{rr} \\ \varepsilon_{zz} \\ \gamma_{rz} \\ \varepsilon_{\theta\theta} \end{bmatrix} = [B_u]\begin{bmatrix} u_r \\ u_z \end{bmatrix} = \begin{bmatrix} \partial/\partial r & 0 \\ 0 & \partial/\partial z \\ \partial/\partial z & \partial/\partial r \\ 1/r & 0 \end{bmatrix}\begin{bmatrix} u_r \\ u_z \end{bmatrix} \tag{3.41}$$

The corresponding statement of static equilibrium is given by Equation 3.42. Note that the radial coordinate r is introduced here to account for the increase in element size with increasing r.

$$-\frac{1}{r}[B_u]^T[r\,\sigma]+[f^B]=[0] \tag{3.42}$$

Expanding the matrices (Equation 3.43) and performing the matrix multiplication produces the expression for static equilibrium of stresses in Equation 3.44.

$$-\frac{1}{r}\begin{bmatrix} -\partial/\partial r & 0 & -\partial/\partial z & 1/r \\ 0 & -\partial/\partial z & -\partial/\partial r & 0 \end{bmatrix}\begin{bmatrix} r\sigma_{rr} \\ r\sigma_{zz} \\ r\sigma_{rz} \\ r\sigma_{\theta\theta} \end{bmatrix} + \begin{bmatrix} f_r^B \\ f_z^B \end{bmatrix} = \begin{bmatrix} 0 \\ 0 \end{bmatrix} \tag{3.43}$$

$$\begin{bmatrix} \dfrac{\partial \sigma_{rr}}{\partial r} + \dfrac{\partial \sigma_{rz}}{\partial z} + \dfrac{\sigma_{rr} - \sigma_{\theta\theta}}{r} \\ \dfrac{\partial \sigma_{zz}}{\partial z} + \dfrac{\partial \sigma_{rz}}{\partial r} + \dfrac{\sigma_{rz}}{r} \end{bmatrix} + \begin{bmatrix} f_r^B \\ f_z^B \end{bmatrix} = \begin{bmatrix} 0 \\ 0 \end{bmatrix} \tag{3.44}$$

The transpose of the strain-displacement matrix, therefore, provides a rather effective means of expressing a statement of equilibrium for any coordinate system. Completing the formulation requires a connection between stress and displacement. The stress terms in Equations 3.38 and 3.42 relate to displacement by the stress–strain matrix $[C]$:

$$[\sigma] = [C][\varepsilon] = [C][B_u][u] \tag{3.45}$$

Thus, the statement of static equilibrium in terms of displacement takes the following form:

$$[B_u]^T [C][B_u][u] - [f^B] = [0] \tag{3.46}$$

To this point, the discussion has been applied to "exact" fields of displacement (e.g., u_x, u_y). We can now consider the FE approximation, for which displacement fields are described by nodal displacements ($\hat{u}_x, \hat{u}_y$). Displacements in the interior of an element are computed from a polynomial interpolation function $h(x, y)$ of order n having the following general form:

$$\begin{bmatrix} u_x \\ u_y \end{bmatrix} = [H_u][\hat{u}] = \begin{bmatrix} h_1 & 0 & h_2 & 0 & \ldots & h_n & 0 \\ 0 & h_1 & 0 & h_2 & \ldots & 0 & h_n \end{bmatrix} \begin{bmatrix} \hat{u}_x \\ \hat{u}_y \end{bmatrix} \tag{3.47}$$

The corresponding strain–displacement matrix is now as follows:

$$[B_u] = \begin{bmatrix} \partial h_1/\partial x & 0 & \ldots & \partial h_n/\partial x & 0 \\ 0 & \partial h_1/\partial y & \ldots & 0 & \partial h_n/\partial y \\ \partial h_1/\partial y & \partial h_1/\partial x & \ldots & \partial h_n/\partial y & \partial h_n/\partial x \end{bmatrix} \tag{3.48}$$

Equation 3.46 may now be invoked to enforce the equilibrium requirement. However, as the displacement field is approximate, the left-hand side of the equation is no longer identically zero. Treatment of this residual force imbalance R varies according to which numerical scheme is adopted. One approach is to set integrated sum of the residuals to zero over n elements in the mesh, which produces Equation 3.49.

$$\sum_n \int_V [B_u]^T [C][B_u][\hat{u}]dV = \sum_n \int_V [H_u]^T [f^B]dV \tag{3.49}$$

It is noted that the transpose of the interpolation matrix $[H_u]^T$ is a weighting function; that is, the infinitesimal body forces within an element are distributed to the nodes in proportion to their proximity to a given node. The transpose of the strain–displacement matrix $[B_u]^T$ acts as both a divergence operator and a weighting function, with contributions from infinitesimal stress gradients also being distributed to the nodes. The right-hand side of Equation 3.49 typically requires additional terms accommodate distributed surface loads f^s and concentrated loads f^C. The integrals are typically evaluated numerically by Gauss integration. The result is a system of equations (Equation 3.50) that, after appropriate imposition of displacement boundary constraints, is solved by the methods of linear algebra.

$$[K_{uu}][\hat{u}] = [R_u]$$

$$[K_{uu}] = \sum_n \int_V [B_u]^T [C][B_u] dV \qquad (3.50)$$

$$[R_u] = \sum_n \int_V [H_u]^T [f^B] dV + \sum_n \int_S [H_u]^T [f^S] dS + \sum_i F_i^C$$

Omitted from this brief overview are numerous details with respect to interpolation functions, numerical integrations, assemblage into a global system of equations, imposition of boundary constraints, and efficient storage and solution techniques for systems of linear equations. Additionally, nonlinearity can be introduced by either a nonlinear constitutive law or large deformations and/or large strains. For details of these topics, the reader is referred to a text on FE methods (e.g., Bathe, 1996).

The constitutive matrix in Equation 3.50 should be formulated on a model-specific basis. For the case of an isotropic, linearly elastic material, the stiffness matrix $[C^e]$ can be expressed in terms of the bulk modulus K_b and shear modulus G as follows:

$$[C^e] = \begin{bmatrix} K_b + 4G/3 & K_b - 2G/3 & K_b - 2G/3 & 0 & 0 & 0 \\ K_b - 2G/3 & K_b + 4G/3 & K_b - 2G/3 & 0 & 0 & 0 \\ K_b - 2G/3 & K_b - 2G/3 & K_b + 4G/3 & 0 & 0 & 0 \\ 0 & 0 & 0 & 2G & 0 & 0 \\ 0 & 0 & 0 & 0 & 2G & 0 \\ 0 & 0 & 0 & 0 & 0 & 2G \end{bmatrix} \qquad (3.51)$$

Much of the FE work applied to anchors focuses on ultimate capacity estimates, which requires introduction of plasticity into the model. A simple model satisfying this requirement is the linearly elastic–perfectly plastic model described earlier. Assuming a von Mises yield criterion, incremental stress can be computed by modifying the elastic stiffness matrix $[C^e]$ in Equation 3.51 as shown in Equation 3.50 (Desai, 1976).

$$[\dot{\sigma}] = \{[C^e] - [C^p]\}[\dot{\varepsilon}] \qquad (3.52)$$

$$[C^p] = \frac{G}{k^2} \begin{bmatrix} s_{xx}^2 & s_{xx}s_{yy} & s_{xx}s_{zz} & s_{xx}s_{xy} & s_{xx}s_{yz} & s_{xx}s_{zx} \\ & s_{yy}^2 & s_{yy}s_{zz} & s_{yy}s_{xy} & s_{yy}s_{yz} & s_{yy}s_{zx} \\ & & s_{zz}^2 & s_{zz}s_{xy} & s_{zz}s_{yz} & s_{zz}s_{zx} \\ & & & s_{xy}^2 & s_{xy}s_{yz} & s_{xy}s_{zx} \\ & \text{sym} & & & s_{yz}^2 & s_{yz}s_{zx} \\ & & & & & s_{zx}^2 \end{bmatrix} \qquad (3.53)$$

where $s_{ij} = \sigma_{ij} - \delta_{ij}\sigma$ are the deviatoric components of stress and σ is mean stress.

The formulation for fluid flow utilizes a gradient operator $[B_p]$ relating potential ϕ to hydraulic gradient has the following form for two-dimensional problems:

$$\begin{bmatrix} i_x \\ i_y \end{bmatrix} = [B_p][\phi] = \begin{bmatrix} \partial/\partial x \\ \partial/\partial y \end{bmatrix} [\phi] \tag{3.54}$$

The FE discretization can proceed exactly along the same lines as that described for mechanical behavior; however, when coupled hydromechanical behavior is considered, the interpolating polynomial for pore pressure in the fluid phase $[H_p]$ should be at least one order less than the interpolating polynomial for displacements. Analysis of a coupled hydromechanical system proceeds by considering equilibrium (Equation 3.37) together with the Terzaghi effective stress principle, $\sigma = \sigma' + \mathbf{I}p$, where σ' is the effective stress, $\mathbf{I}$ is the identity matrix, and p is the pore pressure. Equation 3.52 expresses the resulting equations in continuum mechanics and matrix formats. Matrix multiplication of the strain–displacement matrix $[B_u]$ times the identity matrix simply produces a matrix $[B_v]$, relating volumetric strains to displacements.

$$\nabla \cdot \sigma' + \nabla \cdot \mathbf{I}p + \mathbf{f}^B = 0 \tag{3.55}$$

$$[B_u]^T C'[B_u][\hat{u}] + [B_v]^T[H_p][\hat{p}] = [H_u]^T[f^B] \tag{3.56}$$

A second set of equations is generated from the conservation of mass principle, as shown in Equation 3.53 for the case of no distributed fluxes, where $[k]$ is the permeability tensor and γ_w is the unit weight of water. The sign convention used here assumes pressure p is positive in tension.

$$\nabla \cdot \mathbf{v} + \dot{\varepsilon} = 0 \tag{3.57}$$

$$-\frac{1}{\gamma_w}[B_p]^T[k][B_p][\hat{p}] + [H_p]^T[B_v]\left[\frac{\partial}{\partial t}\hat{u}\right] = [0] \tag{3.58}$$

The occurrence of time derivatives in this equation introduces the need for a time-stepping algorithm. A common approach is the finite difference form shown in Equation 3.59, where t is the time of the previous step and $t + \Delta t$ is time of the current step. The time derivative of displacements with respect to time is thus taken as proportional to a weighted average for the second spatial derivatives of pore pressures within the time step. The weight factor α is selected by the analyst, with $\alpha = 0$ producing an explicit time-stepping procedure and $\alpha = 1$ producing a fully implicit procedure. Any $\alpha \geq 1/2$ will be unconditionally stable.

$$\frac{\partial u}{\partial t} \approx \frac{u_{t+\Delta t} - u_t}{\Delta t} = \frac{1}{\gamma_w}\{\alpha \nabla \cdot [k]\nabla p_{t+\Delta t} + (1-\alpha)\nabla \cdot [k]\nabla p_t\} \tag{3.59}$$

3.5.2 Coupled poro-elastic analysis

A fully coupled analysis (Equation 3.53) is generally necessary for any process involving consolidation around the pile or anchor, an example being the simulation of the setup process

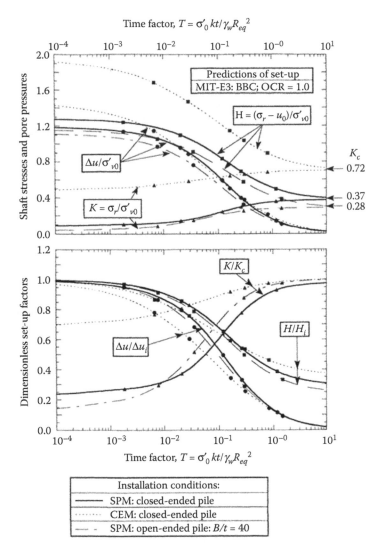

Figure 3.23 Coupled consolidation around piles. (Adapted from Whittle, AJ, 1992, *Proceedings of the Conference on Offshore Site Investigation and Foundation Behaviour*, London, 28, pp 607–643.)

following pile or caisson installation in clay. Figure 3.23 shows a FE simulation of setup around the shaft of solid and open-ended piles following installation by Whittle (1992). The initial fields of effective stresses and pore pressure were computed using a strain path analysis with the MIT-E3 soil model in a K_0-normally consolidated clay—K_0 in the intact soil being 0.48 in this case. The upper figure shows the dissipation process in terms of absolute stresses, from which it is apparent that the details of the disturbance process strongly influence setup stresses, which may be expressed in terms of a horizontal earth pressure coefficient $K = \sigma'_r/\sigma'_{v0}$. Predicted stresses at the end of consolidation show $K_c = 0.72$ (an increase over the *in situ* stress state) for cylindrical cavity expansion disturbance, compared to $K_c = 0.37$ and 0.28 calculated for two-dimensional simulations of solid and open-ended pile penetration. Also noteworthy is the substantial decrease in total stresses occurring during the consolidation process, which is a consequence of soil nonlinearity and coupling effects. If a simplified linear uncoupled analysis (discussed in the next section) were applied to this problem, total stress would remain constant during consolidation, resulting in a severe overestimate of setup stresses. Thus, rigorous predictions of installation disturbance due to pile penetration requires both (1) realistic simulation of the strain histories in the soil adjacent to the pile during penetration and (2) a consolidation analysis that accounts for the effects of both soil nonlinearity and coupling between the solid and liquid phases. Nevertheless, it should be stated that in many instances the analysis has the less ambitious objective of simply estimating the time required for setup to occur. In this case, one is only concerned with the relative degree of setup K/K_c, shown in the lower portion of the figure. The predictions show the setup curves for the various disturbance simulations to fall into a relatively narrow band, suggesting that even relatively crude simulations of disturbance can suffice for predicting setup times, with the important proviso that the disturbance simulation should accurately reflect the volume of soil (R_{eq} equals the radius of an equivalent solid pile having the same cross-sectional area as the open-ended pile) displaced by pile installation.

3.5.3 Uncoupled consolidation

Following the notion of a simplified setup analysis focused solely on the time required for setup to occur leads to consideration of an uncoupled analysis of pore water pressure dissipation. Sills (1975) showed that, for a linear elastic isotropic soil having constant moduli and permeability, Equations 3.55 and 3.57 uncouple under the following conditions: an irrotational displacement field, infinite soil extent, and constant body forces during consolidation. Under these conditions, the uncoupled equation for the fluid phase takes the following form:

$$c\nabla^2(p - p_0) = \frac{\partial p}{\partial t}$$
$$c = (K_B + 4G/3)\frac{k}{\gamma_w} = \frac{Mk}{\gamma_w}$$

(3.60)

Equation 3.60 is a well-known diffusion (or heat) equation that can also be solved by FE techniques. The consolidation coefficient c depends on the elastic constants K_B = bulk modulus and G = shear modulus. One may also recognize that this combination of elastic constants is equivalent to the constrained modulus M (equal to $\sigma_{zz}/\varepsilon_{zz}$, where $\varepsilon_{rr} = \varepsilon_{\theta\theta} = 0$). Thus, the elastic stiffness coefficient governing uncoupled consolidation corresponds to that measured in a conventional one-dimensional consolidation test. It is emphasized that all elastic constants under discussion here refer to drained loading conditions.

The case of radial one-dimensional consolidation around a cylindrical pile shaft actually satisfies the conditions established by Sills (1975) with regard to an irrotational displacement field, infinite soil extent, and constant body forces. Therefore, analysis of setup around a pile shaft in terms of an uncoupled analysis is appropriate for a linearly elastic medium. The question remains as to the validity of the uncoupled analysis in a nonlinear medium and in the vicinity of the pile tip. Figure 3.24 shows a comparison of uncoupled to nonlinear coupled predictions for consolidation around the tip and shaft of a solid pile following installation. A strain path analysis with the MCC constitutive model was used to simulate installation disturbance. The use of the MCC model was carried forward in the nonlinear coupled consolidation analysis. The *in situ* elastic constrained modulus for the undisturbed soil in these simulations was $M/\sigma'_{v0} = 106$. This value was adjusted downward to $M/\sigma'_{v0} = 45$ for the uncoupled analysis to achieve a match with the coupled nonlinear dissipation curve on the pile shaft. The curves in Figure 3.24 indicate that the uncoupled analysis can reliably predict the shape of the consolidation curve in the zone of radial consolidation surrounding

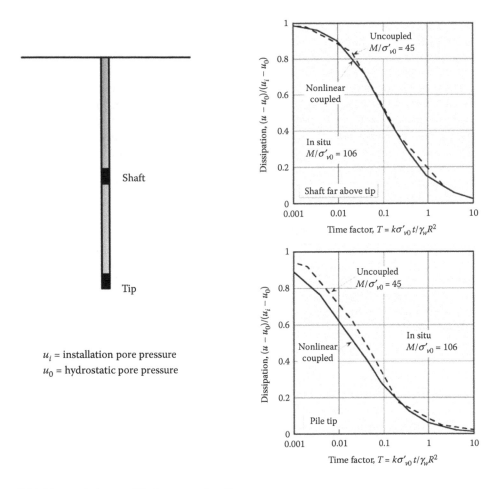

Figure 3.24 Uncoupled consolidation around a pile.

the pile shaft. In the zone of two-dimensional consolidation surrounding the pile tip, the nonlinear coupling effects are noticeable during the early stages of consolidation. However, during the later stages of consolidation, the uncoupled curve tends to merge with the non-linear coupled curve.

The analyses presented above tend to support the overall validity of an uncoupled analysis for predicting pore pressure dissipation around a pile shaft, in the sense that the uncoupled analysis can reasonably capture the shape of the dissipation curves. However, the difficulty remains in selecting an appropriate coefficient of consolidation for the analysis. The use of an elastic constrained modulus M for the intact soil roughly corresponds to using the coefficient of consolidation for an intact over-consolidated (elastic unload–reload) soil, c_{oc}. The analyses show this value to be high by a factor of 2–2.5. It is also noted that the normally consolidated coefficient of consolidation c_{nc} (approximately 1/5 of c_{oc} in this case) is too low. The ambiguity in selecting an appropriate value of c suggests that it should be obtained from either a numerical calibration to a coupled analysis (the procedure illustrated here) or calibration to experimental model tests or field measurements that come close to matching the conditions of the case under consideration. An example of the latter is the piezocone holding test, which essentially provides a small scale simulation of the processes of pile penetration and postinstallation setup.

3.5.4 Collapse loads

Chapter 4 will show that the upper and lower bound estimates of the collapse loads are possible for a number of important cases of anchor loading for conditions involving a single mode of loading, such as pure loading normal to the long axis of a plate anchor, and pure moment loading. However, many cases arise where anchor ultimate capacity under combined loading conditions is required, for which relatively few analytical solutions have been developed. In these cases, recourse to numerical solutions is required. Other complexities such as soil heterogeneity, proximity to a free surface, and complex anchor geometry can also dictate the use of numerical analyses. Under undrained loading—the most common case considered for this class of problem—analysis as a mechanical system uncoupled to the pore fluid equations (e.g., Equations 3.37 and 3.50) adequately describes the system. Displacement-based FE methods have been the most widely used analytical tool for this purpose to date, although alternative methods exist, which will be discussed subsequently. Collapse load analyses are frequently made under the assumption that ultimate load capacity is independent of the elastic properties of the soil (Chen, 1975). This may be true if a kinematically admissible rigid-plastic collapse mechanism is the critical collapse mechanism, as is the case for shallowly embedded anchors and deeply embedded strip anchors (Figure 3.14). However, as noted earlier, cases arise where ultimate capacity cannot be evaluated independently of the elastic properties of the soil. Notable cases for which ultimate capacity of an anchor depends on E/s_u include deeply embedded square and circular anchors (Wang et al., 2010) and tip resistance of a long caisson or pile. Thus, although the principle that ultimate load capacity is independent of elastic behavior can be useful, care is required to ensure that it actually applies to the problem under consideration. Displacement-based FE formulations require at least one finite elastic parameter, irrespective of whether it influences collapse loads. Typically, the simplest possible model—linearly elastic—is selected to satisfy this requirement.

Figure 3.25 shows plastic strain contours from a collapse load analysis of a strip anchor with aspect ratio $L/t = 7$. The predicted collapse mechanism that is remarkably similar to the upper and lower bound mechanisms presented earlier in Figures 3.6 and 3.9. The bearing factor from this analysis was $N_{nmax} = 11.98$, compared to an upper bound estimate $N_{nmax} = 12.1$ from Equation 3.20. This level of accuracy was achievable by adopting the extremely fine mesh shown in Figure 3.26. Increasing the mesh density often suffices as a means of achieving accuracy, but can be impractical for three-dimensional problems where the computational effort becomes exorbitant. In certain cases, accurate solutions are elusive even with extreme levels of mesh refinement. Particularly, difficult cases arise in connection with translation of a thin plate parallel to its long axis. Abnormal stress distributions arise from the displacement discontinuity that takes place at the edges of the rigid structure penetrating the softer material (Van Langen and Vermeer 1991). To ensure accurate modeling of the singularity at the anchor tip, Rowe and Davis (1977) introduced potential rupture lines near the edge of the anchor. This approach attempts to overcome the inhibition of free plastic flow inherent in the usual stiffness formulation of the FE method by permitting the formation of velocity discontinuities in the regions of high stress and velocity gradient near the tip of the anchor plate. Van Langen and Vermeer (1991) introduced zero-thickness interface elements to enhance the unrealistic stress nonuniformity on surfaces with singular points. Using the same general concept, Nouri et al. (2011) insert interfaces—contact pairs rather than elements—placed in the soil at the corners of the plate to accommodate displacement discontinuities.

The analysis discussed above assumed full bonding between the soil and the anchor surface. More realistic modeling of the interface utilizes a contact surface with a specified adhesion factor α defined as the ratio of adhesion at the interface to the soil shear strength. Full adhesion ($\alpha = 1$) actually involves less constraint than full bonding, since slip can occur

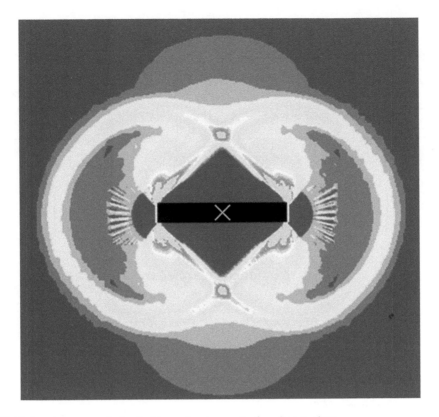

Figure 3.25 Finite element analysis of ultimate load capacity for plate anchor.

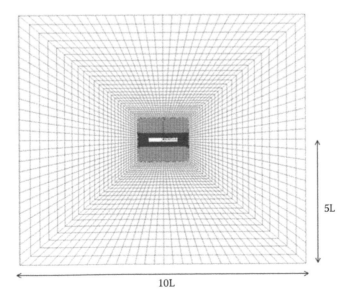

Figure 3.26 Finite element model of strip anchor.

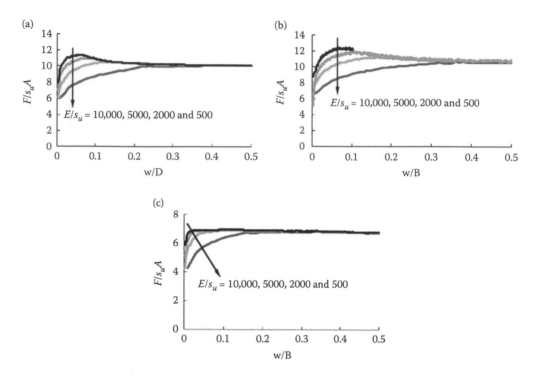

Figure 3.27 LDFE analyses on plates. (a) Circular anchor, (b) square anchor, and (c) strip anchor. (From Wang D, Hu Y, and Randolph MF, 2010, *ASCE J Geotech Geoenviron Eng*, 136(2), 355–365, with permission from ASCE).

even when full shearing resistance is mobilized, but results reported by Murff et al. (2005) show the effects to be relatively minor.

Large deformation finite element (LDFE) analyses have been applied to anchors as described by Song et al. (2008) in two dimensions and Wang et al. (2010) in three dimensions. The remeshing and interpolation technique with small strain (RITSS) method performs a series of small strain analysis increments followed by remeshing and interpolation of stresses and material properties from the old to the new mesh. Among other interesting results, their analyses shed light on the effect of soil rigidity E/s_u on ultimate load capacity (Figure 3.27). In the case of a normally loaded deeply embedded strip anchor, ultimate capacity is essentially independent of E/s_u. By contrast, LDFE analyses on deeply embedded square and circular anchors show computed ultimate loads for a soil rigidity $E/s_u = 500$ (a realistic value for normally consolidated clays) to be about 15% less than those computed for a nearly rigid soil $E/s_u = 100,000$. Interestingly, LDFE calculated load capacities are often even less than lower bound values. As small displacement FE and lower bound analyses are both based on the undeformed geometry of the soil, the small displacement FE calculation will equal or exceed the lower bound. This is not the case for LDFE calculations based on the deformed configuration of the soil, where changes in geometry—which are influenced by E/s_u—can impact the computed ultimate load capacity.

3.5.5 Optimization methods

An alternative to displacement-based FE calculations of collapse loads are numerical limit analyses using linear programming methods with FE discretization. Although the technique was proposed early on by Lysmer (1970) and Bottero et al. (1980), the excessive computation time for the optimization using simplex algorithms hindered adoption of the method. The use of more efficient optimization methods (Sloan, 1988a) permit practical implementation of the method for lower bound (Sloan, 1988b) and upper bound (Sloan and Kleeman, 1995) formulations. Ukritchon et al. (1998) apply the lower and upper bound methods to obtain collapse loads for surface strip footings under combined loading. Lyamin (1999) extends the lower bound approach to three-dimensional geometries. In regard to application to anchors, the method was applied to strip anchors by Merifield et al. (2001) and to circular and rectangular anchors by Merifield et al. (2003). Details of the derivations and resulting equations are contained in the references noted above.

In a two-dimensional analysis, the predominant body of the mesh comprises three-node triangular elements, with triangular and rectangular extension elements placed at infinite boundaries. In contrast to displacement-based FEs, nodes are unique to the elements and stress discontinuities can occur along edges shared by elements. In the lower bound formulation, stress is the nodal variable and the stress field varies linearly within an element. Constraints imposed on the lower bound collapse load include satisfying equilibrium and the external stress boundary conditions, and not exceeding the yield criterion. Static admissibility is ensured if equilibrium is satisfied within each element and if shear and normal tractions are in equilibrium along the discontinuities between elements. To retain linear constraints the nonlinear (circular) yield criterion is approximated by an interior polygon. Merifield et al. (2003) extend the lower bound analysis to three dimensions using tetrahedral elements. The lower bound is obtained by maximizing collapse loads subject to the constraints described above.

The upper bound formulation also uses three-node triangular elements with velocity as the nodal variable. Imposed constraints ensure that the velocity field satisfies strain

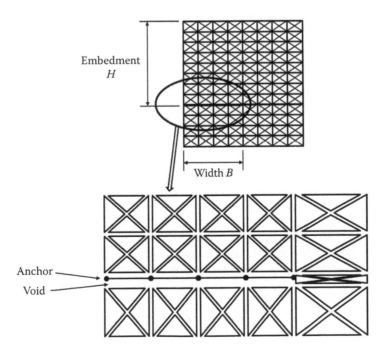

Figure 3.28 Typical mesh for upper bound limit analysis.

compatibility, the flow rule, and the kinematic boundary conditions. Jumps in tangential velocity are permitted at the edges of the elements. As with the lower bound analysis, the yield criterion is approximated as a polygon, except that it circumscribes the parent yield surface to ensure a rigorous upper bound. A collapse load is computed by equating the external rate of work performed by the anchor to the internal rate of energy dissipation in the continuously deforming regions within the elements and the velocity jumps at the element edges. The upper bound is obtained by minimizing collapse loads subject to the constraints. Figure 3.28 shows the type of upper bound mesh used by Merifield et al. (2001) for analysis of a shallowly embedded horizontal plate anchor. In this case, a void beneath the plate simulates an immediate breakaway condition where no suction is assumed to be mobilized beneath the plate.

REFERENCES

Atkinson JH and Bransby PL, 1977, *The Mechanics of Soils: An Introduction to Critical State Soil Mechanics*, McGraw-Hill, London, 375p.
Aubeny C and Grajales F, 2015, Strain path analysis of setup time around piles and caissons, *Proceedings of the ASME 2015 34th International Conference on Ocean, Offshore and Artic Engineering OMAE2015*, St John's Newfoundland.
Aubeny CP, 1992, Rational interpretation of in situ tests in cohesive soils, Doctor of Philosophy Thesis, Massachusetts Institute of Technology, Cambridge.
Aubeny CP, Whittle AJ, and Ladd CC, 2000, Effects of disturbance on undrained strengths interpreted from pressuremeter tests, *ASCE J Geotech Geoenvon Eng*, 126(12), 1133–1144.
Baligh MM, 1985, *Fundamentals of Deep Penetration I: Soil Shearing and Point Resistance.* (Tech. Report R85-9, Order No. 776), Department of Civil Engineering, MIT, Cambridge, MA.
Bathe K-J, 1996, *Finite Element Procedures*, Prentice Hall, Upper Saddle River, New Jersey.
Bottero A, Negre R, Pastor J, and Turgeman S, 1980, Finite element method and limit analysis theory for soil mechanics problems, *Comput Meth Appl Mech Eng*, 22, 131–149.
Bransby MF and O'Neill MP, 1999, Drag anchor fluke–soil interaction in clay, *International Symposium on Numerical Models in Geomechanics (NUMOG VII)*, Graz, Austria, pp 489–494.
Chen WF, 1975, *Limit Analysis and Soil Plasticity.* Elsevier Publishing Co., Amsterdam, The Netherlands.
Chin CT, 1986, Open-ended pile penetration in saturated clays, Doctor of Philosophy thesis, Massachusetts Institute of Technology, Cambridge.
Davis EH and Christian JT, 1971, Bearing capacity of anisotropic cohesive soil, *J Soil Mech Found Div Am Soc Civ Eng*, 97(5), 753–769.
Dsimulia, 2012, ABAQUS Version 6.8, User Manual.
Hill R, 1950, *The Mathematical Theory of Plasticity*, Oxford University Press, London.
Kavvadas M, 1982, *Non-linear consolidation around driven piles in clays*, ScD Thesis, Dept. of Civil Engineering, MIT, Cambridge, MA.
Levadoux J-N and Baligh MM, 1980, *Pore Pressures in Clays due to Cone Penetration*, Research Report R80-15, Order No. 666, Dept of Civil Engineering, MIT, Cambridge, MA.
Lyamin AV, 1999, *Three-Dimensional Lower Bound Limit Analysis using Nonlinear Programming*, PhD thesis, Department of Civil, Surveying and Environmental Engineering, University of Newcastle, NSW, Australia.
Lysmer J, 1970, Limit analysis of plane problems in soil mechanics, *J Soil Mech Found Div*, ASCE, 96(4), 1311–1334.
Malvern LE, 1969, *Introduction to the Mechanics of a Continuous Medium*, Prentice-Hall, Inc., Englewood Cliffs, New Jersey.

Martin CM and Randolph MF, 2001, Applications of the lower and upper bound theorems of plasticity to collapse of circular foundations, *Proceedings of the 10th International Conference International Association of Computer Methods and Advances in Geomechnics*, Tucson, 2, pp 1417–1428.

Merifield RS, Lyamin AV, Sloan SW, and Yu HS, 2003, Three-dimensional lower bound solutions for stability of plate anchors in clay, *J Geotech Geoenviron Eng*, ASCE, 129(3), 243–253.

Merifield RS, Sloan SW, and Yu HS, 2001, Stability of plate anchors in undrained clay, *Geotechnique*, 51(2), 141–153.

Meyerhof GG, 1951, The ultimate bearing capacity of foundations, *Geotechnique*, 2(4), 301–332.

Murff JD, 1980, Vane shear testing of anisotropic, cohesive soils, *Int J Numer Anal Meth Geomech*, 4, 285–289.

Murff JD, Randolph MF, Elkhati, S, Kolk HJ, Ruinen RM, Strom PJ, and Thorne CP, 2005, Vertically loaded plate anchors for deepwater applications, *Proceedings of the International Symposium on Frontiers in Offshore Geotechnics*, ISFOG05, Balkema, Perth, Australia, pp 31–48.

Nouri H, Biscontin G, and Aubeny C, 2011, Numerical simulation of plate anchors under combined translational and torsional loading, *13th International Conference of the International Association for Computer Methods and Advances in Geomechanics*, Melbourne.

Pestana JM and Whittle AJ, 1999, Formulation of a unified constitutive model for clays and sands, *Int J Numer Anal Meth Geomech*, 23, 1215–1243.

Potts DM and Zdravkovic L, 2001, *FiniteElement Analysis in Geotechnical Engineering*, Thomas Telford, London, 427p.

Prager W, 1959, *An Introduction to Theory of Plasticity*. Addison-Wesley, Reading, Massachusetts.

Rao SS, 2002, *Applied Numerical Methods of Engineers and Scientists*, Prentice Hall, New Jersey.

Roscoe KH and Burland JB, 1968, On the generalized stress-strain behavior of 'wet' clay, *Engineering Plasticity*, Ed. J Heyman, (ed.), Cambridge University Press, Cambridge, pp 535–609.

Rowe RK and Davis EH, 1977, *Application of the Finite Element Method to the Prediction of Collapse Loads*. Research Report R310. Sydney: University of Sydney.

Schofield AN and Wroth CP, 1968, *Critical State Soil Mechanics*, McGraw-Hill, London, 310p.

Sills G, 1975, Some conditions under which Biot's equation of consolidation reduce to Terzaghi's equation, *Geotechnique*, 25(1), 129–132.

Sloan SW and Kleeman PW, 1995, Upper bound limit analysis using discontinuous velocity fields, *Compo Meth Appl Mech Eng*, 127, 293–314.

Sloan SW, 1988a, Lower bound limit analysis using finite elements and linear programming, *Int J Numer Anal Meth Geomech*, 12(1), 61–77.

Sloan SW, 1988b, A steepest edge active set algorithm for solving sparse linear programming problems, *Int J. Numer Anal Meth Geomech*, 26(12), 2671–2685.

Song Z, Hu Y, and Randolph MF, 2008, Numerical simulation of vertical pullout of plate anchors in clay, *ASCE J Geotech Geoenviron Eng*, 134(6), 866–875.

Strang, G, 1986, *Introduction to Applied Mathematics*, Wellesley-Cambridge Press, Wellesley, Massachusetts.

Teh CI and Houlsby GT, 1988, Analysis of the cone penetration test by the strain path method, in *Numerical Methods in Geomechanics*, Swoboda (ed.), Balkema, Rotterdam, pp 397–402, ISBN 90619809X.

Ukritchon B, Whittle AJ, and Sloan SW, 1998, Undrained limit analyses for combined loading of strip footings on clay, *ASCE J Geotech Geoenviron Eng*, 124(3), 265–276.

Van Langen H and Vermeer PA, 1991, Interface elements for singular plasticity points, *Int J Numer Anal Meth Geomech*, 15(5), 301–315.

Wang D, Hu Y, and Randolph MF, 2010, Three-dimensional large deformation finite element analysis of plate anchors in uniform clay, *ASCE J Geotech Geoenviron Eng*, 136(2), 355–365.

Whittle, AJ, 1992, Assessment of an effective stress analysis for predicting the performance of driven piles in clays, *Proceedings of the Conference on Offshore Site Investigation and Foundation Behaviour*, London, 28, pp 607–643.

Whittle AJ and Aubeny CP, 1991, Pore pressure fields around piezocone penetrometers installed in clays, *Comput Meth Adv Geomech.*

Whittle AJ and Aubeny CP, 1991, Pore pressure fields around piezocone penetrometers installed in clays, in *Proceedings of the 7th International Conference on Computer Methods and Advances in Geomechanics*, Cairns, Australia, G. Beer, J.R. Booker, and J.P.C. Carter (eds.), AA Balkema, Rotterdam, pp 285–290.

Whittle AJ and Kavvadas MJ, 1993, Formulation of MIT-E3 constitutive model for overconsolidated clays, *J Geotech Eng*, ASCE, 120(1), 173–198.

Zakeri A, Liedtke E, Clukey EC, and Jeanjean P, 2014, Long-term axial capacity of deepwater jetted piles, *Geotechnique*, 4(12), 966–980.

Chapter 4

Fundamental studies

While relevant studies unique to specific anchor types are introduced in subsequent chapters of this book, a number of studies elucidating the basic aspects of anchor behavior apply to broad categories of anchor systems. This chapter summarizes the basic findings from these studies. Issues covered include the effects of anchor geometry, mode of loading, combined loading, and proximity to a free surface. This chapter first covers plate sections relevant to DEAs, VLAs, PDPAs, SEPLAs, and DEPLAs, followed by cylindrical sections relevant to piles, caissons, and most dynamically installed piles.

4.1 PLATE ANCHORS

Characterizing the yield locus of a plate anchor generally requires load capacity estimates for six loading degrees of freedom: three force and three moment components. Setting the origin of an x–y–z coordinate system at the centroid of the plate (Figure 4.1), there will be one normal component of force N and two shear components F_{sx} and F_{sy}. Eccentric loading can introduce two possible moment loads, M_x and M_y, and one torsion load T. In addition to the six basic conditions of "pure" or uniaxial loading, interactions under various conditions of combined loading should be defined. Figure 4.1 illustrates some combinations that should be considered—additional combinations such as F_{sx}–F_{sy} and M_x–M_y not shown in the figure may also have to be evaluated. Analytical solutions occasionally exist for simple plate geometries for conditions of pure loading. However, with the exception of the shear–torsion combinations, the definition of the yield locus normally requires computationally intensive probes employing finite element techniques.

It is noteworthy that pure normal loading typically produces the greatest anchor pullout capacity, which could raise a question as to why other loading modes should be investigated. There are two basic reasons: (1) eccentricity and inclined loading are often deliberately introduced into anchor design to achieve desirable behavior, typically to promote a tendency of the anchor to dive more deeply when overloaded and (2) unintended load combinations can occur during partial failure of the mooring system. As the design basis for a mooring system typically considers the possibility of failure of one or more mooring lines, the anchor should be capable of resisting loads from unintended directions.

The bearing factors for pure loading as well as the various interaction relationships depend on (1) the details of the plate geometry such as shape, aspect ratio, and thickness and (2) the depth of embedment, in particular, the reduction in load capacity owing to the proximity of a free surface. The following discussion will first consider deeply embedded plate anchors, followed by a discussion of free surface effects.

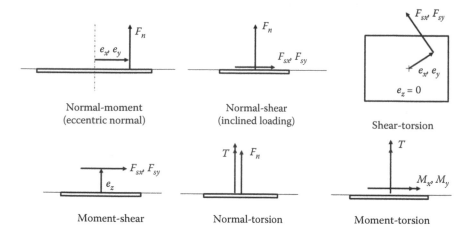

Figure 4.1 Loading modes on plate anchors.

4.1.1 Strip anchors

Revisiting the upper bound analysis of pure normal loading on a strip anchor of finite thickness, provided that the computed bearing factors are insensitive to deviations from 45° in the optimal wedge angle suggest a simplified form as follows:

$$N_{nmax} = \frac{F_{nmax}}{s_u L} + 3\pi + 2 + \frac{2t}{L}\left(\alpha + \frac{(1+\alpha)}{\sqrt{2}}\right) \qquad (4.1)$$

Bransby and O'Neill (1999) employ essentially the same upper bound approach for a strip anchor subjected to pure tangential loading (Figure 4.2). In this case, the optimized wedge angles are generally greater than 75°. Considering the general values of adhesion α in the analysis leads to Equation 4.2. In this case, the upper bound estimate appears to consistently overestimate the estimates edge resistance when compared to numerical estimates. Therefore, Murff et al. (2005) suggest a simplified adjusted form by assuming an end bearing factor $N_e \approx 7.5$, which produces Equation 4.3.

$$N_{smax} = \frac{F_{smax}}{s_u L} = 4\left\{\frac{t}{L}\left[(\pi - \beta) + \frac{\tan \beta}{2}\right] + \frac{1}{2}[\alpha + (1+\alpha)\cos \beta]\right\} \qquad (4.2)$$

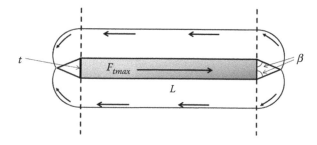

Figure 4.2 Tangentially loaded strip anchor.

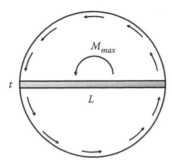

Figure 4.3 Rotational slip mechanism.

$$N_{smax} = 2\alpha + 2N_e \frac{t_f}{L_f} \approx 2\alpha + 15 \frac{t_f}{L_f} \tag{4.3}$$

Finally, Bransby and O'Neill (1999) suggest the slip mechanism in Figure 4.3 for a strip anchor subjected to pure moment loading. An upper bound analysis for this mechanism produces the bearing factor for pure moment loading N_{mmax} shown by Equation 4.4:

$$N_{mmax} = \frac{M_{max}}{L^2 s_u} = \frac{\pi}{2}\left[1 + \left(\frac{t}{L}\right)^2\right] \tag{4.4}$$

Evaluation of anchor load tcapacity under conditions of combined loading generally requires probing of the yield locus employing finite element techniques. In an API/Deepstar study (Andersen et al., 2003), such finite element analyses were conducted for the case of a strip anchor with a fluke length-to-thickness ratio $L/t = 7$ and 10, with subsequent independent analyses reported by Yang et al. (2010) for $L/t = 7$. The analyses assumed full bonding between the soil and the anchor surface. Full bonding may be considered to roughly correspond to full adhesion ($\alpha = 1$), although it actually imposes a slightly greater level of restraint with computed full bonding collapse loads being on the order of 4% greater than those for full adhesion. The computed bearing factors for pure normal, shear, and moment loading shown in Table 4.1 agree well with to the values from Equations 4.2 through 4.4.

Figure 4.4 shows the load capacity interactions for a strip anchor with $L/t = 7$. Bransby and O'Neill (1999) and O'Neill et al. (2003) propose a convenient functional form of the interaction relationship:

$$f = \left[\left(\frac{F_n}{F_{nmax}}\right)^q + \left[\left(\frac{M}{M_{max}}\right)^m + \left(\frac{F_s}{F_{smax}}\right)^n\right]^{\frac{1}{p}}\right] - 1 = 0 \tag{4.5}$$

Least squares fits of Equation 4.5 to the finite element solutions yield the interaction coefficients m, n, p, and q listed in Table 4.1. Subsequent calculations can be sensitive to the shape of the yield locus, particularly where kinematic behavior is concerned. Accordingly, assessing the sensitivity of the predicted behavior to variability in the yield locus parameters is advisable.

Table 4.1 Bearing and interaction factors for strip anchors

Capacity factor	Yang et al. (2010)	API/deepstar (2003)
N_{nmax}	11.98	11.58
N_{smax}	4.39	4.49
N_{mmax}	1.64	1.74
M	1.56	1.40
n	4.19	3.49
p	1.57	1.32
q	4.43	4.14

4.1.2 Circular disks

Application of the method of characteristics to axisymmetric geometries (in a cylindrical $r - \theta - z$ system) requires modification of the equations shown previously for the plane strain case. The directions of the characteristic lines are actually identical to those for the plane strain case, with r and z replacing x and y. However, the mean stress gradient relationships are altered. Martin and Randolph (2001) show the following revised equation for the case of a soil profile with a vertical strength gradient $\partial s_u / \partial z$ and unit weight γ:

$$d\sigma_m \pm 2s_u d\theta = \left(\mp \frac{\partial s_u}{\partial z} + \frac{s_u \cos 2\theta + \omega s_u}{r} \right) dr + \left(\gamma - \frac{s_u \sin 2\theta}{r} \right) dz \tag{4.6}$$

The sign of the parameter $\omega = \pm 1$ is selected as negative if the expected direction of plastic flow is outward and positive if inward.

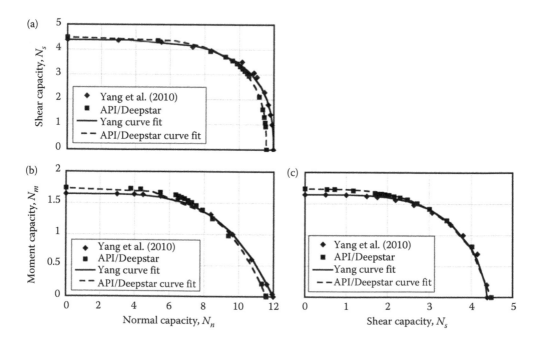

Figure 4.4 Load interactions for strip anchor. (a) Normal-shear, (b) normal-moment, and (c) shear-moment.

For upper bound solutions, a so-called "consistent" velocity field can be selected such that the principal stress rates coincide with the principal stress directions obtained from the lower bound method of characteristics solution. Velocity components along the characteristics become

$$v_r \sin(\theta \pm \pi/4) + v_z \cos(\theta \pm \pi/4) = \frac{-udr}{2r\sin(\theta \pm \pi/4)}$$

(4.7)

Martin and Randolph (2001) demonstrate that for the case of smooth ($\alpha = 0$) circular disk their lower and upper bound solutions produce identical values of $N_{nmax} = F_{nmax}/A = 12.42$. Similarly, for the case of a rough ($\alpha = 1$) disk, the lower and upper bound analyses produce an exact solution $N_{nmax} = 13.11$.

In regard to moment capacity of a circular disk with diameter L, Yang (2009) considers a spherical slip mechanism to obtain the upper bound estimate of rotational bearing resistance:

$$N_{mmax} = \frac{M_{max}}{s_u L^3} = \frac{\pi^2}{8}$$

(4.8)

4.1.3 Eccentric normal loading on rectangular plates

Considering now the capacity and interaction characteristics of a finite length rectangular plate, for the time being the discussion will be restricted to the normal–moment loading shown in Figure 4.5. The convention adopted here will define the long and short dimensions of the plate as L and W, respectively. If the x–z plane is the intended plane of the anchor line force, this is often termed "in-plane" loading, as opposed to "out-of-plane" loading that occurs for eccentricity e_y unequal to zero. The angle β in Figure 4.5 defines the out-of-plane eccentricity of loading. The relevant bearing factors are now as follows:

$$N_n = \frac{F_n}{LWs_u}$$

(4.9)

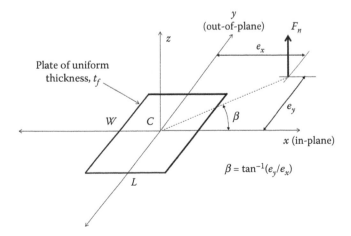

Figure 4.5 Eccentric normal loading on plate.

$$N_{mx} = \frac{M_x}{L^2 W s_u} = \frac{F_n e_x}{L^2 W s_u} \tag{4.10}$$

$$N_{my} = \frac{M_y}{L^2 W s_u} = \frac{F_n e_y}{L^2 W s_u} \tag{4.11}$$

Table 4.2 shows the uniaxial bearing factors for square and rectangular plates from finite element probes by Yang et al. (2010). The reported bearing factor for pure normal loading on a square plate, $N_{nmax} = 12.5$, exceeds the Merifield et al. (2003) value of 11.9 obtained from their lower bound finite element analysis procedure. The moment bearing factors for a square plate, $N_{mxmax} = N_{mymax} = 1.9$, exceed the moment capacity for a circular disk (Equation 4.8) by approximately 20%. Some of this difference may be due to the tendency of finite element calculations to exceed actual values. For rectangular plates, the ratio of strong-axis to weak-axis moment resistance is predicted to be about $N_{mymax}/N_{mxmax} = 1.25$.

Figure 4.6 shows the results of a full finite element probe of normal–moment interactions for square and rectangular $(L/W = 2)$ plates. The resultant moment bearing factor in this figure is defined as $N_m = \sqrt{N_{mx}^2 + N_{my}^2}$. For square plates, the eccentricity angle β has virtually no influence on the interaction normal–moment relationship. The moment–moment $(M_x–M_y)$ interaction for the square plate visually appears to approach a circular relationship—subsequent curve fits show a nearly circular relationship. The increased aspect ratio associated with the rectangular plate introduces a directional preference to moment resistance as evidenced by the divergence in the normal–moment interaction curves for different eccentricity angles β. The moment–moment $(M_x–M_y)$ interaction predictably tends toward an elliptical shape owing to the directional preference in moment resistance for rectangular plates.

Characterization of normal–moment interactions in a functional form can proceed by adapting the O'Neill–Bransby equation, as shown in Equation 4.12. The shear force term is omitted for now, but an additional moment term is added to reflect the two components of moment loading. Table 4.2 shows the coefficients for fitting Equation 4.12 to the curves in Figure 4.6.

$$f_{nm} = \left(\frac{F_n}{F_{nmax}}\right)^q + \left[\left(\frac{M_x}{M_{xmax}}\right)^{mx} + \left(\frac{M_y}{M_{ymax}}\right)^{my}\right]^{\frac{1}{p}} - 1 = 0 \tag{4.12}$$

Table 4.2 Normal–moment bearing factors for rectangular plates

Symbol	W/L = 1	W/L = 2
N_{nmax}	12.5	12.35
N_{mxmax}	1.9	1.70
N_{mymax}	1.9	2.15
m_x	1.91	2.47
m_y	1.91	1.86
p	1.56	1.93
q	3.26	3.20

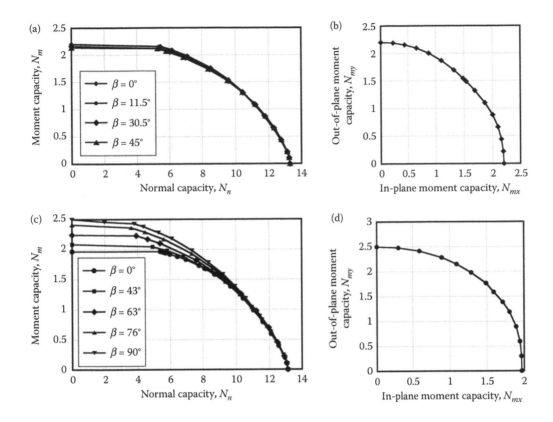

Figure 4.6 Normal–moment interactions for square and rectangular plates. (a) Normal–moment, (b) M_x–M_y, (c) normal–moment, and (d) M_x–M_y.

While the yield locus formulation shown above frequently proves convenient for analytical modeling, an alternative characterization of interaction effects in terms of anchor pullout capacity versus load eccentricity frequently proves useful for practical analysis and evaluation of experimental measurements. This is achieved by simply substituting $M_x = F_n e_y$ and $M_y = F_n e_x$ and solving for F_n in Equation 4.12. Figure 4.7 shows the resulting interactions. Comparison to pullout tests in a laboratory test bed in kaolin (Gilbert et al., 2009) confirms the general trend of the analytical models.

4.1.4 Shear–torsion on plates

Pure sliding occurs when the line of action of F_s passes through the plate centroid (Figure 4.8), while pure torsion occurs when F_s acts at an infinite distance from the plate. The relevant equations can be simplified by defining the aspect ratio of the plate $R_a = W/L$ and the thickness ratio as $R_t = t/L$. The bearing capacity factor for pure sliding depends on the direction of the load ϕ as follows:

$$N_{smax} = 2\alpha + 2[(N_e + R_a)\sin\phi + (N_e R_a + 1)\cos\phi]R_t/R_a \qquad (4.13)$$

As noted for the case of a strip anchor, evidence suggests an end bearing coefficient $N_e \approx 7.5$. A load angle $\phi = 0$ with an infinite aspect ratio reproduces the strip footing formula Equation 4.3.

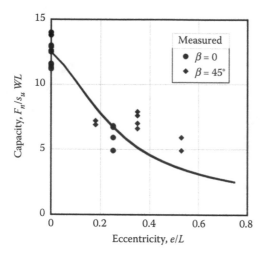
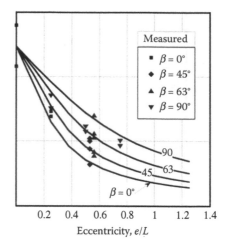

Figure 4.7 Normal load capacity versus eccentricity.

The bearing factor for pure torsion is given by

$$
N_{tmax} = \frac{\alpha R_a^2}{6} \left\{ \frac{\sin\psi}{\cos^2\psi} + \ln\left[\tan\left(\frac{\pi}{4} + \frac{\psi}{2}\right)\right] \right\} + \frac{\alpha}{6R_a} \left\{ \frac{\cos\psi}{\sin^2\psi} - \ln\left[\tan\left(\frac{\psi}{2}\right)\right] \right\}
$$
$$
+ \frac{1}{2} N_e R_t \left(R_a + \frac{1}{R_a} \right) + 2R_t
$$

(4.14)

where $\psi = \tan^{-1}(L/W)$. Torsion bearing factors for infinitely thin square and rectangular (2:1 aspect ratio) plates are $N_{tmax} = 0.765$ and 1.19, respectively. The development of this yield function assumes no interaction between sliding and bearing resistance on the edges of the plate, which will tend to overestimate the effects of a finite plate thickness on torsional resistance.

General eccentric in-plane loading of a finite length plate (Figure 4.8) can be obtained by considering a plate rotating about a center of rotation (x_0, y_0) at angular velocity $\dot{\beta}$. The

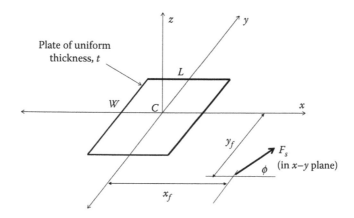

Figure 4.8 Shear–torsion on a plate.

least upper bound for F_s is obtained by equating the external rate of work to internal energy dissipation:

$$F_s = \dot{D}/(\rho + e)\dot{\beta} \tag{4.15}$$

A least upper bound occurs when F_s is minimized with respect to the distance to the center of rotation:

$$F_s = \frac{1}{\dot{\beta}}\frac{\partial \dot{D}}{\partial \rho} \tag{4.16}$$

Thus, once an internal dissipation and its derivative are established, a complete shear–torsion yield locus can be established by varying ρ from zero to infinity in a selected direction. This analysis is formulated in terms of the force F_s being perpendicular to the moment arm e. In actual systems, the load is not necessarily physically attached to the anchor in this fashion (moment arm CA in Figure 4.9). However, a statically equivalent arm CB can always be constructed. A question may also arise as to how one can be assured that the optimal center of rotation is always aligned normal to the direction of the applied force, that is, along line OC. Considering a center of rotation along a line that is not normal to the direction of F_s is equivalent to applying a component of velocity v_n normal to the direction of the applied force (Figure 4.9). The added velocity component will increase the internal rate of energy dissipation while having no impact on the rate of external work. Therefore, v_n equal to zero will always produce the minimum collapse load for a given value of ρ.

In the case of a vanishingly thin plate, the dissipation function and associated minimum collapse load are as follows:

$$\dot{D} = 2\alpha s_u \dot{\beta} \int_{-W/2}^{W/2} \int_{-L/2}^{L/2} \sqrt{(\rho\cos\phi + x)^2 + (\rho\sin\phi + y)^2}\, dx\, dy \tag{4.17}$$

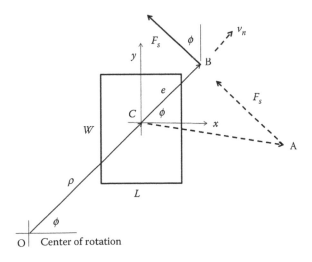

Figure 4.9 System of shear–torsion loads on plate.

$$F_s = 2\alpha s_u \int\limits_{-W/2}^{W/2} \int\limits_{-L/2}^{L/2} \frac{(\rho\cos\phi + x)\cos\phi + (\rho\sin\phi + y)\sin\phi}{\sqrt{(\rho\cos\phi + x)^2 + (\rho\sin\phi + y)^2}} \, dx \, dy \tag{4.18}$$

The eccentricity e corresponding to the selected value of ρ is calculated from Equation 4.15. For the special case of $\phi = 0$, further simplification is possible since Equation 4.17 can be integrated to obtain the following closed-form expression:

$$F_s = 2\alpha s_u \left[a_1^2 \ln\left|\frac{b_1 + W/2}{b_1 - W/2}\right| - a_2^2 \ln\left|\frac{b_2 + W/2}{b_2 - W/2}\right| + W(b_1 - b_2) \right]$$

$$a_1 = \rho_{opt} + L/2$$

$$a_2 = \rho_{opt} - L/2 \tag{4.19}$$

$$b_1 = \sqrt{a_1^2 + W^2/4}$$

$$b_2 = \sqrt{a_2^2 + W^2/4}$$

Figure 4.10 presents the sliding–torsion interaction for a thin plate for various aspect ratios.

As described earlier for other load interactions, shear–torsion interactions can be expressed in an empirical functional form:

$$f_{st} = \left[\left(\frac{F_s}{F_{smax}}\right)^n + \left(\frac{T}{T_{max}}\right)^k \right] - 1 = 0 \tag{4.20}$$

For the infinitely thin square plate, the fitting parameters are $n = 2.02$ and $k = 1.45$; whereas, for the rectangular plate $(W/L = 2)$ case, they are $n = 1.81$ and $k = 1.32$. The shear load here is $F_s = \sqrt{F_{sx}^2 + F_{sy}^2}$, a circular interaction that is exact for an infinitely thin plate. If a finite thickness is considered, then the F_{sx}–F_{sy} interaction must be evaluated by finite element analyses.

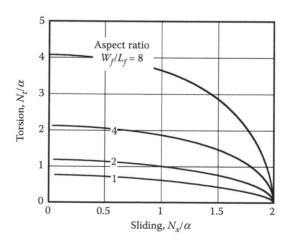

Figure 4.10 Shear–torsion interaction.

4.1.5 General combined loading

For general combined loading, Yang et al. (2010) propose to unify Equations 4.12 and 4.20 into the following equation:

$$f = \left(\frac{F_n}{F_{nmax}}\right)^q + \left\{ \left[\left(\frac{M_x}{M_{xmax}}\right)^{m1} + \left(\frac{M_x}{M_{xmax}}\right)^{m2} \right] + \left[\left(\frac{F_s}{F_{smax}}\right)^n + \left(\frac{T}{T_{max}}\right)^k \right]^s \right\}^{\frac{1}{p}} - 1 = 0 \qquad (4.21)$$

The interaction coefficient s is introduced to characterize the moment–shear interaction. Based on finite element normal–torsion (F_n–T) and moment–torsion (M–T) probes, they suggest $s = 4.80$ and 2.84 for infinitely thin square and rectangular $W/L = 2$ plates, respectively. This equation is formulated from the standpoint of determining how the primary pullout capacity (F_n) is reduced owing to load demand imposed by all other modes of loading—moment, shear, and torsion. It is not clear whether this is the most effective grouping of the load combinations. For example, in their analysis of six degrees of freedom loading on mat foundations, Feng et al. (2014) choose to reduce the interactions into a resultant shear–moment (F_s – M) space.

4.1.6 Free surface effects

An anchor is considered shallowly embedded if the failure mechanism extends to the free surface (Figure 4.11). Early studies on shallowly embedded plate anchors include the work of Rowe and Davis (1982), with more recent studies by Merifield et al. (2003), Song et al. (2008), and Wang et al. (2010). The proximity of a plate anchor to a free surface requires several additional considerations. First, the overall shearing resistance to pullout is reduced. Second, separation (breakaway) of the plate from the soil is possible if tension cannot mobilize on the underside of the plate. Third, if separation occurs, then the soil self-weight contributes to load capacity. Merifield et al. (2003) provide the general form of

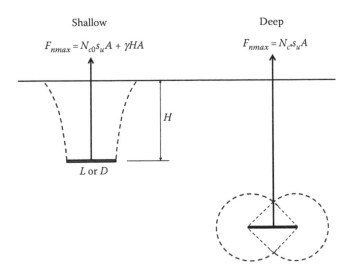

Figure 4.11 Shallowly embedded anchor.

the bearing factor equation for vertical pullout resistance to a horizontally oriented plate anchor:

$$N_{c\gamma} = N_{c0}(H/L) + \frac{\gamma'H}{s_u} \leq N_c^*$$ (4.22)

The bearing factor $N_{c\gamma}$ denotes the total resistance derived from shear strength and soil self-weight. N_{c0} is the shearing resistance factor for a weightless soil, which is a function of embedment depth H. The combined resistance $N_{c\gamma}$ cannot exceed the resistance N_c^* corresponding to a condition of no separation (no breakaway).

Assuming that breakaway does occur, Merifield et al. (2003) performed lower bound finite element analyses for circular disks and rectangular plates of various aspect ratios. Figure 4.12 summarizes the findings. A circular plate has slightly greater pullout capacity than a square plate, with maximum "deep" embedment behavior occurring at embedment ratios H/L greater than about 7. As the aspect ratio W/L increases, the reduction in bearing resistance owing to the free surface effect becomes increasingly significant, with strip anchors having roughly half the capacity of square or circular anchors when comparable embedment depths are considered.

Wang et al. (2010) conducted large displacement three-dimensional finite element analyses for conditions of no breakaway in the zone of tension. As shown in Figure 4.13, a reduction in capacity occurs owing to proximity to the free surface, but it is not nearly as dramatic as that which occurs for the case of breakaway. The obvious significance of the development of separation at the soil–anchor interface beneath the anchor led Wang et al. to investigate the conditions for which breakaway occurs. At sufficiently large depths, the soil overburden pressure will force the soil into contact with the base of the anchor plate even if no tension is assumed to develop at the base of the anchor. Based on their finite element studies as well as previous studies along similar lines performed by Song et al. (2008) for strip anchors, they proposed the following equation for estimating the embedment depth beyond which breakaway will not occur:

$$H_s/L = 6.2s_u/\gamma'L$$ (4.23)

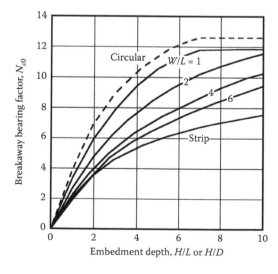

Figure 4.12 Shallow bearing factors for immediate breakaway conditions.

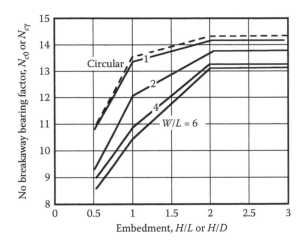

Figure 4.13 Shallow bearing factors for no breakaway.

The complexity of anchor behavior at shallow embedment depths is illustrated by considering the case of a plate anchor with aspect ratio $W/L = 2$ with unit weight ratio $\gamma'/s_u L = 2$. Figure 4.14 shows the capacity to be drastically reduced when breakaway occurs. Soil self-weight mitigates the breakaway effect somewhat, as shown by the comparison of N_{c0} to $N_{c\gamma}$; however, breakaway still results in a substantial reduction in pullout capacity. Finally, the dashed line denotes the depth below which no separation occurs (Equation 4.23), illustrating the desirability of embedding the anchor below the breakaway depth when possible.

4.1.7 Inclined shallow anchors

The pullout capacity of deeply embedded anchors is essentially independent of orientation of the anchor. By contrast, the collapse mechanism and associated pullout capacity of

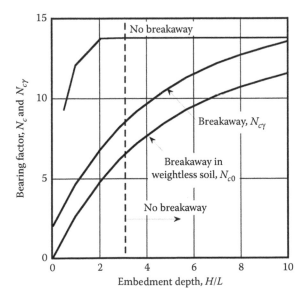

Figure 4.14 Anchor capacity at shallow depths.

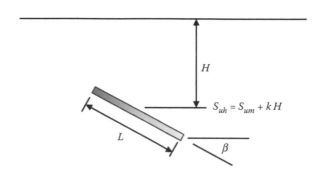

Figure 4.15 Failure mechanisms for shallowly embedded plates.

shallowly embedded anchors are affected by proximity to a free surface, as illustrated by the finite element studies of Yu et al. (2011) in Figures 4.15 and 4.16. The discussion that follows considers an anchor embedded at depth H measured to the center of a plate oriented at an angle β measured from horizontal (Figure 4.15). The seminal studies by Rowe and Davis (1982) similarly considered shallowly embedded strip anchors at horizontal and vertical orientations, $\beta = 0°$ and $90°$. Subsequent investigation into the pullout capacity of inclined plate anchors at shallow depths include experimental studies by Das and Puri (1989), upper and lower bound solutions of Merifield et al. (2001), and finite element studies by Yu et al. (2011).

The bearing factor for shallowly embedded inclined plate anchors can be expressed in the same framework embodied in Equation 4.22, with $N_{c\gamma}$ being the overall bearing factor that includes the effects of soil self-weight, N_{c0} being the bearing factor for a weightless soil with gapping, and N_c^* being the bearing factor for the case of no gapping. Figure 4.17

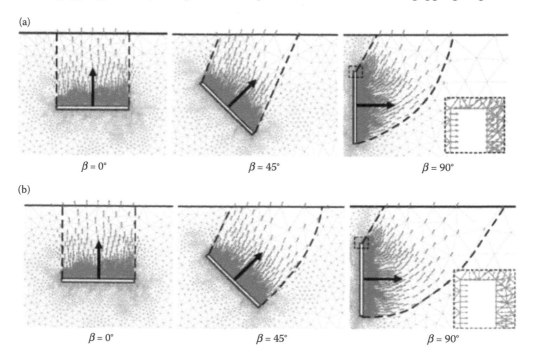

Figure 4.16 Schematic representation of shallowly embedded inclined plate anchor. (a) Smooth interface and (b) rough interface.

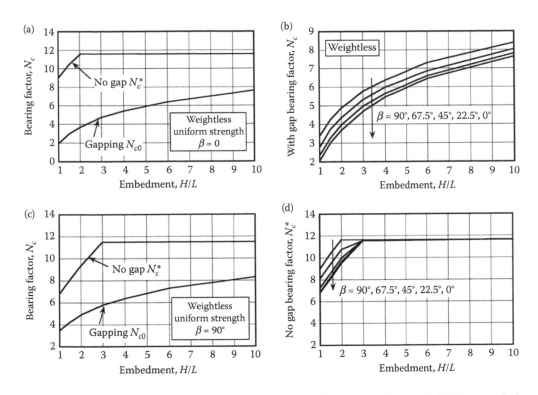

Figure 4.17 Bearing factors for horizontal and vertical plate anchors in a uniform soil. (a) Horizontal plate, (b) vertical plate, (c) gapping $\beta = 0$–$90°$, and (d) no gap $\beta = 0$–$90°$.

shows bearing factors computed by Yu et al. (2011) as a function of embedment depth H/L and plate inclination angle β. All curves shown in this figure are for a uniform soil strength profile and a smooth ($\alpha = 0$) plate. Yu et al. compute a maximum bearing factor for the no-gapping case $N^*_{cmax} = 11.59$, which slightly exceeds the limit analysis value of 11.48. The analyses show the effects of plate orientation to be greatest at shallow embedment depths, with the bearing factors for a vertical plate ($\beta = 90°$) exceeding those of a horizontal plate by up to 70% at an embedment $H/L = 1$. The differences diminish with increasing embedment. N_{c0} values for intermediate β values are also shown in Figure 4.17.

The bearing factors shown in Figure 4.17 correspond to uniform soil strength profiles. Actual seabed strength profiles typically vary with depth. Yu et al. (2011) consider the effects of a linearly varying undrained shear strength profile, with s_{um} being the strength at the mudline and k being strength gradient. The effects of strength inhomogeneity are expressed in terms of a dimensionless strength parameter kL/s_{ub}, where s_{ub} is the strength at the middle of the plate (Figure 4.15). The bearing factor N_{c0} is adjusted for inhomogeneity effects by means of a factor $s_k = N_{ck}/N_{c0}$, where N_{ck} is the bearing factor corresponding to gapping in a weightless soil having a strength gradient k, and N_{c0} corresponds to a uniform soil strength profile. A similar inhomogeneity correction factor is defined for the no-gapping bearing factor $s^*_k = N^*_{ck}/N^*_{c0}$. Figure 4.18 shows values of s_k and s^*_k as computed by Yu et al. (2011). The parameter s_k (describing resistance with gapping) is seen to decrease nearly linearly with increasing strength gradient kL/s_{ub}. A similar trend occurs for the parameter s^*_k (describing resistance with no gapping) for a vertical plate anchor. By contrast, the inhomogeneity parameter s^*_k for a horizontal plate shows a more complex trend of increasing for shallow embedment depths ($H/L = 1$) and decreasing at greater embedment depths ($H/L > 3$).

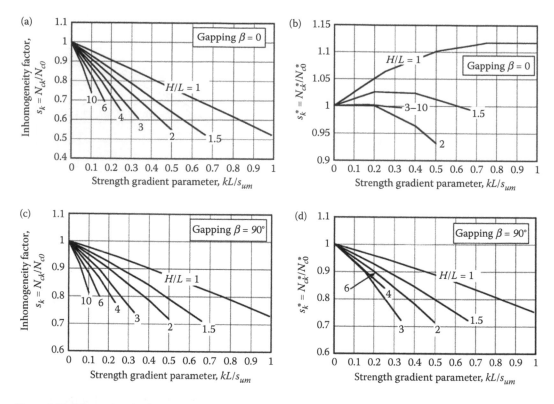

Figure 4.18 Effect of soil inhomogeneity on bearing factors. (a) Horizontal plate w/gap, (b) vertical plate w/gap, (c) gapping $\beta = 0$–$90°$, and (d) no gap $\beta = 0$–$90°$.

4.2 CYLINDRICAL ANCHORS

The other broad category of anchors is cylindrical geometries exemplified by piles and caissons. The discussion that follows focuses on ultimate load capacity using the methods of plasticity theory. In contrast to plate anchors, the performance of these types of anchors can be significantly influenced by elastic effects. This is particularly the case for long, slender piles for which it will be later shown that it is just about impossible to mobilize the full lateral load capacity. Nevertheless, even in cases where elastic behavior dominates performance, an understanding of ultimate resistance is important since most elastic ("$P - y$ spring") models used in practice are scaled to ultimate resistance. Consideration of elastic effects will be introduced in later chapters. As was seen to be the case for plate anchors, combined loading effects (for example, lateral–axial) can be important. Since they manifest their effects differently for piles and caissons, the issue of combined loading will also be postponed for lateral chapters covering these individual topics.

4.2.1 Lateral loading

Randolph and Houlsby (1984) considered the case of lateral resistance on a horizontally translating cylinder in a region far from a free surface (Figure 4.19). The zone of interest is assumed to be free of the influence of a free surface implying a plane strain condition. The analysis applies to undrained loading such that the soil may be considered frictionless with a cohesive strength c equal to the undrained shear strength s_u. The maximum shearing resistance at the

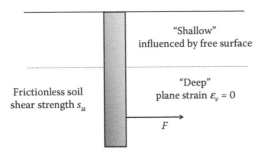

Figure 4.19 Translating cylinder.

soil–pile interface equals the adhesion $a = \alpha s_u$. They first develop a lower bound solution using the method of characteristics. Randolph and Houlsby demonstrate that one set of characteristics should be linear. Figure 4.20 illustrates the orientation of these characteristics relative to the pile boundary. If the full adhesion a mobilizes at the soil–pile interface, it follows from Mohr's circle of stress that the direction of maximum shear stress occurs at an angle $\pi/4 + \Delta/2$, where $\Delta = \sin^{-1}(a/c)$, from the normal to the pile boundary. This constraint defines the orientation of the characteristics originating along line AB in Figure 4.21. This condition prevails till the region BCD is reached. Characteristics originating from opposing sides of the cylinder should intersect at right angles, which leads to the rigid wedge comprising two sets of linear characteristics shown in Figure 4.22. Noting that a singularity occurs at Point A, the third region of linear characteristics comprises the circular fan AEF. The linear characteristics provide the necessary information for completely defining the shear and normal tractions along paths AB and BD. Integration along these paths leads to the following expression for collapse load:

$$N_p = \frac{P}{s_u D} = \pi + 2\Delta + 2\cos\Delta + 4[\cos(\Delta/2) + \sin(\Delta/2)] \qquad (4.24)$$

For the case of zero adhesion ($\Delta = 0$), the lower bound estimate of the bearing factor becomes $N_p = 9.14$, while for full adhesion ($\Delta = \pi/2$) $N_p = 11.94$. The relationship between bearing factor versus adhesion (Figure 4.23) shows slight curvature, but linear interpolation between the perfectly smooth and rough cases is reasonable.

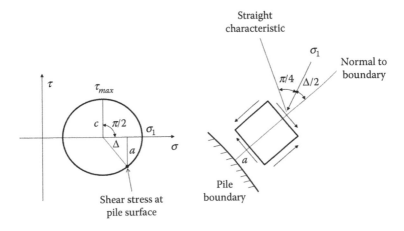

Figure 4.20 Translating cylinder boundary condition.

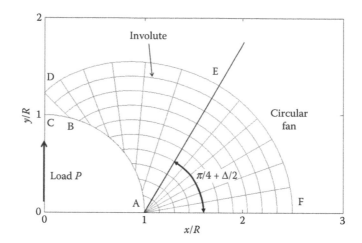

Figure 4.21 Typical characteristic net.

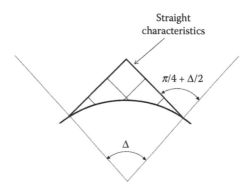

Figure 4.22 Wedge region of characteristic net.

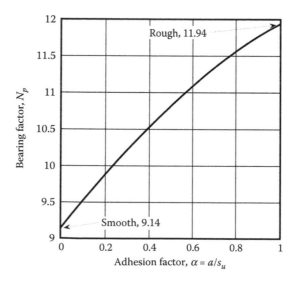

Figure 4.23 Bearing factors for laterally translating cylinder.

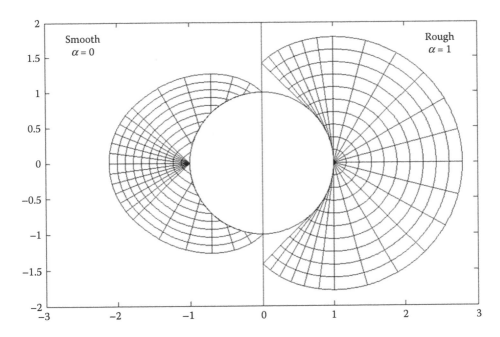

Figure 4.24 Characteristic nets for smooth and rough cylinders.

The second set of characteristics is circular in the fan region AEF and a form of involute in region ABDE. To develop an upper bound estimate of collapse load, Randolph and Houlsby use a velocity field from the field of characteristics. Their original analysis produced an upper bound estimate of collapse load that exactly matched the lower bound equation in Equation 4.24. However, the velocity field used in the original Randolph–Houlsby analysis actually has a small region of conflict for adhesion factors $\alpha < 1$. Subsequent refinements (Martin and Randolph, 2006) have narrowed the gap between upper and lower bound estimates to the extent that Equation 4.24 can be considered accurately to within 1%; that is, the current least upper bound estimate for the case of a smooth ($\alpha = 0$) cylinder is $N_p = 9.20$ versus the method of characteristics lower bound value of 9.14. Thus, Equation 4.24 can be considered sufficiently accurate for practical purposes. Figure 4.24 shows the characteristic nets for the smooth and rough cases.

4.2.2 Axial loading

Martin and Randolph (2001) and Martin (2001) present lower and upper bound collapse load estimates for vertically loaded caissons (Figure 4.25). In the latter study, finite element calculations were also performed. Strictly speaking, the analyses were not specifically directed toward uplift loading. Nevertheless, their conclusions still apply so long as "suction" (actually negative differential pressure) can be transmitted through the soil plug to the base of the caisson, such that a reverse end bearing collapse mechanism can develop. Figure 4.26 summarizes the findings for the cases perfectly smooth and rough ($\alpha = 0$ and 1) skirts, from which the following important observations can be made:

- For the smooth case, the only source of uplift resistance is reverse end bearing; thus, the increase in the end bearing factor is unambiguously seen to increase from $N_c = 6.05$ at zero embedment depth to $N_c = 9.2$ at embedment depths greater than L/D greater than 1–1.5.

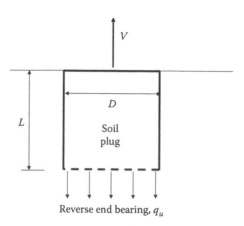

Figure 4.25 Vertically loaded circular caisson.

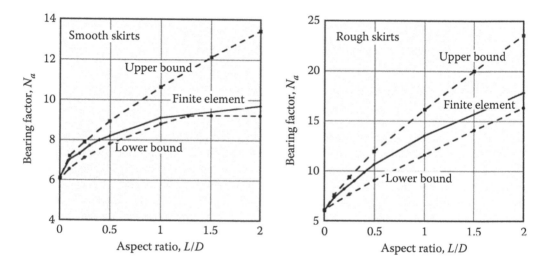

Figure 4.26 Vertical bearing capacity factors in homogeneous soil.

- When a rough skirt is considered, the increase in resistance above that mobilized by end bearing is consistent with a simple addition of side friction to reverse end bearing.

The latter observation provides a strong justification for the equation commonly used in practice for predicting total vertical load capacity:

$$V = \frac{\pi D^2}{4} N_c s_u + \alpha \pi D L \tag{4.25}$$

4.2.3 Free surface effects for lateral loading

Murff and Hamilton (1993) evaluated the free surface effects for a laterally loaded cylindrical pile by considering the surface failure wedge mechanism in Figure 4.27. At the free surface, a passive failure mechanism intersects the ground surface at a distance r_0 from the pile and extends to a depth z_0. Below this depth, the Randolph–Houlsby flow-around condition

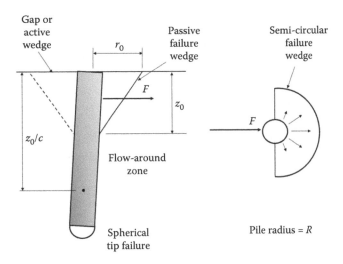

Figure 4.27 Failure mechanism for laterally loaded cylindrical pile.

is assumed. Spherical slip mechanism is assumed at the pile tip. The pile rotates about a center of rotation at depth z_0/c, which can occur below the pile tip. Soil resistance within the flow-around zone is assumed to be independent of the center of rotation. The parameters r_0, z_0, and c are taken as optimization parameters that are varied to minimize the collapse load F. Although a general upper bound approach is used here, the analysis is not a strict upper bound since a kinematically admissible velocity fields are not ensured at the interface of the flow-around zone and the surface wedge or the interface of the flow-around zone and the tip failure mechanism.

For a pile of radius R, the passive failure wedge deforms internally according to the radial velocity distribution described by Equation 4.26.

$$v_r = v_o \left(\frac{R}{r} \right)^\zeta \left(1 - \frac{cz}{z_0} \right) \cos\theta \qquad (4.26)$$

This introduces a fourth optimization variable, ζ, into the analysis. Enforcing the incompressibility constraint and compatibility at the interface between the wedge and the undeforming soil leads to the following expression for the vertical velocity distribution.

$$v_z = g(r,\theta) \left(z - \frac{cz^2}{2z_0} \right) + h(r,\theta) \qquad (4.27)$$

where

$$g(r,\theta) = v_o \frac{(\zeta - 1)R^\alpha}{r^{1+\alpha}} \cos\theta$$

$$h(r,\theta) = v_o z_o \cos\theta \left(\frac{R}{r} \right)^\zeta \left(\frac{r_o - r}{r_o - R} \right) \left\{ -\frac{1}{r_o - r} + \left(\frac{c}{r_o - R} \right) + \frac{1-\zeta}{r} \left[1 - \frac{c}{2} \left(\frac{r_o - r}{r_o - r} \right) \right] \right\}$$

Strain rates are evaluated from the spatial derivatives of the radial and vertical velocity terms in Equations 4.26 and 4.27. In general, all six strain rate components are nonzero. Energy dissipation terms are evaluated using the normal methods of plasticity theory. While the analysis is also applicable to an anisotropic medium, Table 4.3 lists the energy dissipation terms for the special case of an isotropic medium with a von Mises yield surface. The first internal energy dissipation term arises from deformation within the passive failure wedge. The second and third are the dissipation terms at the respective interfaces between the passive wedge and the rigid soil and the passive wedge and the pile. Below the surface wedge, the energy dissipation is simply the product of the Randolph–Houlsby flow-around resistance times the pile velocity at the depth in question. In cases where the center of rotation occurs within the pile (Figure 4.28), a fifth component of energy dissipation occurs. If the pile does not yield, then a kick-back region with a reverse flow-around resistance develops. Alternatively, if the pile itself yields, then the pile below the center of rotation remains

Table 4.3 Dissipation and work terms for laterally loaded pile analysis

Location	Equation
(1) Deforming wedge	$\dot{D}_1 = 2 \displaystyle\int_{r=R}^{r=r_0} \int_{z=0}^{z=z_0[(r_0-r)/(r_0-R)]} \int_{\theta=0}^{\theta=\pi/2} \sqrt{2\dot{\varepsilon}_{ij}\dot{\varepsilon}_{ij}}\, r\, d\theta\, dz\, dr$
(2) Wedge-rigid soil	$\dot{D}_2 = 2v_0 \sqrt{1+\left(\dfrac{z_0}{r_0-R}\right)^2} \sqrt{1+\left(\dfrac{R-r_0}{z_0}\right)^2} \displaystyle\int_0^{z0} s_u \dfrac{R^\varsigma\left(1-c\dfrac{z}{z_0}\right)}{\left[r_0 - \dfrac{z}{z_0}(r_0-R)\right]}\, dz$
(3) Wedge–pile interface	$\dot{D}_3 = 2\displaystyle\int_{z=0}^{z=z0}\int_{\theta=0}^{\theta=\pi/2} \alpha s_u \sqrt{v_z^2 + v_c^2}\, d\theta\, dz$ v_z from Equation 4.27
	$v_c = (1 - cz/z_0)\sin\theta$
(4) Flow-around above z_0/c	$\dot{D}_4 = 2\displaystyle\int_{z_0}^{z_0/c} RN_p s_u v_o \left|1 - \dfrac{cz}{z_0}\right| dz \quad z_0/c < L$
	$\dot{D}_4 = 2\displaystyle\int_{z_0}^{L} RN_p s_u v_o \left|1 - \dfrac{cz}{z_0}\right| dz \quad z_0/c > L$
(5a) Flow-around below z_0/c	$\dot{D}_5 = 2\displaystyle\int_{z_0/c}^{L} RN_p s_u v_o \left|1 - \dfrac{cz}{z_0}\right| dz \quad z_0/c < L$
	$\dot{D}_5 = 0 \quad z_0/c > L$
(5b) Plastic hinge in pile	$\dot{D}_5 = \left|M_p\right| \dfrac{v_0}{z_0/c} \quad z_0/c < L$
(6) Pile tip	$\dot{D}_6 = \dfrac{v_0 R_2^3 c}{z_0} \displaystyle\int_{\varphi=0}^{\varphi=2\pi} \int_{\omega=0}^{\omega=\sin^{-1}\left(R/\sqrt{R^2+R_l^2}\right)} s_u \sqrt{\cos^2\omega + \sin^2\omega\sin^2\varphi}\, \sin\omega\, d\omega\, d\varphi$
(7) External work by wedge	$\dot{W}_g = 2\displaystyle\int_{r=R}^{r=r_0}\int_{z=0}^{z=z_0[(r_0-r)/(r_0-R)]}\int_{\theta=0}^{\theta=\pi/2} v_z \gamma'\, r\, d\theta\, dz\, dr$ v_z from Equation 4.27

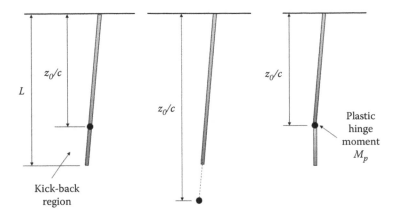

Figure 4.28 Possible pile collapse mechanisms.

vertical while energy dissipation occurs at the plastic hinge at a rate equal to the plastic moment capacity times the angular velocity of the pile. The final component of energy dissipation occurs at the pile tip assuming a spherical slip mechanism. The tip resistance is effectively integrated in a spherical coordinate $(\rho - \omega - \phi)$ system with dimensions defined by Figure 4.29. For long slender piles, the tip resistance term is typically minor. However, it can be a significant component of resistance for suction caissons. If a gap is assumed such that no active wedge forms, the final term required for the energy balance is the external work from the weight of the uplifting passive wedge $\dot{W}_g$. The collapse load F is now computed by equating internal energy dissipation to external work:

$$F v_0 = \sum_{i=1}^{5} \dot{D}_i - \dot{W}_g \qquad (4.28)$$

This equation presumes a sign convention that produces negative external work for an uplifting wedge, such that soil self-weight contributes to load capacity. If no gap is assumed then the first three dissipation terms will be doubled, but the work performed by the active wedge will counteract that of the passive wedge, in which case soil self-weight does not contribute to load capacity F.

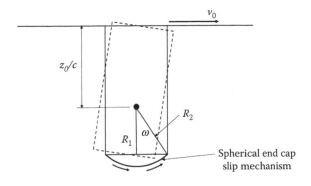

Figure 4.29 Tip resistance definition sketch.

The Murff–Hamilton analysis is well suited to complexities such as stratified soil profiles and, with appropriate modification of the dissipation functions, strength anisotropy. However, the following discussion focuses on the development of a bearing function $N_p(z)$ that transitions from the deep flow-around Randolph–Houlsby solution to the reduced resistance at the surface for the case of a horizontally translating pile. As shown in Figure 4.30, if the collapse load F is computed for successively longer pile lengths L, the normalized incremental increase in load capacity with respect to pile length is simply the lateral bearing resistance factor at the depth in question $N_p(z)$. Defining a strength gradient parameter $\lambda = s_{um}/kD$, bearing functions can be developed in terms of this parameter.

Based on this analysis Murff and Hamilton proposed a bearing resistance function having the following form:

$$N_{pg}(z) = N_1 - N_2 \exp(-\eta/D) \tag{4.29}$$

$$\eta = 0.25 + 0.05\lambda \qquad \text{for } \lambda \le 6$$
$$\eta = 0.55 \qquad \text{for } \lambda \ge 6$$

The subscript "g" emphasizes that this bearing factor applies to the case of zero suction on the trailing side of the pile such that a gap will form. The parameter N_1 is simply the Randolph–Houlsby flow-around bearing resistance factor. As previously discussed, it is a function of surface roughness α and varies from approximately 9.2–11.94. From Coulomb theory, the bearing at the free surface, $N_p(0) = (N_1 - N_2)$, ranges from 2 to 2.82 for surface roughness α ranging from 0 to 1, assuming a gap develops. Thus, the parameter N_2 defines the increase in bearing resistance from the free surface to the flow-around zone. Finally, the parameter η defines the depth interval for the transition from free surface to flow-around conditions. Equation 4.29 considers only the soil shearing resistance contribution to load capacity. Since gapping occurs, the soil self-weight contribution to resistance should be added, subjecting to the constraint that total resistance does not exceed the flow-around resistance:

$$N_{p\gamma}(z) = N_p(z) + \gamma z/s_u D \le N_1 \tag{4.30}$$

If a wedge develops behind the caisson, then the work performed by the leading surface wedge is offset by that performed by the trailing wedge; thus, soil self-weight contributes

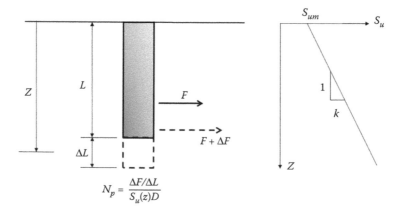

Figure 4.30 Development of lateral bearing resistance profile.

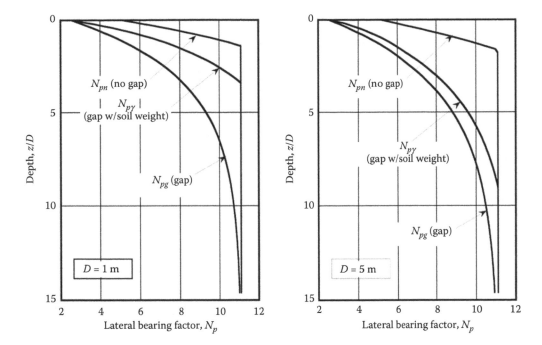

Figure 4.31 Lateral bearing resistance versus depth.

nothing to resistance. However, since shearing resistance mobilizes in both the active and passive surface wedges, the bearing resistance is a doubled subject to the constraint that total resistance does not exceed the flow-around resistance:

$$N_{pn}(z) = 2N_p(z) \le N_1 \tag{4.31}$$

We can now consider an illustrative case for a typical normally consolidated soil strength profile with mudline strength $s_{um} = 2$ kPa, strength gradient $k = 1.6$ kPa/m, soil submerged unit weight $\gamma' = 6.5$ kN/m³, and adhesion factor $\alpha = 0.7$. When the soil self-weight effect is not considered, Figure 4.31 shows gapping to have a very significant effect on the bearing resistance profile. However, the inclusion of soil self-weight can substantially offset much of the loss in resistance owing to the gapping. Furthermore, the relative proportions of soil resistance from shear strength and self-weight will depend on the size of the pile. Thus, the beneficial effect of soil self-weight is considerably more significant for smaller diameter piles.

REFERENCES

Andersen KH, Murff JD, and Randolph MF, 2003, Deepwater anchor design practice-vertically loaded drag anchors, *Phase II Report to API/Deepstar*, Norwegian Geotechnical Institute, Norway, Offshore Technology Research Center, USA and Centre for Offshore Foundation Systems, Australia.

Bransby MF and O'Neill MP, 1999, Drag anchor fluke–soil interaction in clays, *International Symposium on Numerical Models in Geomechanics (NUMOG VII)*, Graz, Austria, pp 489–494.

Das BM and Puri VK, 1989, Holding capacity of inclined square plate anchors in clay, *Soils Found* 29(3), 138–144.

Feng X, Randolph MF, Gourvernec S, and Wallerand R, 2014, Design approach for rectangular mud-mats under fully three-dimensional loading, *Geotechnique* 64(1), 51–63.

Gilbert RG, Lupulescu C, Lee CH, Miller J, Kroncke M, Yang M, Aubeny CP, and Murff JD, 2009, Analytical and experimental modeling of out-of-plane loading of plate anchors, *Offshore Technology Conference*, Houston, Texas, USA 4–7 May 2009, OTC 20115, pp 1–15.

Martin, 2001, Vertical bearing capacity of skirted circular foundations on Tresca soil, *Proceedings of the 15th International Conference on Soil Mechanics and Geotechnical Engineering*, Istanbul, Vol. 1, pp 743–746.

Martin CM and Randolph MF, 2001, Applications of the lower and upper bound theorems of plasticity to collapse of circular foundations, *Proceedings of the 10th International Conference International Association of Computer Methods and Advances in Geomechnics*, Tucson, Vol. 2, pp 1417–1428.

Merifield RS, Lyamin AV, Sloan SW, and Yu HS, 2003, Three-dimensional lower bound solutions for stability of plate anchors in clay, *J Geotech Geoenviron Eng, ASCE*, 129(3), 243–253.

Merifield RS, Sloan SW, and Yu HS, 2001, Stability of plate anchors in undrained clay, *Geotechnique*, 51(2), 141–153.

Murff JD and Hamilton JM, 1993, P-Ultimate for undrained analysis of laterally loaded piles, *ASCE J Geotech Eng*, 119(1), 91–107.

Murff JD, Randolph MF, Elkhatib S, Kolk HJ, Ruinen RM, Strom PJ, and Thorne CP, 2005, Vertically loaded plate anchors for deepwater applications, *Proceedings of the International Symposium on Frontiers in Offshore Geotechnics*, IS-FOG05, Perth, pp 31–48.

O'Neill MP, Bransby MF, and Randolph MF, 2003, Drag anchor fluke–soil interaction in clays, *Canad Geotech J*, 40, 78–94.

Randolph MF and Houlsby GT, 1984, The limiting pressure on a circular pile loaded laterally in cohesive soil, *Geotechnique*, 34(4), 613–623.

Rowe RK and Davis EH, 1982, The behavior of anchor plates in clay. *Geotechnique*, 32(1), 9–23.

Song Z, Hu Y, and Randolph MF, 2008, Numerical simulation of vertical pullout of plate anchors in clay, *ASCE J Geotech Geoenviron Eng*, 134(6), 866–875.

Wang D, Hu Y, and Randolph MF, 2010, Three-dimensional large deformation finite element analysis of plate anchors in uniform clay, *ASCE J Geotech Geoenviron Eng*. 136(2), 355–365.

Yang M, Murff JM, and Aubeny CP, 2010, Undrained capacity of plate anchors under general loading, *ASCE J Geotech Geoenviron Eng*, 136(10), 1383–1393.

Yang M, Murff JD, Aubeny CP, Lee CH, Lupulescu C, and Gilbert RB, 2009, *Out of Plane Loading of Plate Anchors Analytical and Experimental Modeling*, Final Project Report for ABS Consulting, OTRC Library Number ABSC/JIP2/09C, College Station, Texas, 109p.

Yu L, Liu J, Kong X-J, and Hu Y, 2011, Numerical study on plate anchor stability in clay, *Geotechnique*, 61(3), 235–246.

Chapter 5

Anchor line mechanics

Anchor line or chain behavior can influence the overall evaluation of anchor performance in several ways. First, soil resistance on the anchor line can contribute to overall anchor pullout capacity. Second, the downward curvature of the anchor line in the soil column generates vertical forces on the anchor, which may have to be considered in the load capacity evaluation of the anchor. Finally, plate anchors tend to orient themselves in a fixed direction relative to the direction of the anchor line tension; thus, the kinematics of plate anchors is largely driven by the behavior of the anchor line. This chapter first discusses the soil resistance forces that act on anchor lines, the line tension that develops in response to these forces, and the geometric configuration of the embedded anchor line.

Anchor line behavior in the water column is governed by well-established catenary equations describing the line tension and geometric configuration of a line loaded by its self-weight. Analysis of anchor line behavior in the soil column provides required information of the water-column catenary analysis in regard to the line tension at the mudline, possible slack in the line on the seabed, and possible uplift of the anchor line at its contact with the seabed. Analysis of anchor line behavior in the soil column can also provide information on the length of line that becomes embedded in the soil. In shallow water moorings, the reduction in the total length of the anchor line can become sufficiently large to affect the catenary analysis in the water column. In deep moorings, the embedded line length is likely to be small relative to the total length of the mooring. However, accurate prediction of the embedded length is still required to ensure that the forerunner in a composite mooring system is sufficiently long; for example, to ensure that the polyester cable in the middle segment does not embed into the seafloor. The section in this chapter on the coupling of the soil column equations to the water column equations addresses some of these issues.

5.1 SOIL RESISTANCE ON ANCHOR LINE

Soil resistance acting on the anchor line or chain may have to be calculated for the two possible conditions shown in Figure 5.1: a slack line resting on the seabed and a line fully embedded in the soil. Anchor penetration in sands is typically small, so the embedded line case is considered for clays only.

5.1.1 Slack line on seabed

The frictional resistance force T_s from the slack portion of the anchor line with length is normally computed from a simple friction equation:

$$T_s = f w L_s \tag{5.1}$$

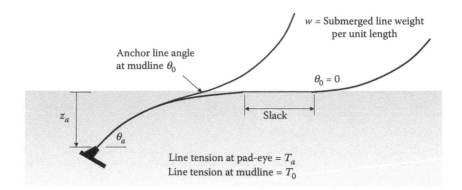

Figure 5.1 Soil resistance on elemental length of anchor line.

where

f = is the coefficient of friction,
w = is the submerged weight per unit length of the line, and
L_s = is the length of the slack portion of the line.

Friction coefficients for various seabed soils are listed in Table 5.1 (Taylor and Valent, 1984).

5.1.2 Embedded line or chain

The forces acting on an embedded line or chain are typically computed in terms of a limit state involving an impending motion of the line. The resulting normal and tangential soil resistance forces per unit length of anchor line (Figure 5.2) are described by Equations 5.2 and 5.3.

$$Q = E_n N_c s_u b \quad \text{Normal to chain} \tag{5.2}$$

$$F = \alpha E_t s_u b \quad \text{Parallel to chain} \tag{5.3}$$

where

b = is the wire diameter, or bar diameter from which chain is fabricated,
N_c = is the plane strain bearing factor for translating cylinder,

Table 5.1 Friction coefficients for anchor line resting on seabed

Line type	Seabed soil	Coefficient of friction, f	
		Static	Sliding
Chain	Sand	0.98	0.74
	Mud with sand	0.92	0.69
	Mud/clay	0.90	0.56
Wire	Sand	0.98	0.25
	Mud with sand	0.69	0.23
	Mud/clay	0.45	0.18

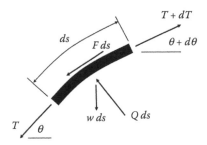

Figure 5.2 Anchor line in soil column.

$E_n =$ is the normal area multipliers for chains (use 1.0 for wire lines),
$E_t =$ is the normal area multipliers for chains (use 3.14 for wire lines),
$s_u =$ is the undrained shear strength, and
$\alpha =$ is the adhesion factor.

Table 5.2 lists the values recommended by Degenkamp and Dutta (1989) for use in Equations 5.2 and 5.3. The deep bearing factor $N_c = 7.6$ is somewhat lower than the theoretical value from Randolph and Houlsby (1984). Moreover, some publications (Aubeny and Chi, 2010) recommend a value more in line with theoretical values, say $N_c = 10$, although this value should arguably be somewhat lower than the Randolph–Houlsby value owing to the interaction effect of the combined normal and tangential loading on the anchor line. The value of α can vary from approximately 0.2–1.0, depending on the conditions of the setup. During installation, the adhesion factor is often considered as the inverse of soil sensitivity $\alpha = 1/S_t$ (DNV, 2002). The ratio of tangential to normal resistance, "friction factor," frequently appears in calculations. This ratio is assigned the symbol μ (Equation 5.4). Table 5.2 shows the μ values recommended by Degenkamp and Dutta (1989). Since it is affected by the adhesion factor α, μ is sensitive to the conditions of the setup.

$$\mu = F/Q \tag{5.4}$$

5.2 ANCHOR LINE TENSION IN SOIL COLUMN

The behavior of an anchor line (or chain) embedded in a cohesive soil is analyzed by considering an elemental length ds having weight per unit length w, subjected to tension T (Figure 5.2). The soil resistance forces acting on the element can be resolved into a normal component

Table 5.2 Recommended soil resistance parameters on anchor line

Parameter	Description	Recommended value Degenkamp and Dutta (1989)
E_n	Area multiplier normal to chain	2.5
E_t	Area multiplier parallel to chain	8.0
N_c	Plane strain bearing factor	5.1 at seabed 7.6 at depth 6b
μ	Relative magnitude of F and Q	0.4–0.6

Q and a tangential component F. Through the summation of forces normal and parallel to the element, Vivatrat et al. (1982) produced the following equations of equilibrium:

$$T\frac{d\theta}{ds} = -Q + w\cos\theta \quad \text{Normal to chain}$$
$$\frac{dT}{ds} = F + w\sin\theta \qquad \text{Parallel to chain} \tag{5.5}$$

In terms of the friction factor $\mu = F/Q$, Equation 5.5 can be expressed as follows:

$$\frac{d}{ds}(Te^{\mu\theta}) = we^{\mu\theta}(\sin\theta + \mu\cos\theta) \tag{5.6}$$

By neglecting the weight of the anchor line, Neubecker and Randolph (1995a,b, 1996) developed a number of very useful closed-form equations for anchor line behavior. Setting the right-hand side of Equation 5.6 to zero shows the line tension T to simply vary exponentially with line angle θ. If anchor line tension and angle are known at the pad eye (T_a, θ_a), then the tension at any other line angle is given by

$$T = T_a \exp[\mu(\theta_a - \theta)] \tag{5.7}$$

The use of Equation 5.7 requires a knowledge of θ at the point of interest. Neubecker and Randolph (1995b) produce a second equation by substituting Equation 5.7 into the normal equilibrium component of Equation 5.5. Assuming the line curvature to be sufficiently small such that $d\theta/ds \cong \sin\theta$, leads to the following differential equation:

$$T_a \exp[\mu(\theta_a - \theta)]\theta\frac{d\theta}{ds} = -T_a \sin\exp[\mu(\theta_a - \theta)]\frac{d\theta}{dz} = -Q \tag{5.8}$$

Integrating both sides from a depth z to the pad-eye depth z_a provides a means of relating line angle θ to depth:

$$\frac{T_a}{1+\mu^2}[e^{\mu(\theta_a-\theta)}(\cos\theta + \mu\sin\theta) - \cos\theta_a - \mu\sin\theta_a] = \int_z^{za} Q(z)\,dz \tag{5.9}$$

Finally, Neubecker and Randolph apply Taylor series expansions to Equation 5.9 to produce a simpler form that is more amenable to routine calculations:

$$T_a\left(\theta_a^2 - \theta^2\right)/2 = \int_z^{za} Q(z)\,dz \tag{5.10}$$

For the common case or relating anchor line conditions at the pad eye to the mudline in a linearly varying soil strength profile having mudline strength s_{um} and strength gradient k, Equation 5.10 becomes

$$T_a\left(\theta_a^2 - \theta_0^2\right)/2 = E_n N_c b z_a(s_{um} + kz_a/2) \tag{5.11}$$

where

z_a = is the depth of the pad eye,
s_{um} = is the soil strength at the mudline,
k = is the soil strength gradient,
E_n = is the area multiplier for chains,
N_c = is the bearing factor for the anchor line,
b = is the anchor line diameter,
T_a = is the anchor line tension at the pad eye, and
θ_0 = is the anchor line angle at the mudline.

Equation 5.11 has reasonable accuracy for zero or small mudline angles θ_0. At larger mudline angles (say greater than 10°), its accuracy deteriorates. In this case, Equation 5.9 is recommended. Even then, Equations 5.10 and 5.11 have utility in providing the initial estimates of angles for the iterative solution of Equation 5.9.

5.3 ANCHOR LINE CONFIGURATION IN SOIL

We may now consider the case where an anchor is embedded at a specified pad-eye depth z_a, tension T_a, and mudline angle θ_0. We further assume a linearly varying soil strength profile with intercept s_{um} and strength gradient k. Following the same reasoning used for developing Equation 5.11, the line angle θ at any depth is governed by the following equation:

$$\theta = \sqrt{\theta_0^2 + \frac{2E_nN_cb}{T_a}(s_{um}z + kz^2/2)} \tag{5.12}$$

To simplify the analysis, we will now introduce the following normalizations:

X = normalized horizontal coordinate = x_e/z_a
Z = normalized vertical coordinate = z_e/z_a
Q_1 = normalized soil resistance due to mudline strength = $E_n N_c b s_{um} z_a/T_a$
Q_2 = normalized soil resistance due to strength gradient = $E_n N_c b k z_a^2/2T_a$

The coordinate x_e is the horizontal distance from the pad eye and z_e is depth below the mudline. We can further approximate θ by dz/dx, so Equation 5.11 is now as follows:

$$\theta \approx \frac{dZ}{dX} = \sqrt{\theta_0^2 + Q_1Z + Q_2Z^2} \tag{5.13}$$

Aubeny and Chi (2014) integrate this analytically to obtain a closed-form equation for the chain configuration in a linearly varying soil profile:

$$X = \sqrt{\frac{1}{2Q_2}} \ln\left[\frac{Q_2 + Q_1/2 + \sqrt{Q_2^2 + Q_1Q_2 + Q_2\theta_0^2/2}}{Q_2Z + Q_1/2 + \sqrt{Q_2^2Z^2 + Q_1Q_2Z + Q_2\theta_0^2/2}}\right] \tag{5.14}$$

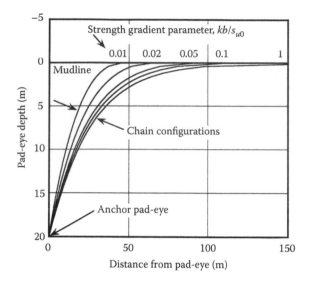

Figure 5.3 Effect of strength profile on anchor line configuration.

For the special limiting case of strength being proportional to depth ($s_{u0} = 0$), Q_1 is zero and Equation 5.14 can be shown to be equivalent to the solution given by Neubecker and Randolph (1995a,b). For the case of a uniform strength profile, Q_2 is zero. Invoking L'Hopital's rule to Equation 5.14 leads to the Neubecker and Randolph for a uniform strength profile:

$$X = \sqrt{\frac{Q_1}{4}} \left[\sqrt{\frac{\theta_0^2}{Q_1} + 1} - \sqrt{\frac{\theta_0^2}{Q_1} + Z} \right] \tag{5.15}$$

Figure 5.3 illustrates the effect of soil strength gradient on anchor line configuration for a series of soil profiles having the same average strength (18 kPa) down to the anchor depth, with strength gradient ratios (kb/s_{u0}) varying from 0.01 to 1, which roughly spans the range from a uniform to triangular strength profiles. The analysis shows that the horizontal distance from the pad eye at which the anchor line daylights at the mudline can vary by a factor of more than two according to variations in the soil strength gradient.

Sometimes required is the arc length of the embedded anchor line, which is computed as the following integral:

$$s_e = \int_0^{z_a} \sqrt{dx_e^2 + dz_e^2}\, dz_e \cong \sqrt{x_{e0}^2 + z_a^2} \tag{5.16}$$

The integral in Equation 5.16 can be evaluated numerically by any convenient method using Equations 5.14 or 5.15 to compute z_e for selected x_e values. Alternatively, noting that the effect of the curvature in the anchor line is often relatively small, the length of the embedded anchor line may be approximated as the distance between the pad-eye depth z_a and the horizontal distance x_{e0} between the pad eye and the point at which the anchor line intersects the mudline.

5.4 COUPLING TO ANCHOR LINE IN WATER COLUMN

Once the anchor line response has been defined in the soil column, it remains to define how the anchor line tension is transmitted through the water column to the water surface. This problem can be addressed from several perspectives. A common problem that arises involves predicting the bollard tension on an anchor-handling vessel (AHV) during the installation of a drag embedment anchor. The following discussion approaches the problem from the bottom up, first computing the line tension at the anchor pad eye, through the soil to the mudline and finally through the water to the AHV. The geometric configuration and system of forces for the mooring line in the water column are governed by widely published catenary equations (e.g., Meriam, 1975). The present discussion focuses on the simplest case involving a mooring line having uniform properties along its entire length, including the portion embedded in the seabed.

As shown in Figure 5.4, the anchor line in the water can be in one of the two states. The first is a slack condition, where a catenary extends from the AHV to the seabed, with some segment of the line resting on the seabed before embedding into the soil. The second state involves uplifting of the anchor line at the seabed, such that $\theta_0 > 0$. Normally, it is advantageous to install a drag anchor under catenary conditions with $\theta_0 = 0$; however, uplift of the anchor line sometimes occurs during the latter stages of embedment.

The slack and uplifting states involve somewhat different calculation sequences, so a useful initial calculation identifies the critical condition of zero slack and zero mudline uplift angle. The critical condition is computed for the anchor embedment depth under consideration by the following steps:

1. Compute the anchor line tension at the pad eye from the soil strength and effective bearing factor of the anchor $T_a = s_u \, N_e \, A_f$.
2. Compute the anchor line angle θ_a at the pad eye for a mudline angle $\theta_0 = 0$ using Equation 5.9.

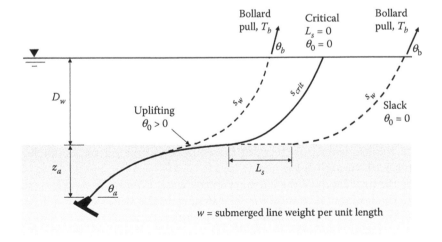

w = submerged line weight per unit length

T_a = pad-eye tension
T_0 = tension at soil exit point
T_a = horizontal tension in water catenary

x_e, z_e = coordinates of embedded line
s_e = arc length of embedded line
x_w, z_w = coordinates of catenary in water
s_w = arc length of catenary in water

Figure 5.4 Interaction with anchor line in water column.

3. Compute the line tension at the mudline T_0 for a mudline angle $\theta_0 = 0$ from Equation 5.7.
4. Compute the arc length of the s_e of the embedded portion of the anchor line from Equations 5.14 through 5.16.
5. Compute the arc length of the critical catenary in the water column $s_{crit} = s_{total} - s_e$.
6. Compute the catenary constant $a = T_0/w$.
7. Compute the height of the critical catenary in the water column z_{crit} by inserting s_{crit} for s in the classical catenary relationship:

$$z_w = \sqrt{s^2 + a^2} - a \tag{5.17}$$

If the water depth D_w is less than z_w, then a slack condition exists and slack should be introduced into the mooring line at the seabed to proceed with the calculation. If the water depth z_w exceeds z_{crit}, then the anchor line is uplifting at the mudline and θ_0 should be increased. Calculations for the slack case involve the following iterative sequence:

1. Introduce a slack distance L_s in the anchor line at the seabed.
2. Compute the frictional resistance T_s along the slack segment from Equation 5.1.
3. Compute the horizontal tension for the catenary in the water column T_w as the sum of T_s and the mudline tension T_0 computed at the point where the anchor line exits the soil.
4. Compute the length of the catenary in the water column, which equals the total line length minus the length embedded in the soil minus the slack length, $s_w = s_{total} - s_e - L_s$.
5. Compute the height of the catenary in the water column from using s_w and T_w in Equation 5.17.
6. Repeat Steps 1–5 with successive increases in slack distance L_s until the computed height of the catenary in the water column equals the water depth $z_w = D_w$.
7. Compute the bollard pull T_b as the line tension at the top of the catenary:

$$T_b = \sqrt{w^2 s_w^2 + T_w^2} \tag{5.18}$$

8. The anchor line angle at the bollard is then $\theta_b = \cos^{-1}(T_w/T_b)$.

Calculations for the uplifting case can proceed as follows:

1. Increase the anchor line angle θ_0 at the seabed.
2. Compute the anchor line angle θ_a at the pad eye for mudline angle θ_0 using Equation 5.9.
3. Compute the line tension at the mudline T_0 for a nonzero mudline angle θ_0 from Equation 5.7.
4. Compute the arc length of the s_e of the embedded portion of the anchor line from Equations 5.14 through 5.16.
5. Compute the arc length of the catenary in the water column $s_w = s_{total} - s_e$.
6. Compute the horizontal tension for the catenary in the water column $T_w = T_0 \cos \theta_0$.
7. Compute the offsets in the origin of the conventional catenary equations owing to the uplift:

$$\begin{aligned} x_0 &= a \sinh^{-1}(\tan \theta_0) \\ z_0 &= a \cosh(x_0/a) \\ s_0 &= a \sinh(x_0/a) \\ a &= T_0 \cos \theta_0 / w \end{aligned} \tag{5.19}$$

8. Compute the height of the catenary above the seabed:

$$x_w = a \sinh^{-1}\left(\frac{s_w + s_0}{a}\right) - x_0$$

$$z_w = a \cosh\left(\frac{x_w + x_0}{a}\right) - z_0$$

(5.20)

9. Repeat Steps 1–8 with successive increases in uplift angle θ_0 until the height of the catenary in the water column equals the water depth $z_w = D_w$.
10. Compute the bollard pull T_b from the following equation:

$$T_b = \sqrt{\left(w\,s_w + T_0\,\sin\theta_0\right)^2 + T_w^2}$$

(5.21)

11. The anchor line angle at the bollard is then $\theta_b = \cos^{-1}(T_u/T_b)$.

5.5 MODIFICATION FOR COMPOSITE MOORING LINES

Mooring systems more commonly comprise multiple line types. Table 5.3 shows typical configurations for catenary and taut leg systems for drag embedded plate anchor systems. Catenary systems utilize DEAs to provide horizontal load resistance, with vertical uplift being resisted by the weight of the chain. A typical system comprises a chain attached to the anchor, with a steel cable middle section followed by a chain segment extending to the water surface. A typical taut leg arrangement comprises a VLA capable of resisting both horizontal loads and uplift attached to a steel cable, with a polyester middle section followed by a chain segment to the water surface.

The classical catenary equations still apply to the three segments of these mooring systems. The computation sequence described in the previous section still applies, but Equations 5.22 through 5.25 are recursively applied to the portion of the mooring line contained within the water column. Figure 5.5 defines the various terms for a 3-segment mooring line. The horizontal component of tension T_x is constant for that portion of mooring line suspended in water. The properties of commercially available chains and cables can be found in mooring references (e.g., Vryhof, 2015).

$$a_i = T_x/w_i$$
$$x_{i-1} = a_i \sinh^{-1}(\tan\theta_{i-1})$$
$$s_{i-1} = a_i \sinh(x_{i-1}/a_i)$$
$$z_{i-1} = a_i \cosh(x_{i-1}/a_i)$$

(5.22)

Table 5.3 Main components of mooring systems

	Mooring system type	
	Catenary	Taut leg
Anchor type	DEA	VLA
Forerunner	Chain	Steel cable
Middle part	Steel cable	Polyester cable
Top part	Chain	Chain or cable

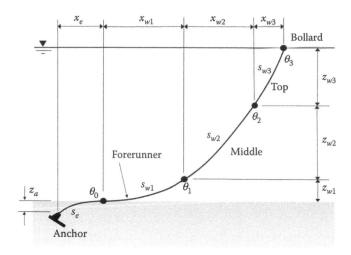

Figure 5.5 Schematic representation of 3-segment mooring line.

$$x_{wi} = a_i \sinh^{-1}[(s_{wi} + s_{i-1})/a_i] - x_{i-1}$$
$$z_{wi} = a_i \cosh[(x_{wi} + x_{i-1})/a_i] - z_{i-1} \qquad (5.23)$$

$$\theta_i = \tan^{-1}\{\sinh[(x_{wi} - x_{i-1})/a_i]\} \qquad (5.24)$$

$$T_i = T_x/\cos\theta_i \qquad (5.25)$$

As an example of the calculation procedure, we consider the case of a drag anchor installation in 1000 m of water in a soft seabed having a mudline strength $s_{um} = 2$ kPa and a
strength gradient $k = 1.6$ kPa/m. Table 5.4 shows the key properties of the mooring system.
The calculations are made for three points in the drag embedment process at pad-eye depths
$z_a = 2$, 5, and 10 m. The prediction of drag embedment trajectory is the subject of another
chapter, but an analysis of these embedment depths to correspond to horizontal drag distances $x_{drag} = 3$, 9.6, and 25 m. Soil embedment calculations are made for a wire cable forerunner of diameter $b = 0.89$ m with a presumed bearing factor for the cable $N_c = 10$ and
friction coefficient $\mu = 0.1$.

Table 5.4 Properties of composite mooring line in example analysis

Component	Property	Value
Anchor	Fluke area	10 m²
	Bearing factor	4.6
Forerunner	Length	600 m
	Submerged unit weight	0.27 kN/m
Middle segment	Length	2,000 m
	Submerged unit weight	0.05 kN/m
Top segment	Length	600 m
	Submerged unit weight	0.27 kN/m

Table 5.5 Results of example mooring line analysis

Component	Variable	Pad-eye depth, z_a		
		2 m	5 m	10 m
Pad eye	Line angle θ_a	13.3°	19.5°	27.6°
	Tension T_a	239 kN	460 kN	828 kN
Forerunner	Embedment distance x_e	20.9 m	41 m	39 m
	Embedment length s_e	21.0 m	42 m	41 m
	Angle at mudline θ_0	0°	0°	7.6°
	Tension T_0 at mudline	245 kN	476 kN	869 kN
	Slack length	378 m	71	0
	Friction in slack segment T_s	32 kN	5.9 kN	0
Middle segment	Horizontal distance x_{w1}	199 m	482 m	545 m
	Vertical distance z_{w1}	19.5 m	65 m	121 m
	Angle θ_1	11.1°	15.2°	17.2°
	Tension T_1	282 kN	500 kN	901 kN
Top segment	Horizontal distance x_{w2}	1690 m	1690 m	1692 m
	Vertical distance z_{w2}	601 m	617 m	611 m
	Angle θ_2	27.5	24.7	22.5°
	Tension T_2	312 kN	530 kN	932 kN
Bollard	Horizontal distance x_{w3}	466 m	508 m	535 m
	Vertical distance z_{w3}	373 m	317 m	271 m
	Angle θ_3	47.9°	38.5°	31.0°
	Tension T_3	413 kN	615 kN	1000 kN

Table 5.5 presents the basic results from the analysis, and Figure 5.6 depicts the details of the solution in the vicinity of the seabed. At the shallower anchor embedment depths, slack exists in the mooring line. At the 10 m embedment depth, the line tension increases to the extent that uplift occurs at the mudline. Comparison of tensions at various points along the mooring line offers some insights into the physical mechanisms of load capacity. Pad-eye tension T_a arising from soil resistance comprises some 60%–85% of total bollard pull (T_3), with the soil resistance contribution increasing with increasing anchor embedment depth. The pullout resistance mobilized along the anchor line between the pad eye and the

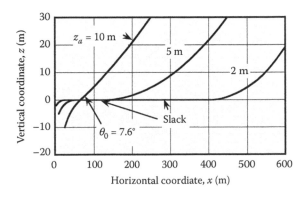

Figure 5.6 Catenary configuration in example analysis.

mudline—the difference between T_a and T_0—accounts for no more than 5% of total capacity in this case. Resistance mobilized from friction in the slack segment of the mooring line accounts for a negligible fraction of overall resistance in this instance. Recognizing that the relative contributions of the various sources of load resistance to overall mooring line capacity can vary for different systems—for example, if chain components are utilized—Table 5.5 nevertheless provides useful insights into general trends that can be expected.

5.6 INTERACTION WITH A STATIONARY ANCHOR

Anchor line behavior, in particular that described by Equation 5.9, should be combined with the anchor load capacity versus load angle relationship to evaluate the behavior of the combined system. The behavior of plate anchors is often more complex, since they tend to reorient themselves in response to changes in line tension and angle. The discussion of plate anchor interaction with anchor lines is therefore covered in later chapters. However, piles and caissons can often be reasonably considered as stationary during loading, at least from the standpoint of interaction with this anchor line. In this case, a description of the combined system is relatively straightforward.

Figure 5.7 shows an example of the line–anchor interaction for a 4-m diameter by 20-m long caisson in normally consolidated clay profile with an anchor chain attached at a depth of 14 m below the mudline. A range of chain angles $\theta_0 = 0\text{–}60°$ is considered. Chapter 7 discusses the prediction of caisson ultimate load capacity versus uplift angle θ_a at the pad eye, but a typical capacity curve is shown in the figure. The chain tension versus uplift angle relationships $(T_a - \theta_a)$ for various θ_0 are obtained from Equation 5.9. As indicated by the simplified approximation to this equation (Equation 5.10), a rough inverse square relationship exists between T_a and θ_a. The applied load magnitude and uplift angle acting on the anchor is defined by the intersection of the respective line tension curves and the anchor capacity curve. For the case of a catenary system with pure horizontal loading at the mudline ($\theta_0 = 0$) the mooring line tension T_a will act at an uplift angle of approximately 14° owing to the downward curvature of the chain in the soil column.

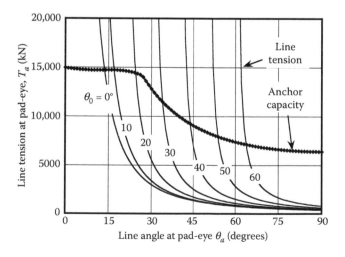

Figure 5.7 Line interaction with caisson.

REFERENCES

Aubeny CP and Chi C-M, 2010, Mechanics of drag embedment anchors in a soft seabed, *ASCE J Geotech Geoenviron Eng*, 136(1), 57–68.

Aubeny CP and Chi C-M, 2014, Analytical model for vertically loaded anchor performance, *ASCE J Geotech Geoenviron Eng*, 140(1), 14–24.

Degenkamp G and Dutta A, 1989, Soil resistance to embedded anchor chains in soft clay, *ASCE J Geotech Eng Div*, 115(10), 1420–1438.

DNV, 2002, *Offshore Standard DNV-RP-E302 Geotechnical Design and Installation of Plate Anchors in Clay*, Det Norsk Veritas, Hovik, Norway, 39p.

Meriam JL, 1975, *Statics*, 2nd Edition, John Wiley and Sons, New York.

Neubecker SR and Randolph MF, 1995a, The performance of embedded anchor chains systems and consequences for anchor design, *Proceedings of the 28th Offshore Technology Conference*, Houston, OTC 7712, pp 191–200.

Neubecker SR and Randolph MF, 1995b, Profile and frictional capacity of embedded anchor chain, *J Geotech Eng Div*, ASCE, 121(11), 787–803.

Neubecker SR and Randolph MF, 1996, The performance of drag anchor and chain systems in cohesive soil, *Mar Geores Geotech*, 14, 77–96.

Randolph MF and Houlsby GT, 1984, The limiting pressure on a circular pile loaded laterally in cohesive soil, *Geotechnique*, 34(4), 613–623.

Taylor R and Valent PJ, 1984, Design guide for drag embedment anchors, Technical Note N-1688, Naval Civil Engineering Laboratory, Port Hueneme, California.

Vivatrat V, Valent PJ, and Ponterio AA, 1982, The influence of chain friction on anchor pile design, *Proceedings of the 14th Annual Offshore Technology Conference*, Houston, Texas, OTC 4178, pp 153–163.

Vryhof Anchors, 2015, Anchor Manual 2015, ISBN/EAN: 978-90-9028801-7.

Chapter 6

Caisson and pile installation and setup

This chapter covers three aspects of caisson and pile installation: (1) the calculations for determining whether the driving forces for installing the pile can overcome the soil resistance forces, (2) the "setup" process involving the recovery of soil strength that was lost owing to disturbance during installation, and (3) the structural integrity of the piles themselves during installation. Installation modes considered include suction (actually a differential water pressure), dynamic penetration from the energy of a free-falling pile through the water column, and conventional pile driving methods.

6.1 SUCTION INSTALLATION IN CLAY

Suction installation in clays entails applying sufficient underpressure to overcome soil resistance. Since the seepage flow rate is negligible, application of the underpressure has essentially no effect on soil shearing resistance. Excessive underpressure can induce instability in the plug of soil inside the caisson, so a second calculation is required to establish the maximum underpressure that may safely be applied. Even in the absence of catastrophic internal soil plug heave, some of the soil displaced by the caisson will flow inside the caisson. This component of heave will reduce the ultimate caisson penetration, typically by a modest amount. The calculations for predicting minimum required underpressure, maximum allowable underpressure, and reduction in penetration depth owing to plug heave are described below. Practical considerations, such as installation tolerances for tilt and twist, are also discussed.

6.1.1 Required underpressure

The underpressure u_{up} required for penetrating a caisson of gross cross-sectional area A is computed from the following limit equilibrium equation:

$$u_{up} = \frac{Q_{tip} + Q_{so} + Q_{si} + Q_{plate} + Q_{ring} - W'}{A} \qquad (6.1)$$

Figure 6.1 defines the various components of soil resistance from the skirt tip, outer skirt wall, inner skirt wall, plate stiffeners, and ring stiffeners. Noting that effective stress methods are sometimes used for evaluating soil resistance to penetration (Andersen et al., 2005), the discussion that follows presents a total stress approach. The analyses are supported by (1) a profile of intact soil undrained shear strength and (2) information on remolded strength either measured directly or calculated based on an estimate of the sensitivity. If the latter

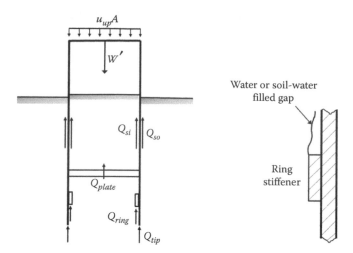

Figure 6.1 Soil resistance to suction penetration in clay.

approach is taken, the adhesion factor is taken as the reciprocal of sensitivity, $\alpha = 1/S_t$. Normally, the soil shear strength in a direct simple shear mode is taken as the operative shear strength in the calculations.

Despite the apparent simplicity of Equation 6.1, evaluation of the individual soil resistance terms requires experience and judgment to predict the effects of remolding as well as the possibility of gapping, extrusion, and displacement of trapped soil. While a fairly extensive database of caisson installation measurements exists, the data typically comprise records of underpressure versus penetration depth. This provides a high degree of certainty in regard to total soil resistance, but leaves considerable uncertainty as to individual components of resistance from the tip, sides and various stiffening elements. Unsurprisingly, variability exists in the guidance for estimating soil resistance. The following discussion presents ranges of bearing factors that more or less encompass the state of practice (Andersen et al., 2005). Situations meriting special attention include stiff clays that are prone to gapping (Andersen and Jostad, 2004) and calcareous soils that exhibit high sensitivity (Erbich and Hefer, 2002). It is also noted that Figure 6.1 does not necessarily cover all possible sources of soil resistance. For example, to facilitate transport some suction caissons feature skids attached to the outer skirt wall along nearly the entire length of the caisson. Diagonal internal plate stiffeners are also used on occasion.

Bearing resistance from the skirt tip is evaluated using bearing capacity theory as follows:

$$Q_{tip} = (N_c s_u + \gamma' z) A_{tip} \tag{6.2}$$

Estimates of the bearing factor N_c vary from 7.5 to 11 among various analysts. The overburden component $\gamma' z$ is essentially a buoyancy term arising from the weight of the soil displaced by the caisson skirt. Bearing resistance from plate stiffeners can be evaluated in a manner similar to that for tip resistance, except that the buoyancy term should be omitted since the soil flows around the top of the stiffener plate. The appropriate bearing factor for plate stiffeners is variously estimated at $N_c = 5-13.5$. The higher values are justified by the fact that reverse end bearing resistance mobilizes at the top of the plate when flow-around occurs. Nevertheless, this resistance may be degraded due to soil remolding. If one or more ring stiffeners are present, the bearing resistance for the lowest ring stiffener is also computed from the conventional bearing capacity formula omitting the buoyancy term.

Outer skirt friction can be estimated by integrating the unit skin resistance $f_s = \alpha s_u$ over the outer surface area of the skirt. In the absence of internal ring stiffeners or below the first stiffener, the unit skin resistance on the inner skirt is computed in the same manner as the outer skirt. In the case of a single internal ring stiffener, zero strength may be assigned to the region above the stiffener (Figure 6.1) where soil is judged to separate from the skirt (Andersen and Jostad, 2004). Alternatively, the lesser of the remolded strength or 1 kPa may be assigned as representative of a soil–water mixture. Where the soil comes back into contact with the skirt, the same shearing resistance is assigned as at the outer wall of the skirt. In normally consolidated clays, the soil generally lacks the strength to support an open void above the stiffener, so full flow-around can be assumed. In over-consolidated clays, various criteria have been developed for estimating the occurrence of an open gap. Andersen and Jostad (2004) use the triaxial compression strength s_{uTC} measured in special tests designed to simulate the extrusion process occurring as an element of soil passes a ring stiffener. By their criterion, a gap may open when $s_{uTC} > 0.5\gamma'z$. Alternatively, Erbich and Hefer (2002) use a criterion for gap formation $s_u/\sigma'_{v0} > 0.4$. In cases of more than one internal ring stiffener, the shearing resistance along the inner skirt wall above the first stiffener is taken as either zero or, assuming that soil from the upper portion of the soil profile is trapped in the compartment between the stiffeners, the remolded shear strength of soil in the upper part of the soil profile.

6.1.2 Plug stability

Negative differential pressure ("suction") inside the skirt above the mudline (Figure 6.2) tends to uplift the plug of soil inside the caisson, creating the potential for an internal heave failure if the soil resistance is insufficient to resist the uplift. The critical under pressure u_c is the maximum value that can be applied without inducing an internal heave failure. The critical under pressure calculation should include all relevant sources of soil resistance, including the inner skirt, plate, and ring stiffeners. Additionally, the reverse end bearing resistance from the soil plug itself, $Q_{reb} = s_{utip} N_{reb} A$, should be considered in the calculation. The

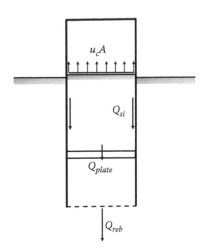

Figure 6.2 Internal soil plug stability.

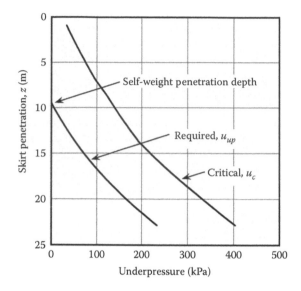

Figure 6.3 Suction installation in clay.

reverse end bearing resistance factor is variously estimated at $N_{reb} = 7–9$. Critical underpressure is then computed from Equation 6.3.

$$u_c = \frac{Q_{reb} + Q_{si} + Q_{plate} + Q_{ring}}{A} \tag{6.3}$$

The factor of safety against internal soil heave in the anchor has been defined in two ways. The first is simply the ratio of critical to required underpressure u_c/u_{up}. It should be noted that soil resistance from the inner skirt wall and the various stiffeners is the same in the required and maximum underpressure calculations; therefore, uncertainty in these sources of resistance should not affect uncertainty in the safety estimate. For this reason, some engineers prefer to define the safety factor in terms of the material safety factor for the reverse end bearing calculation; that is, the ratio of the maximum to the required reverse end bearing resistance. An acceptable minimum factor of safety is generally taken as 1.5, although some will accept 1.3. The maximum recommended caisson penetration depth is governed by maintaining an adequate safety margin against internal heave failure.

Figure 6.3 shows an example calculation of a suction installation of a 4.25-m diameter caisson in soft clay. Self-weight penetration accounts for nearly half of the penetration. For this case, the end of installation, the safety factor against internal plug failure, is $u_c/u_{up} = 1.7$.

6.1.3 Plug heave

Soil displaced by the penetrating caisson entering inside the caisson causes upward movement in the internal soil plug. The heave in this case is a simple consequence of the soil having to accommodate the volume of the caisson, as opposed to the stability related heave discussed in the previous section. When the top plate of the caisson encounters the top of the soil plug, further penetration of the caisson is not possible; therefore, the effective length of the caisson used in load capacity calculations must be reduced by the estimated magnitude of heave. Under self-weight penetration, 50% of the displaced volume of soil is typically assumed to enter the caisson. Under suction penetration, more soil may be drawn inside the caisson. Practice varies on predicting heave under suction penetration, with the percentage of displaced soil volume assumed to enter the caisson varying from 50% to 100%. Caisson tips are sometimes beveled to promote the movement of soil outside the caisson to minimize plug heave.

6.1.4 Practical considerations

Seepage inflow during suction installation is negligible in clay soils, so required pumping and penetration rates are established by a simple flow rate calculation, Q_{pump} = penetration rate times caisson area. A "standard" pump used for this application can have a capacity of approximately 55 m³/hour. For a long aspect ratio, caisson typically used in soft clays, say 4-m diameter by 25-m length, installation would require approximately 6 hours. Larger pumps are available if required.

Temporary mooring systems require provisions for anchor extraction by overpressure u_{op}, typically supplemented by a crane tension. Required overpressure calculations parallel that for required underpressure for installation, except that the caisson submerged weight acts against extraction and a higher adhesion factor α—likely corresponding to a value approaching the full setup value—must be applied. Provision for extraction of anchors for permanent moorings is also frequently required in the event of an unsuccessful installation. In this case, the setup time is less, say 10 days, but the increase in α owing to partial setup must still be considered. During the latter stages of extraction, the overpressure may induce a blowout failure in the soil. This can be avoided by maintaining the overpressure below $u_{op} < s_u N_c$, where s_u is shallow undrained shear strength and $N_c = 6.2$ is the bearing factor for a circular footing. Overpressure can be limited by applying more crane tension.

Tilt and twist will normally occur during installation. Tilt can effectively increase the load angle on the caisson and twist will induce torsional loading, both of which will tend to reduce the load capacity estimates, as discussed in Chapter 7. For tilt and twist angles on the order of ±5° the effects are typically minor. Installation costs are affected by specified tolerances; therefore, limits beyond what is necessary to avoid significant load capacity reduction may not be economical.

6.2 SUCTION INSTALLATION IN SAND

Although superficially similar to suction installation in clay, installation in sand involves two fundamentally different mechanisms of penetration. Firstly, the direct contribution of side resistance to overall penetration resistance is relatively small. Side resistance does, however, have a significant indirect impact on penetration resistance, as drag-down on the soil from friction on the sides of the caisson increases the effective stress level near the tip, substantially increasing tip resistance. Secondly, installation in pervious soil permits upward water flow in the interior of the caisson. This upward flow reduces effective stress in the soil, substantially degrading tip resistance. Absent this degradation, suction penetration is practically impossible in sands. Interestingly, for much of the penetration process, the soil in the interior of the caisson approaches a state of zero effective stress. Although this raises obvious concerns with regard to the potential for heave and boiling of the interior soils, both model tests and field experience show that serious loosening does not occur in the soil, at least for the case of dense sands. Model and field data indicate heaves that are well within tolerable levels, with Erbrich and Tjelta (1999) reporting model tests showing an average of approximately 4% heave.

Reliable prediction of underpressure required for suction installation in sands therefore requires (1) assessment of soil resistance with adequate accounting for side–tip resistance interaction and (2) prediction of upward hydraulic gradients with concomitant reductions in effective stress inside the caisson. The basic equation for installation in sands follows that for installation in clays, that is, a summation of soil tip and side resistance minus the submerged weight of the caisson:

$$P_f = Q_{tip} + Q_{side} = q_{tip} A_{tip} + f_{sav} A_{wall} - W_{sub} \tag{6.4}$$

where P_f is the total force required for penetration, Q_{tip} is the annular tip resitance, Q_{side} is the side frictional resistance, f_{sav} is the average unit side friction, and W_{sub} is the submerged weight of the caisson. The following paragraphs discuss methods for implementing this equation.

6.2.1 Soil resistance for no seepage flow

Andersen et al. (2008) describe two approaches for evaluating soil resistance to caisson penetration: a theoretical bearing capacity approach and an empirical approach centered about cone penetration test (CPT) tip resistance measurements. Table 6.1 shows the respective equations for predicting tip and side resistance. The bearing factors N_q and $N\gamma$ are from classical bearing capacity theory. The term q includes the usual effective overburden stress ($\gamma' z$) plus, following Clausen (1998), the added stress generated from side friction. From trial-and-error back calculation, Andersen et al. (2008) use a skirt wall friction factor $\alpha_f = 1$.

The side friction calculation follows the β-method widely used in pile design for cohesionless soils, which is also adopted in API guidelines. The coefficient K governs the horizontal stress on the sidewalls of the caisson, which in turn drives the frictional resistance calculation. Andersen et al. (2008) report that the API (1993) recommended $K = 0.8$ and a soil–caisson interface angle $\delta = 0.9\phi'$ provides good agreement between penetration predictions and measurements. A higher value of K may be considered if nearby structures can generate lateral restraint or if the sand is over-consolidated. In contrast to the case for suction caissons in clays, ring shear tests conducted at NGI show no significant differences in δ between painted and unpainted surfaces. Andersen et al. also concluded that the use of a drained friction angle (ϕ') measured in triaxial compression, as opposed to plane strain, gives the best agreement between prediction and measurement.

When internal stiffeners are present, soil resistance is normally computed in the same manner as for tip resistance. An alternative to the theoretical approach described above is

Table 6.1 Penetration resistance from bearing capacity theory

Component of resistance	Equation
Tip	$q_{tip} = 0.5\,\gamma'\,t_w\,N\gamma + q\,N_q$ = unit tip resistance
	$N_q = \exp(\pi \tan \phi')\,\tan^2(45 + \phi'/2)$
	$N\gamma = 1.5\,(N_q - 1)\,\tan \phi'$
	$q = \gamma'\,z\,(1 + \alpha_f K \tan\delta)$ = adjusted effective overburden stress
	$A_{tip} = \pi D\,t_w$ = annular area of pile tip
Side	$f_{sav} = 0.5\,K\gamma'\,z\,\tan\delta$ = average unit wall friction
	$A_{wall} = 2\pi D\,z$ = inner and outer wall surface area

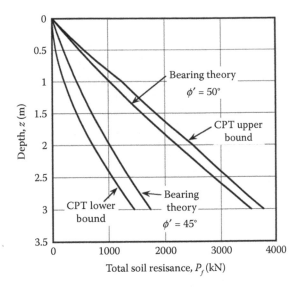

Figure 6.4 Penetration resistance for no-flow condition.

an empirical model based on CPT tip resistance q_c. If both skirt tip resistance and sidewall friction are correlated to q_c, then the model takes the following form:

$$P_f = k_{tip}q_cA_{tip} + A_{wall}\int k_{side}q_c dz \qquad (6.5)$$

where k_{tip} and k_{side} are empirical constants. From back-analysis of laboratory model and field prototype measurements, Andersen et al. (2008) report the following parameter ranges and combinations: $k_{tip} = 0.01$–0.55 in combination with $k_{side} = 0.0015$, and $k_{tip} = 0.03$–0.6 in combination with $k_{side} = 0.001$. Figure 6.4 shows penetration resistance predictions for a 3.25-m diameter caisson having a 0.016-m wall thickness in dense sand. The CPT-based predictions are based on actual CPT measurements in North Sea dense sand. The theory-based predictions show a high sensitivity to friction angle, with comparable sensitivity in the CPT-based approach to the lower and upper bounds of the tip and side resistance parameters. In the absence of clay layers in the soil profile, this high sensitivity is not quite as serious as it may appear, since it will be primarily reflected in the caisson self-weight penetration calculation, which is based on conditions of no seepage flow. When underpressure application with seepage occurs, much of the soil resistance is eliminated as described in the following paragraphs.

6.2.2 Pore water flow

The differential head induced by suction inside the caisson induces seepage flow. Erbrich and Tjelta (1999) performed steady-state seepage analyses for various conditions of caisson embedment z/D and soil disturbance, as illustrated in the example in Figure 6.5. Their analyses consider two conditions with respect to the spatial extent of disturbance: (1) in a 0.07D vertical strip adjacent to the interior sidewall of the caisson and (2) in the entire soil plug inside the caisson. Disturbance was considered to manifest itself in the form of increased soil permeability in the disturbed zone that may be expressed in terms

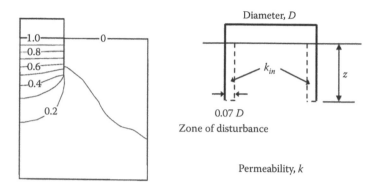

Figure 6.5 Equipotential lines during suction installation. (From Andersen KH, Jostad HP, and Dyvik R, 2008, *J Geotech Geoenviron Eng*, 134(1), 106–116, with permission from ASCE).

of a permeability ratio k_{in}/k, where k_{in} is the disturbed soil permeability and k is the permeability of the surrounding soil. The seepage flow is upward inside the caisson, which decreases effective stress. In the limiting state of a "quick" condition, effective stress approaches zero, with a concomitant reduction of internal skin friction and tip resistance. The converse situation occurs outside the caisson: downward hydraulic gradients increase effective stress and frictional resistance on the outer wall of the caisson. Examination of the equipotential lines in Figure 6.5 shows that the upward hydraulic gradients inside the caisson are greater than the downward gradients in the exterior; thus, a net reduction penetration resistance occurs. Increased permeability inside the caisson owing to soil disturbance increases the downward external gradients, thereby increasing the resistance to penetration.

The critical hydraulic gradient associated with quick condition is calculated from classical soil mechanics theory, $i_{cr} = \gamma'/\gamma_w$, where γ' is submerged soil unit weight and γ_w is the unit weight of water. The occurrence of a quick condition can be assessed in terms of either an average internal gradient i_{avg} between the skirt tip and the caisson base plate or a local exit gradient i_E at the interior sand surface. A measure of applied differential head relative to critical hydraulic gradient can be expressed in terms of a "suction number" S_N defined as follows:

$$S_N = \left(\frac{H}{z}\right)\left(\frac{\gamma_w}{\gamma'}\right) = \frac{\Delta u}{\gamma' z} \tag{6.6}$$

where H is the differential water head and z is the skirt penetration depth. The critical suction number $S_{N,cr}$ denotes the condition of the maximum differential head associated with a quick condition. Andersen et al. (2008) followed the same methodology as Erbrich and Tjelta but for a greater range of geometries. Figure 6.6 shows critical suction number S_{Ncr} versus skirt penetration depth z/D assuming (1) a thin zone (0.07D) of increased permeability adjacent to the inner skirt wall and (2) the exit gradient i_E at the inner sand surface controls the onset of a quick condition. The dashed lines show $S_{N,cr}$ assuming average gradient along the inner skirt wall controls, based on the earlier study by Erbrich and Tjelta (1999).

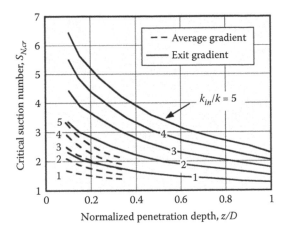

Figure 6.6 Critical suction number versus skirt penetration depth (From Andersen KH, Jostad HP, and Dyvik R, 2008, *J Geotech Geoenviron Eng*, 134(1), 106–116, with permission from ASCE).

6.2.3 Estimating soil disturbance effects

Before proceeding to an analysis of penetration under conditions of water flow, an estimate is required of the ratio of permeability inside versus outside of the caisson, k_{in}/k. Owing to shear strains from the penetrating skirts, some dilation of the soil with a concomitant increase in permeability will occur. Andersen et al. (2008) estimate the volumetric strain owing to shearing as follows:

$$\varepsilon_{vol} = -4\varepsilon_r \sin\psi/(1-\sin\psi) \tag{6.7}$$

where ε_{vol} is the volumetric strain, ε_r is the radial strain, and ψ is the dilatancy parameter. Andersen et al. (2008) estimate radial strain and dilatancy from the following relationships:

$$\varepsilon_{vol} = -2t_w/D \tag{6.8}$$

$$\phi' = 32.6° + 0.66\psi \tag{6.9}$$

where t_w and D are the caisson wall thickness and diameter, respectively. Change in void ratio can be computed from volumetric strain, $\Delta e = \varepsilon_{vol}(1 + e_0)$. If site-specific data relating void ratio to permeability are lacking, a common proportionality relationship between permeability k and void ratio e is the following (USACE, 1993):

$$k = Ce^3/(1+e) \tag{6.10}$$

where C is a constant. Since only the relative change in permeability with increasing void ratio is necessary for the analysis, knowledge of the constant C is not strictly necessary.

The ratio k_{in}/k used in conjunction with the relationship between and skirt penetration depth and critical suction number S_{Ncr} ($=\Delta u_{cr}/\gamma'z$) permits direct prediction of the maximum allowable underpressure versus depth, as shown in the example in Figure 6.7.

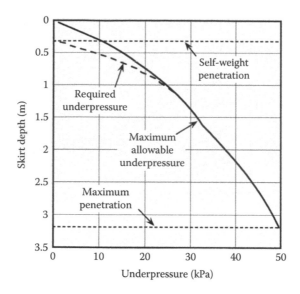

Figure 6.7 Penetration prediction example.

6.2.4 Side resistance and maximum penetration depth

If a quick condition develops, tip resistance and internal skin friction approach zero. Considering the increase in effective stress owing to downward flow outside the caisson and assuming $i_{cr} = 1.0$ (Andersen et al., 2008) leads to the following expression for external skin friction:

$$F_{s\_out} = 0.5K\gamma'z^2 \tan\delta\pi DS_{n,cr} \tag{6.11}$$

Combining this with the expression discussed above for computing maximum allowable underpressure leads to the following condition for continued penetration of the caisson:

$$W' > (2Kz/D\gamma' \tan\delta - 1)S_{n,cr}\gamma'z \; \pi D^2/4 \tag{6.12}$$

6.2.5 Required underpressure

With maximum allowable underpressure and maximum penetration depth determined, the remaining required calculation is a prediction of the underpressure required to advance the caisson at all stages of penetration. The required underpressure is of course zero when the submerged weight of the caisson exceeds the soil resistance under no-flow conditions from Equation 6.4 or 6.5. To estimate the underpressure required to advance the caisson beyond the self-weight penetration depth, Andersen et al. (2008) performed a detailed analysis of penetration records from the laboratory model and field prototype suction installations in sand. From these data, they obtained the plot in Figure 6.8 by relating normalized penetration resistance ($P_{f,Flow}/P_{f,No\ Flow}$) to normalized suction number ($S_N/S_{N,cr}$) for various normalized skirt penetration depths (z/t). The chart is used for determining the intersection of a line having intercept and slope as defined below:

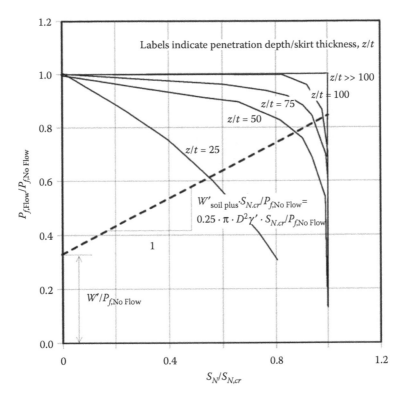

Figure 6.8 Required underpressure for suction installation in sand (From Andersen KH, Jostad HP, and Dyvik R, 2008, Penetration resistance of offshore skirted foundations and anchors in dense sand, *J Geotech Geoenviron Eng*, 134(1), 106–116, with permission from ASCE).

$$\text{Intercept} = W'/P_{f,\text{NoFlow}}$$
$$\text{Slope} = W'_{\text{soil plug}} S_{N,cr}/P_{f,\text{NoFlow}}$$

(6.13)

The submerged weight of the soil plug $W'_{\text{soil plug}}$ is simply $\pi D^2/4\gamma'$. Once the intersection is identified, $S_N/S_{N,cr}$ from Figure 6.8, it is used in conjunction with $S_{N,cr}$ from Figure 6.6 and Equation 6.6 to compute required underpressure.

Figure 6.7 shows the results of a suction installation analysis for a relatively small caisson ($D = 3.25$ m, $t_w = 0.015$ m, $W' = 140$ kN) in a dense sand with friction angle $\phi' = 45°$ and submerged unit weight $\gamma' = 10$ kN/m³. The self-weight penetration of about 0.3 m proves to be minor relative to the maximum penetration. Beyond a penetration depth of approximately 1 m, the required underpressure curve merges into the maximum allowable underpressure prediction. Thus, the sand inside the caisson approaches a quick condition throughout much of the penetration process. This is in contrast to caisson penetration in soft clays, where a substantial safety margin is maintained between the applied underpressure and the maximum allowable underpressure for plug stability. Although penetration under incipient quick conditions raises concerns regarding possible piping and channeling of the soil inside the skirt compartment, repeated successful suction installations of caissons in dense sand indicate no significant detrimental consequences. The no-flow soil resistance calculation in Figure 6.4 for the same caisson and soil considered in Figure 6.7 indicates a soil resistance of approximately 1000–1300 kN, which translates into approximately

120–160 kPa of required underpressure at a depth of 2.5 m. Figure 6.7 shows a required underpressure for the with-flow condition of 43 kPa at this depth, approximately 25%–33% of the no-flow required underpressures.

The substantial advantages of upward seepage inside the skirt compartment can only be developed if no clay layers occur that can interrupt the seepage. Thus, uniform sand profiles represent the ideal conditions for suction installation in sand. Nevertheless, more recent developments have shown that suction installation in sand profiles containing clay interbeds is possible by introducing measures that include one and two-way cyclic penetration and water injection at the skirt tip (Tjelta, 2015), although installation is more difficult and requires more detailed knowledge of site conditions than installation in uniform profiles.

6.2.6 Practical considerations

Since seepage flow is significant, the calculation of the pumping rate Q_0 required to empty the skirt compartment in a reasonable time interval must be augmented by the pore water flow from the soil Q_{pwf}. The calculation of pore water flow rate can be made directly in terms of Darcy's law:

$$Q_{pwf} = kiA \tag{6.14}$$

Since most of the suction penetration process takes place at a critical hydraulic condition, taking $i = i_{cr} = 1.0$ is a reasonable estimate of the governing hydraulic gradient i. Normally, Q_{pwf} does not dominate the calculation, but it can be sufficiently significant (say $Q_{pwf} = 0.1$ Q_0) to noticeably affect the predicted installation times.

Planning for temporary mooring systems and contingency plans for failed installations require estimates of required overpressures u_{op} for extracting the caisson. Tip resistance in the form of reverse end bearing is not expected to develop. However, the weight of the caisson W' acts against extraction. Furthermore, in contrast to the case of penetration, hydraulic gradients increase rather than decrease the side resistance along the inner skirt. The governing equation for average unit side resistance f_s is the following:

$$f_s = 0.5 \, (\gamma' + \gamma_w i_{avg}) z \, K_1 \tan\delta \tag{6.15}$$

where K_1 is lateral earth pressure coefficient, δ is soil–caisson interface friction angle, and i_{avg} is hydraulic gradient, with sign convention positive for downward flow. Along the interior of the caisson, the average hydraulic gradient is simply that fraction of the differential head that is dissipated inside the caisson, which we will define as η. If disturbance is neglected, we can estimate $\eta_{in} = 0.6$ based on the equipotential contours in Figure 6.5. Perhaps more precisely, it can be taken as the inverse of the suction number derived for average gradients, $\eta_{in} = 1/S_{Ncr}$. Negative gradients along the exterior wall may be estimated assuming the remainder of the differential head dissipates uniformly along the external skirt, although the equipotential contours in Figure 6.5 suggest that may a bit optimistic. According to engineering judgment, the selected η_{out} should be between 0 and $1 - \eta_{in}$. The following equation relates overpressure to average hydraulic gradient on the surface under consideration:

$$i_{avg} = \eta u_{op} / \gamma_w z \tag{6.16}$$

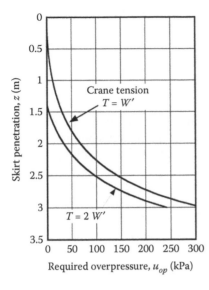
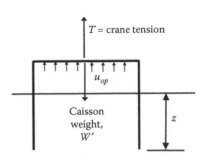

Figure 6.9 Caisson overpressure extraction in sand.

At the least, a crane must apply sufficient line tension T to overcome the submerged weight of the caisson W' to lift it off the seabed following extraction. Additional line tension may be applied to maintain overpressures u_{op} to acceptable levels and to avoid the requirement for overpressure during the latter stages of retrieval, where the skirt depth is shallow and flow channels may form that short-circuit the overpressure. Accounting for an arbitrary level of line tension, the required overpressure is as follows:

$$u_{op} = \frac{\gamma' z K_1 \tan \delta A_s + W' - T}{A - \eta' K_1 \tan \delta A_s}$$

$$A = \pi D^2 / 4$$
$$(6.17)$$
$$A_s = \pi D z$$
$$\eta' = (\eta_{in} - \eta_{out})/2$$

Figure 6.9 shows estimated overpressure requirements for caisson extraction for the same example problem considered in Figures 6.4 and 6.7. Predictions are presented for applied crane tensions equal to the submerged weight and twice the submerged weight of the caisson, and the distribution of head loss along the inner and outer skirt is computed based on $\eta_{in} = 0.6$ and $\eta_{out} = 0$. Depending on the applied crane tension T, overpressure required for extraction is on the order of two to three times the underpressure required for insertion.

6.3 DYNAMIC INSTALLATION

Dynamically installed piles, sometimes termed "torpedo piles" penetrate the seabed by kinetic energy acquired by free falling through the water column. Ultimate pullout capacity for such anchors depends heavily on achieving sufficient embedment depth, so much of the installation analysis entails developing reliable penetration predictions, making a careful accounting of all sources of soil resistance to penetration. Two approaches to penetration prediction have been employed: limit equilibrium methods and energy methods. Both methods have been validated and calibrated to laboratory model (centrifuge) tests, field tests, and sophisticated large displacement finite element simulations.

6.3.1 Limit equilibrium

Much of the work on dynamic anchor penetration adopts the limit equilibrium framework proposed by True (1974), which employs Newton's second law in conjunction with the individual forces resisting penetration described in Figure 6.10. The resulting dynamic equilibrium equation is expressed in terms of pile mass m_p, pile submerged unit weight W_b, soil buoyant resistance, bearing resistance, side frictional resistance, and inertial drag resistance. Resistance arising from soil shear strength is affected by the strain rate dependence of this parameter. The terms R_{fb} and R_{fs} characterize the strain rate dependence effect to produce the following equation for acceleration $\ddot{u}$:

$$m_p\ddot{u} = W_b - F_\gamma - R_{fb}(F_{pt} + F_{pr} + F_{ft} + F_{fr}) - R_{fs}(F_{ps} + F_{fs}) - F_d \qquad (6.18)$$

While the limit equilibrium calculation is conceptually simple, a decision is required for treating the region in the wake of the pile after it has fully penetrated into the seabed. In soft soils, full flow-around is a reasonable assumption, so the buoyant force simply becomes the pile volume times γ_b. Stiff soils can be sufficiently strong to support an open void above the pile. In this case, the void volume must be included in the displaced soil volume.

Capacity arising from soil shearing resistance parallels that for conventional piles, except for consideration of strain rate effects owing to the high velocities involved and inclusion of resistance terms associated with the fins. The bearing resistance terms (F_{pt}, F_{pr}, F_{ft}, F_{fr}) take the familiar form $N_c s_u A_p$, where N_c is the dimensionless bearing factor, s_u is the soil shear strength at the depth under consideration, and A_p is the cross-sectional area. Kim et al. (2015) recommend bearing factors $N_c = 13.6$ and $N_c = 7.5$ for the pile and fins, respectively. Bearing resistance mobilizes beneath the tip and fins. However, after full embedment,

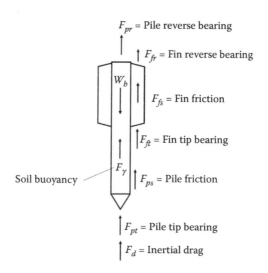

Figure 6.10 System of forces on dynamically penetrating pile.

reverse end bearing resistance may mobilize in the wake of the pile. In stiff soils, open voids are likely to develop, while in soft soils full flow-around is expected. Kim et al. (2015) recommend including reverse end bearing in the calculation when $s_{um}/\gamma'D < 0.2$ and $k\,D/s_{um} > 1$. The basic calculation for side friction resistance also parallels conventional analyses, except for consideration of strain rate effects discussed in the next paragraph. In terms of adhesion α, average shear strength s_u, and surface area A_s, the frictional force is $F_s = \alpha s_u A_s$. Other appurtenances, such as pad eyes and chains will create additional sources of bearing and skin resistance that must be considered.

Undrained shear strength is well known to depend on strain rate $\dot{\gamma}$. Kim et al. (2015) adopt the Herschel and Bulkley approach for characterizing strain rate dependency. Omitting consideration of remolding effects, the equation takes the following form:

$$s_u = \left[1 + \eta \left(\frac{\dot{\gamma}}{\dot{\gamma}_{\text{ref}}}\right)^{\beta}\right] \frac{s_{u,ref}}{1 + \eta} \tag{6.19}$$

The terms η and β are material parameters characterizing the rate dependence, and $s_{u,ref}$ is the undrained shear strength measured at strain rate $\dot{\gamma}_{ref}$. This equation is essentially a "microscopic" characterization of the strain rate effect at a point. More directly applicable to dynamic pile installation is a "macroscopic" multiplier R_f on the soil resistance force representing the net effect of a nonuniform strain rate field surrounding the pile. For this purpose, Kim et al. (2015) adopt the following equation:

$$R_f = \left[1 + \eta \left(\frac{nv/D}{\dot{\gamma}_{ref}}\right)^{\beta}\right] \frac{1}{1 + \eta} \tag{6.20}$$

In this case, v is the instantaneous velocity of the pile, D is the pile diameter, and n is a multiplier that depends on the type of soil resistance, end bearing versus side friction. For end bearing resistance, the relationship is fairly simple, with the total resistance force increasing with velocity in essentially the same manner as shear strength increases with increasing strain rate; that is, $n = 1$. For side friction resistance, the relationship is more complex, with n for side resistance being about an order of magnitude higher than that for bearing resistance (Einav and Randolph, 2005; Chow et al., 2014; Steiner et al., 2014). An order of magnitude difference in n for bearing versus side friction effectively implies that the rate multiplier factor for side resistance is twice that of bearing resistance: $R_{fs} = 2R_{fb}$. Based on data collected from various clays, Kim et al. (2015) adopted the following rate parameters: $\eta = 1.0$ and $\beta = 0.1$.

The inertial drag resistance is calculated as described by Hossain et al. (2014), in a manner similar to that employed in classical fluid mechanics:

$$F_d = C_d \rho_s A_p v^2 / 2 \tag{6.21}$$

where C_d is the drag coefficient, ρ_s is the soil bulk density, and A_p is the projected pile cross-sectional area. Kim et al. (2015) recommend a drag coefficient $C_d = 0.24$. They report that in soft clays inertial drag dominates resistance during the early stages (30%–40% of final penetration), after which soil shearing (end bearing and friction) becomes dominant.

6.3.2 Analytical solution

Calculation of dynamic pile installation normally requires numerical integration owing to the complex geometry of the pile as well as the complex form of the various components of soil resistance. However, if strain rate dependence and inertial drag effects are neglected, analytical solution of dynamic installation is possible for simplified cylindrical piles. In spite of the modest simplifications required, closed form analytical solutions are very useful for developing intuitive insights into the process of dynamic penetration. The following analysis considers dynamic installation in a soil with a strength profile that has a mudline strength s_{um} and strength gradient k (Figure 6.11). The effects of a tip cone and tail fins are neglected here, such that the analysis considers a cylinder of length L_f and diameter D. The dominant soil resistance forces considered in the analysis include end bearing resistance, side friction and buoyancy. The components of soil resistance can be grouped according to whether the resistance is dependent or independent of tip depth z, as itemized in Table 6.2 and illustrated in Figure 6.11. At a minimum, the analysis should be considered in two stages: Stage 1 where the penetration depth is less than the pile length L_f, and Stage 2 where the pile exceeds the pile length and becomes enveloped in the surrounding soil. Various assumptions can be made regarding how the soil flows around the pile after the penetration depth exceeds the pile length. In this analysis, full flow-around is assumed such that (1) reverse bearing mobilizes at the top of the pile and (2) the buoyant force is simply the volume of the pile times the submerged unit weight of the soil. In all cases except one, the soil resistance can be expressed as a linear function of depth. The exception is the side frictional resistance during Stage 1, which varies with the square of the penetration depth. To preserve linearity, the side resistance is approximated as a linear function.

Invoking Newton's second law produces the differential equation shown below. The coefficients A_i represent the combined depth-dependent force terms divided by pile mass m_p, while B_i represent the depth-independent terms. The subscript i denotes the stage of penetration.

$$\ddot{z} = B_i - A_i z \tag{6.22}$$

The governing equation is an ordinary, nonhomogeneous second-order differential equation. The solution has a harmonic form, as shown below. The solution can proceed recursively through Stages 1 and 2. Initial conditions are set to an initial tip depth at the mudline $z_0 = 0$ and the initial impact velocity v_0.

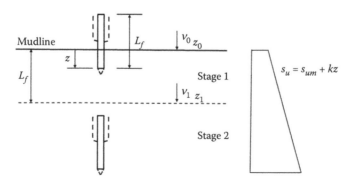

Figure 6.11 Analytical solution for dynamic pile penetration.

Table 6.2 Soil resistance terms for dynamic penetration

Stage 1	Stage 2
Depth-dependent force terms	Depth-dependent force terms
$A_1 = (a_{t1} + a_{s1} + a_{b1})/m_p$	$A_2 = (a_{t2} + a_{r2} + a_{s2})/m_p$
Tip bearing: $a_{t1} = k\, N_{ct}\, A_p$	Tip bearing: $a_{t2} = k\, N_{ct}\, A_p$
Side friction: $a_{s1} = \alpha(s_{um} + kL_f/2)\,\pi D$	Reverse bearing: $a_{r2} = k\, N_{cr}\, A_p/S_t$
Buoyancy: $a_{b1} = \gamma_b A_p$	Side friction: $a_{s2} = \alpha k\, A_s$
Depth-independent force terms	Depth-independent force terms
$B_1 = (W_b - b_{t1})/m_p$	$B_2 = (W_b - b_{t2} - b_{r2} - b_{s2} - b_{b2})/m_p$
Tip bearing: $b_{t1} = s_{um}\, N_{ct}\, A_p$	Tip bearing: $b_{t2} = s_{um}\, N_{ct}\, A_p$
	Reverse bearing: $b_{r2} = (s_{um} - k\, L_f)\, N_{cr}\, A_p/S_t$
	Side friction: $b_{s2} = \alpha\,(s_{um} - k\, L_f/2)\, A_s$
	Buoyancy: $b_{b2} = \gamma_b\, A_p\, L_f$
Pile properties:	*Soil resistance parameters:*
W_b = submerged weight of pile	k = undrained shear strength gradient
A_p = pile area = $\pi D^2/4$	α = adhesion factor
A_s = pile surface area = $\pi D\, L_f$	S_t = soil sensitivity
L_f = pile length	γ_b = soil submerged unit weight
m_p = pile mass	N_{ct} = tip bearing factor
	N_{cr} = reverse end bearing factor

$$z_i = \frac{B_i}{A_i} + \left(z_{i-1} - \frac{B_i}{A_i}\right)\cos\sqrt{A_i}\,t + \frac{v_{i-1}}{\sqrt{A_i}}\sin\sqrt{A_i}\,t$$

$$v_i = -\sqrt{A_i}\left(z_{i-1} - \frac{B_i}{A_i}\right)\sin\sqrt{A_i}\,t + v_{i-1}\cos\sqrt{A_i}\,t$$

(6.23)

Figure 6.12 illustrates the solution for the case of a pile of diameter $D = 1.07$ m, length $L_f = 17$ m, and submerged weight $W_b = 779$ kN. The analysis assumes bearing factors $N_{ct} = N_{cr} = 13.5$, adhesion factor $\alpha = 0.33$, and soil unit weight $\gamma = 16$ kN/m³. The analyses consider impact velocities $v_0 = 15\text{--}20$ m/second in soft and stiff soil profiles. The soft soil has mudline strength $s_{um} = 1$ kPa with strength gradient $k = 1$ kPa/m. The stiff soil has $s_{um} = 10$ kPa and $k = 3$ and kPa/m. The predicted history of velocity shows an initial stage of acceleration to maximum velocities that can exceed impact velocity by 25% or more. Eventually, the increased soil resistance leads to increasing levels of deceleration until zero velocity occurs. Kim et al. (2015) actually performed large displacement finite element analyses for the cases considered here. They predicted pile penetrations in soft clay of 57 and 62 m for impact velocities of 15 and 20 m/second, respectively. In the stiff clay, they predicted penetrations of 28 and 33 m for impact velocities 15–20 m/second. The analytical model is in reasonable agreement with the more rigorous analysis for penetration in the soft clay, but it underpredicts penetration into the stiff clay. It should also be noted that additional resistance terms, such as fin resistance may be introduced into this analysis, so long as the resistance force is a linear function of depth z.

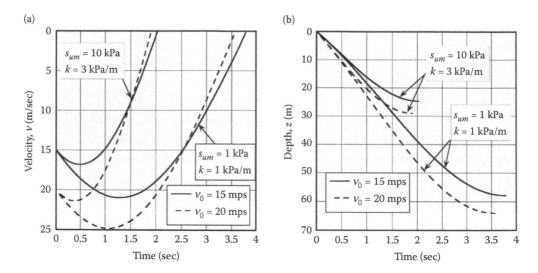

Figure 6.12 Predicted time history of velocity and penetration. (a) Velocity and (b) embedment depth.

6.3.3 Total energy method

O'Loughlin et al. (2013) propose a simplified approach for predicting dynamically installed pile penetration depth in terms of the total kinetic and potential energy of the pile at the seabed. The potential energy is computed relative to the final tip embedment depth $d_{e,t} = z_{tmax}$; thus, the energy equation takes the following form:

$$E_{total} = \frac{1}{2}m_p v_0^2 + W_b d_{e,t} \tag{6.24}$$

As defined earlier, m_p is the pile mass, v_0 is the impact velocity, and W_b is the submerged pile weight. In the modified form developed by Kim et al. (2015), penetration in a soil profile with a uniform strength gradient k, tip embedment depth correlates to total energy through a power law relationship:

$$\frac{d_{e,t}}{D_p} = a\left(\frac{E_{total}}{kA_s D_p^2}\right)^b \tag{6.25}$$

The equivalent pile diameter D_p considers the projected area of the pile, including fins, and A_s is the total surface area of the pile. O'Loughlin et al. (2013) selected $a = 1.0$ and $b = 1/3$ for use in Equation 6.25. Subsequent large displacement finite element predictions by Kim et al. (2015) indicate that an improved fit is achieved using $a = 4.16$ and $b = 0.376$ (Figure 6.13). Since E_{total} is a function of z_{tmax}, iteration is required to estimate pile embedment depth.

6.4 DRIVING INSTALLATION

Driven piles may be installed by impact or vibratory hammers. Impact hammers lift a ram by air, steam, or diesel, and can be single acting or double acting. A single-acting system lifts

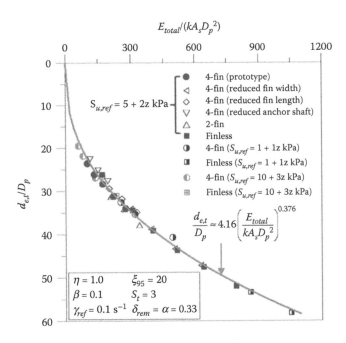

Figure 6.13 Dynamic pile penetration by the total energy method. (Reprinted from *Ocean Eng*, 108, Kim et al., Numerical investigation of dynamic installation of torpedo anchors in clay, 820–832, Copyright 2015, with permission from Elsevier.)

a ram that free falls to strike the pile. Double-acting hammer additionally apply air or steam to drive the ram downward. Underwater hydraulic hammers clamped to the pile head can be used for pile installation in deep water. Vibratory hammers advance the pile by fluidizing cohesionless soils in conjunction with an applied downward load.

Offshore pile installation initially adapted onshore technology to drive piles from the water surface, largely utilizing steam hammers. A "follower" was introduced for permitting transmission of the stress wave from a hammer at the water surface to the submerged pile. The added length of pile added to the pile weight and the longer path of the stress wave transmission generated additional energy losses. The working distance under the barge crane and the weight of the hammer assembly imposed further limitations on the follower length. In water depths up to approximately 300 m, limitations on the use of surface hammers were generally surmountable. However, as development proceeded into deeper water, the requirement for underwater hammers became more pressing. Early development of underwater driving technology is reported by Cox and Christy (1976), who report successful installation of a 0.6-m diameter, 94-m long pile with a HBM-500 underwater hydraulic hammer at an offshore Louisiana test site. This test was performed for supporting the design effort for the Cognac fixed platform in 320 m of water, where twenty four 2.1-m diameter, 122-m long piles were later successfully installed. Other pioneering developments in underwater hammer technology are reported by Jansz (1977), Heerema (1980), and Aurora (1984). Prior to 1980, the power pack for operating underwater hammers was situated on a surface barge. Subsequent developments included a slender hammer that could pass through the same guides that the pile has passed and underwater power packs (van Luipen, 1987).

Impact hammers generate a force pulse that causes the pile to displace downward; the inelastic component of displacement is termed "set." The process is inherently one of the elastic wave propagations and is not adequately described in terms of a quasi-static analysis (appropriate to suction pile installation) or rigid body dynamics (appropriate to dynamic pile installation). Accordingly, wave equation analysis is widely applied to pile driving problems. A number of questions can be answered from a wave equation analysis; common ones are listed by Lowery (1993):

1. *Hammer size.* A larger size than necessary incurs needless expense, while an under-sized hammer can lead to inability to reach the design penetration depth.
2. *Optimal cushion selection.* The cushion must limit driving stresses while maintaining the maximum permanent set per hammer blow.
3. *Soil resistance.* Static soil resistance to pile penetration can be estimated from a wave equation analysis. Resistance at the time of penetration is not the same as pile capacity, since processes such as setup can significantly affect long-term capacity. However, long-term capacity can be predicted by estimating the setup factor from soil mechanics principles or by restriking the pile after a specified setup period.
4. *Pile design.* Since driving stresses follow naturally from the wave equation analysis, minimum pile wall thickness to avoid overstress are determined from the analysis. Additionally, pile geometry can significantly affect drivability. For example, a greater wall thickness can be as effective as a larger hammer in achieving greater penetration.
5. *Effects of accessories.* Various accessories such as the anvil and capblock can absorb a significant (up to 50%) of the hammer energy. Wave equation analyses can support optimized accessory design to minimize energy losses.

Owing to complexity of the soil resistance mechanisms, wave equation analysis of pile driving is largely based on computer programs. Although many of these programs have been validated through extensive experience, developing an intuitive understanding of elastic wave propagation is advisable prior to attempting an analysis. The next section discusses the major points of wave propagation in an elastic rod, simplifying, or neglecting consideration of the effects of soil resistance. This discussion is followed an example computer program for pile driving, considering both the complexities of actual driving systems as well as soil resistance.

6.4.1 Elastic wave propagation

The governing equation for wave propagation is derived by considering the element of pile of cross-sectional area A and elastic modulus E shown in Figure 6.14. The soil resistance is modeled as springs that act independently along the pile shaft. This assumption of

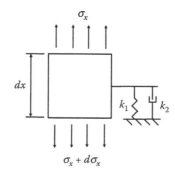

Figure 6.14 Development of one-dimensional model of wave propagation.

noninteracting springs is not strictly correct, but it is made for mathematical convenience. The soil springs are velocity-dependent with unit resistance to particle displacement u and velocity $\dot{u}$ described by spring constants k_1 and k_2, respectively. Invoking Newton's second law produces the following equation:

$$\frac{\partial^2 u}{\partial x^2} - \frac{\pi D k_1}{AE} u - \frac{\pi D k_2}{AE} \frac{\partial u}{\partial t} = \frac{1}{c^2} \frac{\partial^2 u}{\partial t^2}$$

$$c = \sqrt{E/\rho}$$

(6.26)

A pulse induced by a hammer blow travels at a wave propagation velocity c, which should not be confused with particle velocity. Equation 6.26 is a second-order hyperbolic partial differential equation. As with the bearing capacity problem discussed earlier in this book, this class of equation lends itself to solution by the method of characteristics, which permits solution as an ordinary differential equation by integrating along the characteristics. Details of the mathematical derivations can be found in a text on partial differential equations (Farlow, 1993). The characteristics can be shown to be $x \pm ct$. A change of variables from (x, t) to (ξ, η) produces the following equation:

$$4u_{\xi\eta} = 4\frac{\partial^2 u}{\partial \xi \partial \eta} = \frac{\pi D}{AE}\left[k_1 u - k_2 c \frac{\partial u}{\partial \xi} + k_2 c \frac{\partial u}{\partial \eta} \right]$$

$$\xi = x - ct$$

$$\eta = x + ct$$

(6.27)

Figure 6.15 depicts the ξ-characteristics. Transforming the independent variables (x, t) to ξ produces an ordinary differential equation along the characteristic. In the case of no soil resistance (setting the right-hand side of Eqaution 6.27 to zero), the displacement magnitude along any given characteristic will be constant. For the purpose of developing insights into the nature of wave propagation along a pile, we will postpone consideration of the effects of soil resistance by considering this special case. Integration of Equation 6.27 shows displacement u to be the sum of two functions: a waveform ϕ traveling in the negative x-direction and a waveform ψ traveling in the positive x-direction:

$$u(\xi,\eta) = \phi(\eta) + \psi(\xi) = \phi(x + ct) + \psi(x - ct)$$

(6.28)

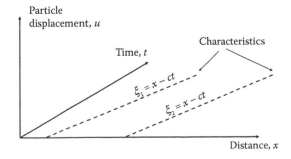

Figure 6.15 Characteristics for one-dimensional wave propagation.

For an infinitely long pile with specified initial displacement $u(x, 0) = f(x)$ and $\dot{u}(x,0) = g(x)$, the expression for particle displacement can be shown to take the following form:

$$u = \frac{1}{2}\left[f(x-ct)+f(x+ct)\right]+\frac{1}{2c}\int_{x-ct}^{x+ct} g(\tau)\,d\tau \tag{6.29}$$

If the initial velocity is zero, the initial displaced shape is seen to be composed of one component $1/2\, f\,(x-ct)$ traveling in the positive x-direction and a second component $1/2\, f\,(x+ct)$ traveling in the negative x-direction:

$$u = \frac{1}{2}[f(x-ct)+f(x+ct)] = \frac{1}{2}[f(\xi)+f(\eta)] \tag{6.30}$$

Considering a downward traveling wave $f\,(\xi)$ and noting that strain at a point is the spatial derivative of displacement, $\varepsilon = \partial u/\partial x$, Equation 6.31 shows stress being proportional to particle velocity:

$$\sigma_x = E\frac{\partial f}{\partial \xi} = E\left(-\frac{1}{c}\right)v = -c\rho v \tag{6.31}$$

This result can be used for analyzing the case of a hammer of mass m impacting at velocity v_0 onto a pile with cross-sectional area A, density ρ, and elastic modulus E. Starting with Newton's second law (Figure 6.16) of motion, Equation 6.31 can be substituted into the velocity term for producing a first-order ordinary differential equation in terms of stress σ. Invoking the initial condition that $v = v_0$ produces the following relationship between stress and impact velocity (Murff, 2000):

$$\sigma = v_0\sqrt{E\rho}\,\exp\left(-A\sqrt{E\rho}/mt\right) \tag{6.32}$$

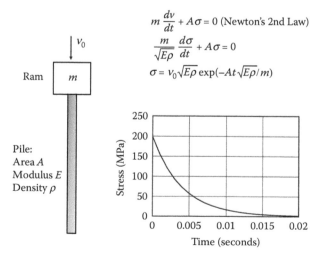

Figure 6.16 Simplified analysis of hammer impact.

Note that stress is linearly proportional to impact velocity v_0. Stress also increases with increasing hammer mass m, but the effect is typically much slighter (Hirsch and Samson, 1966). Thus, for the same amount of hammer energy, a heavier hammer with a shorter stroke (drop height) will induce less stress in the pile than a lighter hammer with a larger stroke.

Equation 6.32 is also useful for getting a sense of the physical length of a pulse induced by a hammer blow. For example, if we consider a 45,000-kg hammer impacting with a velocity 16.5 m/second onto a 1.8-m diameter steel ($E = 200$ MPa, $\rho = 7850$ kg/m³) pile with a 0.05-m wall thickness, Equation 6.32 predicts the stress history at the pile head shown in Figure 6.16. The velocity of wave propagation for steel is computed as $c = 5080$ m/second. Taking the duration of the pulse as the time at which the stress σ decays to 10% of its initial value, the duration of the pulse will be approximately $t_p = 0.01$ seconds. The corresponding physical length of the pulse will then be approximately $c t_p = 5080$ m/second times 0.01 second, or 50.8 m. In most actual situations, a cushion is typically used, which will have the effect of reducing stress levels and increasing the length of the pulse.

Finally, we now consider the reflection of waves for limiting cases of free and fixed conditions at the pile tip (Figure 6.17). The former corresponds to pile penetration in a soft soil with minimal tip resistance, while the latter corresponds to a condition where a stiff bearing stratum is encountered. Reflection at the tip can be considered in terms of a waveform $f(x - ct)$ traveling down the pile (positive x-direction) and a waveform $g(x + ct)$ moving up the pile. At a free tip the stress, which is proportional to the derivative of displacement with respect to x, is zero. Assigning a coordinate $x = 0$ to the pile tip leads to the following expression:

$$\sigma = E\frac{\partial u}{\partial x} = E\left[\frac{\partial f}{\partial x} + \frac{\partial g}{\partial x}\right]$$
$$f'(-ct) + g'(ct) = 0 \qquad\qquad (6.33)$$
$$g'(ct) = -f'(-ct)$$

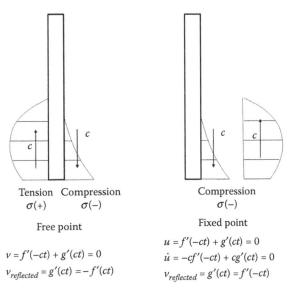

Figure 6.17 Wave reflection at pile tip.

Since f is a compression wave, the reflection g is a tension wave. As shown in Figure 6.17, if the wavelength is short relative to the pile length, the trailing end of the compression wave f offsets the leading end of the tension wave g, such that critical tension does not occur. However, in long piles, the two waveforms completely bypass by one another, so the pile is subjected to the full tension from the g-wave. This is not normally a concern in steel piles, but can cause failure in concrete piles having low tensile strength.

The second limiting case is that of a fixed boundary at the tip, where the particle displacement and velocity are always zero. In this case, Equation 6.34 applies and the downward and reflected waves f and g are both compressive. The stresses are now additive such that a doubling of stress occurs at the fixed boundary. It may be noted that the reflected compression wave will eventually reach the pile head that is a free boundary, if the hammer is no longer in contact with the pile head. Thus, a tension wave may be re-reflected downward at this point. However, in most real situations, damping in the pile and surrounding soil will likely have reduced the magnitude of the stress wave by this point. Nevertheless, cracking has been reported in concrete piles owing to this behavior (Hirsch and Samson, 1966).

$$u = f(-ct) + g(ct) = 0$$
$$\dot{u} = -cf'(-ct) + cg'(ct) = 0 \qquad (6.34)$$
$$f'(-ct) = g'(ct)$$

6.4.2 Wave equation analysis of piles

Full analysis of pile installation requires consideration of soil resistance, which, at a minimum, must consider the elastoplastic response of the soil on the sides and tip of the pile. While relatively sophisticated Laplace transforms and Fourier series solutions have been applied to this problem (Warrington, 1997), most programs in practice (Goble and Rausche, 1976; Hirsch et al., 1976; Randolph, 1992) utilize a relatively simple spring-lumped mass approach illustrated by Figure 6.18. Soil resistance is modeled as a series of spring–dashpot–slider elements along the sides and at the pile tip. Damping behavior is typically empirically based, although Randolph and Simons (1986) utilize an improved approach based on elastodynamic theory, where the effects of radiation damping are considered.

As an example of the general method of analysis, the following discussion presents a wave equation analysis developed by Lowery (1967), which is a modified version of the method developed by Smith (1960), ultimately formulated in the program *Microwave* (1993). The program provides guidance on modeling the many possible variants of pile driving systems. Figure 6.18 shows a schematic representation of a spring–mass analogy to an actual system. The pile driving system (ram, anvil, cushion, pile cap) has many variants, so the system shown is not strictly representative of all systems. Furthermore, elements of the system such as the ram, shown as a single mass element in the figure, may have to be subdivided into smaller elements in some instances. Such variations are easily incorporated into the model. In the portion of the pile extending above the ground surface, the model is comprised solely of mass and spring elements. Not shown in the figure are dashpots that can be added in parallel to the spring elements to simulate internal damping in the pile itself. Beneath the ground surface, spring–slider–dashpot elements are attached to the mass elements for simulating soil resistance to driving. Soil side and tip resistance are described in a similar fashion. The basic output from the analysis is the permanent set (penetration) resulting from a single hammer impact. Repeating for various stages generates the rate of penetration profile illustrated in Figure 6.19. In some instances, the presentation of the output is more conveniently

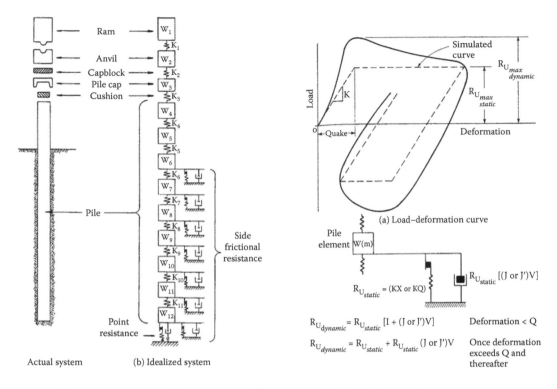

Figure 6.18 Spring–mass model for pile driving. (Courtesy, professor Lee Lowery).

represented in terms of soil resistance that must be overcome, also illustrated in Figure 6.19 (after Bender et al., 1969).

Soil load–deformation behavior at each element along the pile shaft is characterized by three parameters. First, the static resistance on the side of the pile is estimated in the usual manner for computing skin resistance; the horizontal effective stress times the

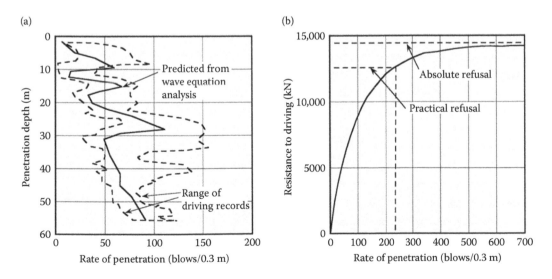

Figure 6.19 Typical output from wave equation analysis (a) rate of penetration versus depth and (b) rate of peneteation versus resistance.

interface friction angle for cohesionless soil and the adhesion factor times undrained shear strength for cohesive soil. End bearing resistance is similarly computed using conventional analyses for static load capacity. Second, yield displacement is specified through a "quake" parameter Q determined empirically through correlation with full-scale pile tests. Finally, side and tip resistance is characterized by viscosity coefficients J and J', respectively. In regard to damping, it should be noted that Randolph and Simons (1986) present an improvement to this formulation, where damping at the tip is characterized entirely as radiation damping and on the sides as combined material and radiation damping.

Key features of the driving system include the hammer, cushion, and other possible accessories. Hammer characteristics include its rated energy, estimated efficiency, mass, stroke, and, in the case of diesel hammers, explosive force. Soil resistance versus penetration resistance curves such as that illustrated in Figure 6.19 are particularly useful for selecting a hammer size capable of overcoming soil resistance with an acceptable range of blow counts. It is noted that simply selecting a larger hammer may not be sufficient for overcoming a certain level of soil resistance. Bender et al. (1969) show examples a larger hammer size being ineffective in overcoming soil resistance when not accompanied by an increase in the pile cross-sectional area. The two key characteristics of the cushion that must be specified are its elastic modulus and its coefficient of restitution, which describes its energy absorption properties. Figure 6.20a shows a parametric study after Bender et al. (1969) illustrating the considerations in an optimal cushion selection, where the rate of penetration is plotted against cushion stiffness for various target levels of soil resistance, designated A through E. Increasing the cushion stiffness clearly reduces the blow count, but continued increase in cushion stiffness at some point ceases to improve drivability. Since a stiffer cushion increases driving stresses, the cushion should not be stiffer than that necessary to achieve an acceptable rate of penetration. Hammer efficiency, which can degrade as a consequence of neglected maintenance, also significantly effects driving performance. Figure 6.20b shows an example parametric study investigating the effect of a 20% reduction in hammer efficiency on the driving resistance–penetration rate relationship. In the low range of driving resistances, the effect of reduced efficiency is relatively minor. However, as resistance to driving increases, reduced hammer efficiency can lead to significant increases in driving time and possibly pile refusal.

(a)

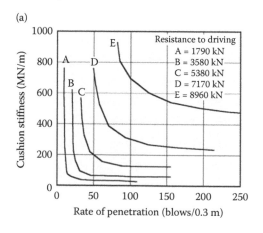

(b)

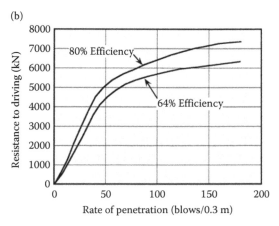

Figure 6.20 Example parametric studies with wave equation analysis (a) effect of cushion stiffness and (b) effect of hammer efficiency.

6.5 SETUP

Pile installation in clays involves intense shearing and remolding of soils in the vicinity of the pile shaft, accompanied by generation of excess pore pressures. Pore pressure dissipation associated with the normal process of consolidation leads to increased effective stress and recovery of much of the strength loss occurring during penetration, a process termed "setup." There are two aspects of setup: the time interval over which strength recovery occurs and the degree to which effective stress and strength recovers. The latter issue is covered in the following chapter in the discussion on axial pile capacity. The discussion below addresses the time-related aspect of the problem.

6.5.1 Predictive framework

A framework for predicting setup time following pile installation requires an estimate of the spatial distribution of excess pore pressures surrounding the pile shaft following installation. Figure 6.21 shows the radial deviatoric and meridional stress components, s_{rr} and s_{rz}, predicted from strain path method (SPM) analyses of solid pile penetration in an elastic–perfectly plastic soil. It can be shown that, to satisfy vertical equilibrium far above the pile tip, the meridional stress component s_{rz} must vary inversely with distance from the pile shaft:

$$s_{rz} = s_{rzB}/r \tag{6.35}$$

where s_{rzB} is the stress at the pile–soil boundary. As will be recalled from Section 3.4.2, the SPM does not necessarily satisfy equilibrium and Figure 6.21 shows that SPM predictions depart substantially from the equilibrium requirement dictated by Equation 6.35. Since the assumption of an irrotational velocity field used in the SPM analysis leads to violation of equilibrium, it is not unreasonable for introducing rigid body rotations to restore equilibrium. By this approach, the strain and stress histories predicted from the SPM are accepted,

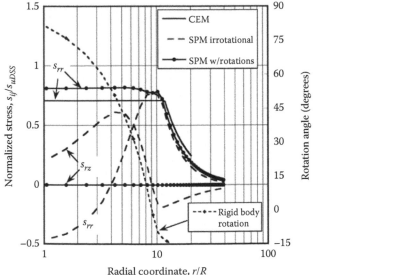
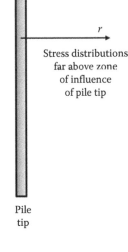

Figure 6.21 Stress field surrounding penetrating pile.

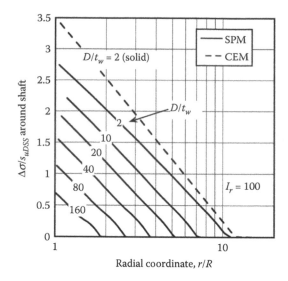

Figure 6.22 Change in mean total stress around shaft during pile penetration.

but rotations are applied to individual soil elements such that Equation 6.35 is satisfied. The resulting velocity field is no longer strictly kinematically admissible; however, satisfying equilibrium is taken as the overriding consideration here. Figure 6.21 shows the outcome of this procedure for the case of a frictionless pile boundary, where s_{rz} must be zero everywhere. The rotations required to satisfy equilibrium approach 80° in the vicinity of the pile shaft, even for the frictionless case. Interestingly, when the rotation correction is applied, the radial component of stress s_{rr} becomes similar to that predicted from the cylindrical cavity expansion method (CEM). The similarity of the corrected SPM to the CEM solution is further illustrated in Figure 6.22, showing the change in mean stress $\Delta\sigma$ in the vicinity of the pile shaft far above the pile tip. The CEM analysis is well known for predicting a semi-logarithmic variation in $\Delta\sigma$ with radial distance from the pile shaft within the zone of plastic yield. The SPM predictions corrected for rigid body rotations conform to this pattern, although the SPM predicts that the mean stress changes $\Delta\sigma$ are some 20% less than those predicted by cavity expansion theory.

The corrected strain path procedure can now be extended to open-ended pile penetration, with diameter-to-wall thickness ratios $D/t_w = 10$–40 being generally representative of driven piles and $D/t_w = 80$–160 being representative of suction caissons. With increasing D/t_w, the $\Delta\sigma - \log r$ distributions show some departure from linearity, but the dominant trend is a qualitative similarity to the solid pile distributions. Decreasing wall thickness also leads to substantial reduction in the magnitude of $\Delta\sigma$ at the pile shaft.

The change in mean total stress $\Delta\sigma$ induced by penetration is one component of excess pore pressure. The second component is the shear-induced pore pressure Δu_s induced by yielding under undrained conditions. Together they comprise the total excess pore pressure u_e that will decay during setup:

$$u_e = \Delta\sigma + \Delta u_s \qquad (6.36)$$

The term Δu_s is determined experimentally from undrained shear strength tests and can be positive or negative according to whether the soil is contractive or dilative. Shear-induced pore pressure is zero in a linearly elastic medium. Therefore, when a linearly elastic–perfectly plastic constitutive model is employed, Δu_s is typically modeled using a step function, being zero prior to yield and jumping to the appropriate experimentally determined value when yield occurs. For the illustrative, analyses that follow shear-induced pore pressure is taken as $\Delta u_s = 0.6\sigma'_0$ in normally consolidated soils, where σ'_0 is the initial mean effective stress in the soil.

As discussed in Section 3.5, the dissipation of excess pore pressures induced by pile installation is a coupled process. However, if the focus is solely on time of consolidation, an uncoupled analysis can provide meaningful results. In the case of radial consolidation around the pile shaft at a distance sufficiently far from the pile tip, the problem reduces to a one-dimensional problem governed by a simple diffusion equation:

$$c\left(\frac{\partial^2 u_e}{\partial r^2} + \frac{1}{r}\frac{\partial u_e}{\partial r}\right) = \frac{\partial u_e}{\partial t} \tag{6.37}$$

The parameter c is coefficient of consolidation, t is time and r is radial coordinate. Solution of Equation 6.37 for the initial excess pore pressure distributions from Figure 6.22 produces the pore pressure dissipation curves shown in Figure 6.23. Predictably, dissipation around thin-walled piles occurs some 1–2 orders of magnitude more rapidly than solid piles.

The analyses presented to this point were confined to a single soil rigidity, $I_r = 100$. Figure 6.24 considers the sensitivity of predicted setup time to variations in soil rigidity, with I_r being varied over a range from 50 to 500. Varying I_r does not result in a simple parallel shift in the setup versus time curves displayed on a semi-logarithmic plot. However, for open-ended piles, the analyses indicate a general trend of setup time increasing in proportion to increases in rigidity index. Setup time around solid piles appears to be somewhat less sensitive to variations in rigidity index. This is consistent with a finding by Houlsby and Teh (1988) that consolidation time around solid piles increases in proportion to the square root of rigidity index.

6.5.2 Suction caissons

For experimental validation of the SPM-based setup predictions, we can first consider laboratory model tests where caissons are inserted and pore pressure decay on the side is measured. Olson et al. (2003) report the results of such tests conducted under single-gravity

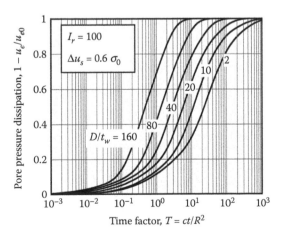

Figure 6.23 Setup around shaft of solid and open-ended piles in normally consolidated clay.

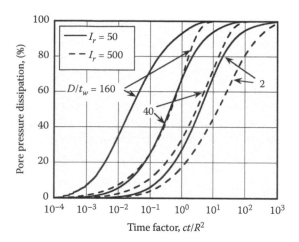

Figure 6.24 Effect of soil rigidity index on setup time.

conditions. The model caissons consist of a 100 mm diameter aluminum tube with a wall thickness of 0.81 mm (diameter to thickness ratio, $D/t_w \approx 125$) and an aspect ratio (L_i/D) of approximately 9. Separate tests were conducted for installation by simulated dead weight penetration and by combined dead weight and suction, but differences in the pore pressure dissipation curves were relatively minor. The soil test bed was a normally consolidated kaolin having liquid limit LL = 54–58 and plasticity index PI = 24. Based on consolidation tests by Pedersen (2001), a coefficient of consolidation representative of normally consolidated conditions is taken as $c_{NC} = 4 \times 10^{-5}$ m²/h (10^{-3} m²/day). The coefficient of consolidation under over-consolidated conditions is not reported; however, based on typical trends (NAVFAC, 1986), it is estimated as 10 times the normally consolidated value for the present comparisons. Figure 6.25 compares SPM-based predictions of excess pore pressure

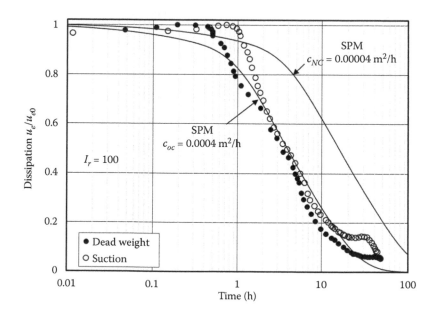

Figure 6.25 Excess pore pressure dissipation around suction Caisson measured in single-gravity tests.

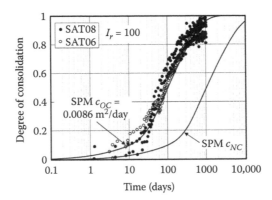

Figure 6.26 Excess pore pressure dissipation around suction caisson measured in centrifuge tests.

dissipation to measurements taken at the mid-depth of the outer wall of the caisson. The comparisons show c_{OC} providing much better agreement with measurements. This observation is consistent with the argument by Levadoux and Baligh (1986) that the reduced effective stress induced by penetration is followed by a time-dependent increase in effective stress during consolidation; thus, the process is dominated by recompression, which is most closely represented by c_{OC}.

Similar tests in normally consolidated kaolin are reported by Cao et al. (2002a,b), this time in centrifuge tests normally consolidated kaolin. The reported coefficient of consolidation in virgin compression is $c_{NC} = 0.9 \times 10^{-3}$ m²/day. As before, the over-consolidated coefficient of consolidation is taken as 10 times this value for the present comparisons. The prototype caisson diameter is $D = 5.17$ m, with wall thickness $D/t_w \approx 80$. The outcome of the comparisons between prediction and measurement (Figure 6.26) is similar to that for the single-gravity tests; namely, the SPM-based predictions are in good agreement with measurements when used in conjunction with the over-consolidated coefficient of consolidation c_{OC}.

Finally, we can consider data reported by Jeanjean (2006) from actual suction caisson installations in Gulf of Mexico (GOM) clay having liquid limit in the range LL = 70–100. Based on data compiled by Bogard (2001) for driven piles in GOM clay—discussed in more detail in the next section—the operative coefficient of consolidation for setup can be taken as c = 0.07 m²/day. In this case, excess pore pressure decay was not directly measured. However, values of the side resistance adhesion factor α inferred from caisson extractions at various times following installation provide a picture of time required for setup (Figure 6.27). Decay of excess pore pressure can be translated into increase in adhesion α according to the following equation:

$$\alpha = \alpha_0 + (\alpha_{eq} - \alpha_0)(1 - u_e/u_{e0}) \tag{6.38}$$

In this equation, α is the adhesion corresponding to the current level of excess pore pressure u_e. The adhesion immediately following pile installation and the adhesion after full equilibrium of pore pressure are denoted by α_0 and α_{eq}, respectively. Figure 6.27 shows scatter in both these values; however, the comparisons to SPM-based setup predictions will consider $\alpha_0 = 0.33$ and α_{eq} ranging from 0.7 to 1.0. Figure 6.27 represents data from caissons having diameters of 5.49 and 3.66 m, with wall thickness $t_w = 0.0381$ m in all cases. The D/t_w ratio is therefore in the range 96–144. Differences in predicted setup curves are

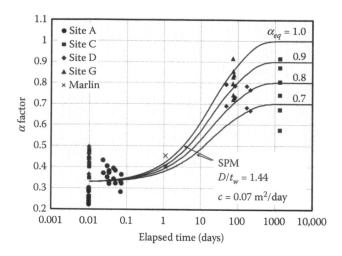

Figure 6.27 Setup times for Gulf of Mexico suction Caisson installations.

minor across this range; therefore, for simplicity, Figure 6.27 shows only the $D/t_w = 144$ predictions. The SPM-based predictions are seen to be reasonably consistent with the measurements. Considering the midrange of the data, with $\alpha_{eq} = 0.8-0.9$, the analyses indicate a setup time to 90% equilibration occurs after approximately 100 days, which is slightly greater than normal expectations (Jeanjean, 2006) for these installations.

6.5.3 Dynamically installed piles

Richardson et al. (2009) report vertical pullout capacity from model tests of dynamically installed piles in a kaolin test bed for various reconsolidation times. The data are reported in terms of net pullout capacity normalized by average shear strength and pile cross-sectional area.

$$F_N = \frac{F - W_b}{s_{uavg} A_p} \tag{6.39}$$

Figure 6.28 shows the normalized capacity versus dimensionless time. For comparison to SPM predictions, normalized capacities immediately after installation and after full consolidation are taken here as $F_{N0} = 25$ and $F_{N100} = 135$. The increase in pile capacity with time can now be predicted by the equation $F_N(T) = F_{N0} + U(T)(F_{N100} - F_{N0})$, where the dissipation $U = (1 - u_e/u_{e0})$ is obtained from the SPM solution in Figure 6.23 for setup around a solid pile. At early times, the SPM appears to be underestimating capacity, possibly owing to thixotropic effects that are not considered in the SPM modeling. Ninety percent setup appears to be occurring at a time factor of about $T_{90} = 200$.

6.5.4 Driven piles

Bogard (2001) compiled times to 50% consolidation following solid pile penetration from a series of GOM sites. His data, when normalized by pile radius, indicate times to 50% setup in the range $t_{50}/R^2 = 270-400$ m²/day. Figure 6.23 provides a nondimensional time factor for solid piles $T_{50} = ct_{50}/R^2 = 23$. Applying this to Bogard's data yields an apparent coefficient of consolidation for the soil in the range $c = 0.057-0.085$ m²/day

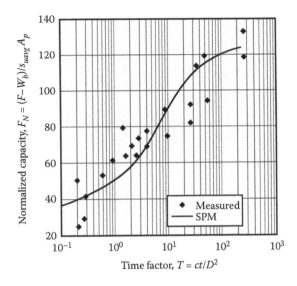

Figure 6.28 Setup around dynamically installed piles.

(average 0.07 m²/day). This relatively high value is consistent with the coefficient of consolidation for a soil in recompression.

Bogard and Matlock (1990) and Bogard (2001) present data on the effects of wall thickness on setup times for driven piles at GOM sites. The data show considerable scatter, but Figure 6.29 shows the trend line of the ratio of the time to 50% consolidation for an open-ended pile to 50% consolidation for a solid pile, $t_{50}/t_{50\text{-}solid}$. In general, the consolidation time for a pile with wall thickness $D/t_w = 24$ is about half and that for wall thickness $D/t_w = 40$ is about one-fifth that of a solid pile. Superimposed over the empirical trend line are the analytical SPM-based predictions. In contrast to the situation for thin-walled

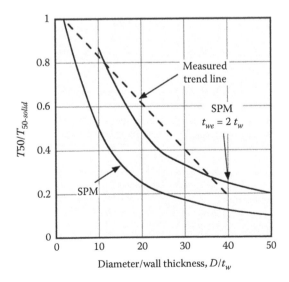

Figure 6.29 Effect of wall thickness on setup time for driven piles in Gulf of Mexico clay.

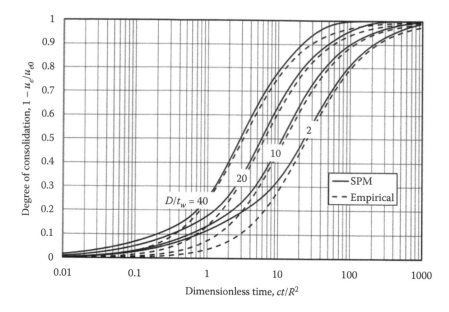

Figure 6.30 Empirical solution for consolidation around driven piles.

suction-installed piles, the SPM appears to significantly overpredict the rate of excess pore pressure dissipation. A possible source of the discrepancy is that the SPM simply overpredicts the amount of soil flowing inside the pile; the additional soil volume pushed outside the pile therefore increases the zone of disturbance in this region, thereby increasing setup time. A simple approach to simulating this "partial plugging" effect is to increase the effective wall thickness of the pile. From trial and error, an effective wall thickness that was two times the physical wall thickness was found to bring SPM predictions into closer agreement with field trends.

From continuous measurements of decay of excess pore pressures at GOM sites, Bogard and Matlock (1990) developed an empirical equation for pore pressure dissipation following installation of driven piles having the following hyperbolic form:

$$U = 1 - u_e/u_{e0} = t/t_{50}/(1.1 + t/t_{50}) \qquad (6.40)$$

Figure 6.30 compares Equation 6.40 to the SPM-based predictions for setup in normally consolidated clay. The agreement is quite good for degrees of consolidation in the range $U = 35\%-90\%$, with the exception of relatively thin-walled piles ($D/t_w = 40$) where noticeable divergence between the theoretical and empirical curves occurs after approximately 85% consolidation.

6.6 STRUCTURAL INTEGRITY

Aside from possible structural damage from driving stresses in driven piles, either driven or suction installed piles can be susceptible to susceptible to extrusion buckling if the stiffness of the surrounding soil exceeds the pile elastic stiffness. Although buckling is commonly analyzed in terms of in situ soil stresses acting on a pile that has been "wished in place" at various stages of embedment, this type of analysis omits a key feature of extrusion buckling—the growth of initial imperfection as a pile is advanced through stiffer soil layers. The two analyses differ at a very fundamental level. For example, the wished-in-place analysis will indicate a reduced risk of buckling in stiff soils, since the greater soil resistance tends to restrain buckling. By contrast, when an extrusion mechanism is considered, stiff soil layers increase the risk of failure since the greater soil stresses drive the extrusion buckling mode. An extrusion buckling analysis will also show sensitivity to conditions at the pile tip. For example, a beveled pile tip will generate inward radial stresses that tend to enhance the extrusion buckling process.

An early case where extrusion buckling was implicated as a likely cause of failure during pile installation was the Goodwyn A platform on the North West Shelf of Australia. In this case, 15 of 20 driven piles were crushed at depths exceeding 80 m below the seabed. Other pile failures have also been attributed to this failure mechanism (Alm et al., 2004). To analyze this buckling mechanism, Barbour and Erbrich (1995) developed a bucket adjusted soil installation loading (BASIL) user element for implementation in the commercial finite element code ABAQUS. The original version of the model comprised user-defined P–y soil springs in conjunction with shell elements simulating the pile. A refined version (Erbrich et al., 2011) characterizes soil resistance in terms of nodal forces that are compatible with computed pile displacements. The original and refined versions model the same underlying P–y mechanisms, the purpose of the modifications being to improve the numerical performance of the model. The analysis assumes that the pile is pushed in place and does not explicitly consider the added potential for damage owing to driving stresses. For suction installation, the destabilizing effect of suction in the interior of the pile can be included in the analysis.

Nonlinear soil springs in the form of a Ramberg–Osgood model are used in the analysis. Based on a series of finite element parametric studies, Erbrich et al. (2011) concluded that

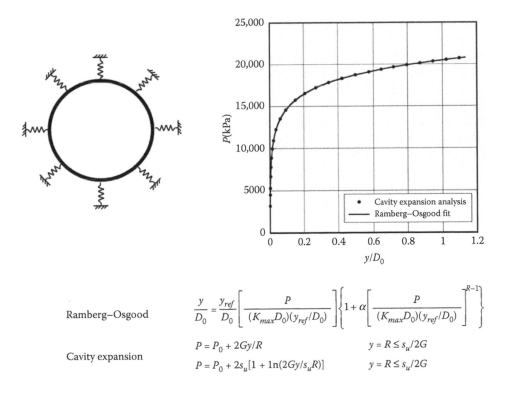

Ramberg–Osgood
$$\frac{y}{D_0} = \frac{y_{ref}}{D_0}\left[\frac{P}{(K_{max}D_0)(y_{ref}/D_0)}\right]\left\{1 + \alpha\left[\frac{P}{(K_{max}D_0)(y_{ref}/D_0)}\right]^{R-1}\right\}$$

Cavity expansion
$$P = P_0 + 2Gy/R \qquad\qquad y = R \le s_u/2G$$
$$P = P_0 + 2s_u[1 + 1n(2Gy/s_uR)] \qquad y = R \le s_u/2G$$

Figure 6.31 Soil spring model for extrusion buckling analysis.

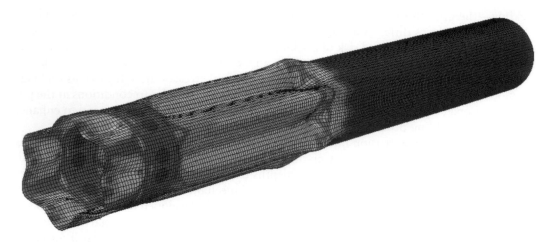

Figure 6.32 Deformed configuration during extrusion buckling.

springs derived from cylindrical cavity expansion theory in an elastic–perfectly plastic (Tresca) medium provide a reasonable approximation to the pressure–displacement response on the sides of the pile. As shown in Figure 6.31, the Ramberg–Osgood model is capable of closely matching the cavity expansion solution. When modeling the soil springs in terms of a cavity expansion solution, the initial cavity diameter under consideration cannot strictly be considered as the pile diameter. Rather, a half-wavelength of the buckle pattern (Figure 6.32) more closely represent the diameter of the expanding "cavity." Only external soil springs acting in compression were found to contribute significantly to response; that is, the influence of the internal soil plug is negligible. Actual discretization of the model is much more refined that the conceptual illustration in Figure 6.31, with shell element dimensions typically being on the order of 0.2 m.

The analysis requires a user-specified initial imperfection that can be expressed in terms of ovality, $D_{max} - D_{min}$. Back-analysis of known buckling failures indicates that the specified initial imperfection must be considerably larger than manufacture tolerances for the BASIL analysis to replicate observed failures. For example, the analysis presented Erbrich et al. (2011) specifies an initial imperfection corresponding to an ovality of 100 mm for a 2.65-m diameter pile (3.77%), which far exceeds a typical manufacturer tolerance. The critical buckling mode may comprise a 2-lobe "peanut" shape, which developed in the analysis by Erbrich et al. (2011). However, other modes can be critical, such as the 6-lobe configuration (exaggerated deformation scale) shown in Figure 6.32. Ring stiffeners can be modeled by specifying a greater wall thickness at the stiffener location. The location of the ring stiffener in Figure 6.32 is easily discernible from zone of low stress levels (denoted by darker shading) in that region. Zones of elevated stress and plastic yield (denoted by lighter shading) are also apparent in the figure.

REFERENCES

Alm T, Snell RO, Hampson K, and Olaussen A, 2004, Design and installation of the Valhall piggyback structures, *Proceedings of the Offshore Technology Conference*, OTC16294, Houston, pp 1–7 (electronic publication).

Andersen KH and Jostad HP, 2004, Shear strength along inside of suction anchor skirt wall in clay, *Proceedings of Paper 16844, Offshore Technology Conference*, Houston, pp 1–13 (electronic format).

Andersen KH, Jostad HP, and Dyvik R, 2008, Penetration resistance of offshore skirted foundations and anchors in dense sand, *J Geotech Geoenviron Eng*, 134(1), 106–116.

Andersen, KH, Murff, JD, Randolph MF, Clukey E, Jostad H., Hansen B, Aubeny C, Sharma P, Erbich C, and Supachawarote C, 2005, Suction anchors for deepwater applications, *Keynote lecture, International Symposium on Frontiers in Offshore Geotechnics*, Perth, Australia, pp 3–30.

API, 1993, Recommended practice for planning, designing and constructing fixed offshore platforms, *API RP 2A*, American Petroleum Institute, Washington, D.C.

Aurora RP, 1984, Experience with driving 84 inch piles with underwater and above water hammers at the South Brae Platform, North Sea, *Offshore Technology Conference*, paper 4803, Houston, pp 237–242.

Barbour RJ and Erbich C, 1995, Analysis of soil skirt interaction during installation of bucket foundations Using ABAQUS, *Proceedings of the ABAQUS Users Conference*, Paris.

Bender CH, Lyons CG, and Lowery LL, 1969, Applications of wave-equation analysis to offshore pile foundations, *1st Annual Offshore Technology Conference*, OTC 1055, Houston, TX, pp I575–I586.

Bogard, D, 2001, Effective stress and axial pile capacity: Lessons learned from empire, OTC 13059, *Offshore Technology Conference*, Houston, TX, pp 1–12 (electronic format).

Bogard D and Matlock H, 1990, Application of model pile tests to axial design, OTC6376, *Offshore Technology Conference*, Houston, TX, pp 271–278.

Cao J, Phillips R, Popescu R, Al-Khafaji, Z, and Audilbert JME, 2002b, Penetration resistance of suction caissons in clay, *Proceedings of 12th ISOPE Conference*, Kyushu, Japan, May 26–31, pp 800–806.

Cao J, Phillips R, Popescu R, Audilbert J, and Al-Khafaji Z, 2002a, Excess pore pressures induced by installation of suction caissons in NC clays, *Proceedings International Conference Offshore Site Investigations and Geotechnics 'Diversity and Sustainability'*, Society of Underwater Technology, London, UK, SUT-OSIG-02-405, pp 405–412.

Chow SH, O'Loughlin CD, and Randolph MF, 2014, Soil strength estimation and pore pressure dissipation for free-fall piezocone in soft clay, *Geotechnique*, 64(10), 817–827.

Clausen CJF, 1998, Fundamentering av plattformer; observasjoner og refleksjoner, *16th Laurits Bjerrum Memorial Lecture*, Norwegian Geotechnical Institute, Oslo.

Cox BE and Christy WW, 1976, Underwater pile driving test offshore Louisiana, *Offshore Technology Conference*, paper 2478, Houston, pp 611–616.

Einav I and Randolph MF, 2005, Combining upper bound and strain path methods for evaluating penetration resistance, *Int J Numer Meth Eng*, 63, 1991–2016.

Erbich C and Hefer P, 2002, Installation of the Laminaria Suction piles—A case history, OTC paper 14240. *Proc. Offshore Technology Conference*, Houston, May 2002, pp 1–14 (electronic format).

Erbich CT, Barbosa-Cruz E, and Barbour R, 2011, Soil–pile interaction during extrusion of an initially deformed pile, *Frontiers in Offshore Geotechnics II*, Gourvenec & White, ed., Taylor & Francis Group, London, ISBN 978-0-415-58480-7.

Erbich CT and Tjelta TI, 1999, Installation of bucket foundations and suction caissons in sand – Geotechnical performance, OTC-10990, *Offshore Technology Conference*, Houston, pp 1–9 (electronic fomat).

Farlow SJ, 1993, *Partial Differential Equations for Scientists and Engineers*, Wiley, New York.

Goble GG and Rausche, F, 1976, *Wave Equation Analysis of Pile Driving-WEAP Program*, Vols. 1–4, FHWA #IP-76-14.1 through #IP-76-14.4.

Heerema EP, 1980, An evaluation of hydraulic vs. steam pile driving hammers, *Offshore Technology Conference*, paper 3829, Houston, pp 321–330.

Hirsch T and Samson TH, 1966, *Driving Practices for Prestressed Piles*, Research Report Number 33-3, Texas Transportation Institute, Texas A&M University, College Station, Texas.

Hirsch TJ, Carr L, and Lowery L, 1976, *Pile Driving Analysis – Wave Equation*, Vol. 14, FHWA HIP-76-13.1 through IP-76-13.4.

Hossain MS, Kim Y, and Gaudin C, 2014, Experimental investigation of installation and pull-out of dynamically penetrating anchors in clay and silt, *J Geotech Geoenviron Eng*, 140(7), 04014026-1:13.

Houlsby GT and Teh CI, 1988, Analysis of the piezocone in clay, *Proceedings International Symposium on Penetration Testing*, Rotterdam, Netherlands, Vol. 1, pp 777–783.

Jansz JW, 1977, North Sea pile driving experience with a hydraulic hammer, *Offshore Technology Conference*, paper 2840, Houston, pp 267–274.

Jeanjean P, 2006, Setup characteristics for suction anchors in Gulf of Mexico clays: Experience from field installation and retrieval, OTC-18005-MS, *Offshore Technology Conference*, Houston, TX, pp 1–9 (electronic format).

Kim Y, Hossain MS, Wang D, and Randolph MF, 2015, Numerical investigation of dynamic installation of torpedo anchors in clay, *Ocean Eng*, 108, 820–832.

Levadoux J-N and Baligh MM, 1986, Consolidation after undrained piezocone penetration. I: Prediction, *J Geotech Eng, American Society of Civil Engineers*, 112(7), 707–726.

Lowery LL, 1967, Dynamic behavior of piling, PhD Dissertation, Texas A&M University, College Station, TX.

Lowery LL, 1993 *Pile Driving Analysis by the Wave Equation*, Wild West Software, Bryan, TX.

Lowery LL, Hirsch TJ Jr, and Samson CH, 1967, *Pile Driving Analysis – Simulation of Hammers, Cushions, Piles and Soils*, Texas Transportation Institute, Research Report 33–9.

Murff JD, 2000, *Marine Foundations, Course Notes*, Texas A&M University, College Station, TX.

NAVFAC, 1986, *Soil Mechanics: Foundations and Earth Structures Design Manual*, U.S. Navy, Naval Facilities Engineering Command.

O'Loughlin CD, Richardson MD, Randolph MF, and Gaudin C, 2013, Penetration of dynamically installed anchors in clay, *Geotechnique* 63 (11), 909–919.

Olson RE, Rauch AF, Luke AM, Maniar DR, Tassoulas JL, and Mecham EC, 2003, Soil reconsolidation following the installation of suction caissons, OTC15263, *Offshore Technology Conference*, Houston, TX, pp 1–9 (electronic format).

Pedersen RC, 2001, Model offshore soil deposit: Design, preparation and characterization, M.S. Thesis, The University of Texas at Austin, May 2001.

Pile Dynamics, Inc., 2003, *GRLWEAP Procedures and Models Version 2003*, 4535 Renaissance Parkway, Cleveland, Ohio 44128, www.pile.com.

Randolph MF, 1992, *IMPACT – Dynamic Analysis of Pile Driving*, Program Manual, Department of Civil and Environmental Engineering, The University of Western Australia.

Randolph MF and Simons HA, 1986, An improved soil model for one-dimensional pile driving analysis, *Proceedings, Third International Conference on Numerical Methods in Offshore Piling*, Nantes, France, pp 3–17.

Richardson MD, O'Loughlin CD, and Randolph MF, 2009, Setup following installation of dynamic anchors in normally consolidated clay. *J Geotech Geoenviron Eng*, ASCE, 135(4), 487–496.

Smith EAL, 1960, Pile driving analysis by the wave equation, *J Eng Mech Div*, ASCE, 86(4), 35–64.

Steiner A, Kopf AJ, L'Heureux J-S, Kreiter S, Stegmann S, Haflidason H, and Moerz T, 2014, In situ dynamic piezocone penetrometer tests in natural clayey soils—A reappraisal of strain-rate corrections, *Can Geotech J*, 51, 272–288.

Tjelta TI, 2015, The suction foundation technology, *Frontiers in Offshore Geotechnics III*, V Meyer, (ed.), CRC Press, Taylor & Francis Group, London.

True DG, 1974, Rapid penetration into seafloor soils, *Proceedings, Offshore Technology Conference*, Houston, OTC2095, pp 607–618.

USACE, 1993, *Seepage Analysis and Control for Dams (CH1)*, U.S. Army Corps of Engineers EM 1110-2-1901.

van Luipen P, 1987, The application of the hydraulic underwater hammer in slender and free riding mode with optional underwater power pack, *Offshore Technology Conference*, paper 5423, Houston, pp 561–568.

Warrington DC, 1997, Closed form solution of the wave equation for piles, Thesis for Master of Science Degree, University of Tennessee at Chattanooga.

Chapter 7

Caisson and pile ultimate capacity

This chapter discusses the ultimate load capacity of caissons and piles. For axially loaded piles as well as short piles and caissons subjected to general loading, ultimate load capacity is a fundamental input for sizing the anchor. For flexible piles, interaction effects should be considered but, even then, most methodologies for estimating stiffness involve scaling to ultimate soil resistance.

Axial resistance of piles and caissons is strongly influenced by installation disturbance. As noted in earlier chapters, analytical methods exist for predicting penetration disturbance. However, such analytical predictions typically rely on computationally intense models of soil constitutive behavior that must be supported by sufficient laboratory testing to provide the required input parameters. Even then, predictions often do not conform to measurements. For this reason, routine design predictions of axial load capacity are generally empirically based predictions. Nonetheless, empiricism has its own pitfalls, including substantial scatter in the database, and uncertainty in extrapolating correlations to soil conditions, pile dimensions, and loading conditions outside of the database that were used for developing the empirical correlations.

Lateral loading of a pile to its ultimate capacity mobilizes soil resistance in a zone that is relatively undisturbed by pile installation. Lateral loading of piles thus lends itself to analytical methods that are both reasonably accurate and suitable for routine design use. Analysis of ultimate capacity under inclined loading requires characterization of the interaction between the combined axial and lateral loads.

7.1 AXIAL CAPACITY IN CLAY

We now consider axial load capacity calculations for driven piles and suction caissons in clay. The basic equations for these two anchor types are identical; however, the correlations developed for slender piles are not generally applicable to caissons, so the cases are considered separately.

7.1.1 Driven piles

Axial capacity of driven piles in clays is computed as the sum of end bearing resistance Q_p and skin friction Q_f:

$$Q_t = Q_p + Q_f \tag{7.1}$$

Point resistance is taken as $Q_p = N_p\, s_u\, A_p$, where A_p is the gross area or annular area according to whether the pile is plugged or unplugged. A bearing factor $N_p = 9$ is widely used and, as will be presented in the following section for suction caissons, it seems to be supported by centrifuge test data.

As discussed in the previous chapter, skin friction on driven piles is influenced by a series of complex processes associated with the intense shearing and distortion occurring during installation, subsequent reconsolidation during setup, and finally shearing to failure during applied axial loading. The complexity and lack of full understanding of these processes leads to a great deal of uncertainty as to the stress state and soil behavior adjacent to the pile wall. Accordingly, practical design largely relies on empirical methods. Current API guidelines for driven piles in clay utilize the "alpha method" (Figure 7.1), which relates skin friction to soil undrained shear strength by a parameter $\alpha = f_s/s_u$. The effects of both radial effective stress and interface friction are embodied in this parameter and, somewhat predictably, considerable scatter occurs in reported values of α (Chow et al., 1997).

Based on the work of Randolph and Murphy (1985), the API (2000) guidelines for estimating α are as follows:

$$\alpha = 0.5(s_u/\sigma'_{v0})^{-1/2} \quad s_u/\sigma'_{v0} < 1$$
$$\alpha = 0.5(s_u/\sigma'_{v0})^{-1/4} \quad s_u/\sigma'_{v0} > 1 \tag{7.2}$$
$$\alpha < 1$$

where σ'_{v0} is vertical effective stress at the depth of interest. These guidelines are supported by a database of over 1000 pile load tests documented by Olsen and Dennis (1982).

The analysis above provides no consideration to pile length effects. Pile length actually appears to affect skin friction, as discussed by Randolph (2003). Owing to compression in the pile itself, relative slip occurs between the soil and pile. As softening occurs, resistance decreases at the top of the pile where the relative slip is greatest. Analysis of this progressive failure mechanism requires a numerical analysis that considers the shape of the load transfer curve. For a simple first-order estimate of the possible reduction in pile capacity owing to progressive failure, Randolph provides the following equation for relating the capacity of an elastic pile Q_{actual} to that computed for a rigid pile Q_{rigid}:

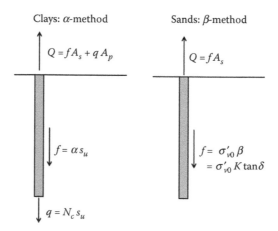

Figure 7.1 Axial capacity of piles.

$$Q_{actual} = R_f Q_{rigid}$$

$$R_f = 1 - (1 - \xi)\left(1 - \frac{1}{2\sqrt{K}}\right) \quad K > 0.25$$

$$K = \frac{\pi D L^2 \tau_{peak}/(EA)_{pile}}{\Delta w_{res}}$$

(7.3)

where EA is the pile stiffness, τ_{peak} is the peak skin friction, ξ is the softening factor, and Δw_{res} is the displacement required to drive the skin friction to its residual value. Ring shear tests show ξ to be in the range 0.5–0.8 and $\Delta w_{res} = 10$–30 mm.

A second mechanism of length effects discussed by Randolph (2003) is based on the measurements of effective radial stresses σ'_{ri} following pile installation showing the ratio of installation stress to cone tip resistance σ'_{ri}/q_c to decrease with depth (Lehane and Jardine, 1994). This effect is not predicted by strain path simulations of installation disturbance (Whittle, 1992). The source of the discrepancy is not well understood but, pending a better understanding of soil behavior under the intense remolding occurring during pile penetration, continued reliance on empirical methods for estimating pile skin friction appears to be justifiable.

7.1.2 Suction caissons

Side resistance for suction caissons in clay is treated within the framework of the α-method in a manner similar to driven piles, although α actual values of used in design of caissons tend to be lower than for piles. Jeanjean (2006) presents an analysis of suction caisson retrieval data from full-scale installations in the normally consolidated GOM clay. The caissons were extracted under vented conditions, so pullout resistance is presumably derived purely from outer and inner side friction with no contribution from reverse end bearing. Retrieval times varied from 0.02 to over 1300 days, permitting a full picture of the setup process. Figure 7.2 shows negligible increase in adhesion factor α beyond 80 days, suggesting full setup essentially occurs beyond this time. The adhesion factor at what may be considered full setup generally falls within a range $\alpha = 0.7$–0.9, which is somewhat lower than that would be expected from a driven pile in similar soils. At 90% consolidation, Jeanjean provides a best-fit estimate from the database as $\alpha = 0.70$–0.75. This is generally consistent with an earlier study by Andersen and Jostad (2002) recommending a lower bound adhesion $\alpha = 0.65$. In this connection, it should be noted that confusion could arise from these recommendations when the full context is not considered. Both the Jeanjean (2006) and Andersen and Jostad (2002) recommendations apply to 90% consolidation, so somewhat greater value can be expected after full setup.

The study cited above provides estimates of adhesion α that represent an average of the external and internal skin friction. Axial loading of suction caissons more typically involves rapid loading that mobilizes external skin friction and reverse end bearing resistance, without mobilization of internal skin friction. This raises the question as to whether significant differences exist between internal and external skin friction. Jeanjean et al. (2006) addresses this issue through a series of centrifuge tests utilizing a double-walled caisson that permits independent measurement of internal and external skin resistance (Figure 7.3). The measurements in fact indicate significant differences between internal and external skin friction. The external friction reaches a peak of about $\alpha = 0.85$ at an upward displacement $0.013D$ and then softens to an ultimate value of about $\alpha = 0.57$. By contrast, the internal skin friction exhibits no peak and tends toward the same value as the ultimate value for external

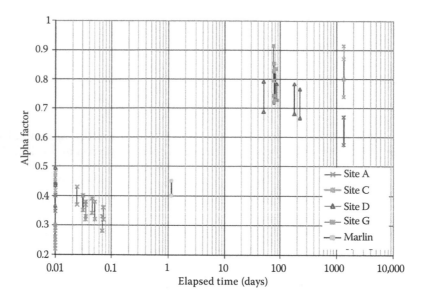

Figure 7.2 Setup following suction caisson installation. (Copyright 2006, Offshore Technology Conference. Reproduced with permission of OTC. Further reproduction prohibited without permission.)

skin friction $\alpha = 0.57$. The average skin friction is seen to be about $\alpha = 0.70$. Thus, the use of an average α to represent external skin friction appears conservative. While the external α-factor in normally consolidated clay for a suction caisson ($\alpha = 0.85$ in this case) remains less than that expected for driven piles, it still exceeds the $\alpha = 0.75$ value representing the average of internal and external skin friction.

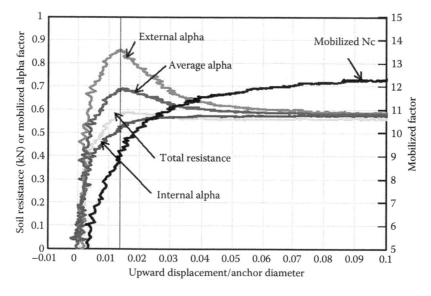

Figure 7.3 Components of uplift resistance. (Copyright 2006, Offshore Technology Conference. Reproduced with permission of OTC. Further reproduction prohibited without permission.)

While the empirical data speaks for itself, one may still question why α is less for suction caissons than for piles. A possible explanation is that suction installation draws additional soil into the interior of the caisson, thereby reducing the overall level of stress from which friction is mobilized on the exterior of the caisson. While this explanation is plausible, it is not well supported by data. Centrifuge tests reported by both Chen and Randolph (2005) and Jeanjean et al. (2006) comparing skin friction on caissons installed by jacking versus suction installation indicate that the mode of installation does not materially affect skin friction. An alternative explanation relates to the small wall thickness of suction caissons relative to driven piles—typical D/t ratios for suction caissons are in the range 144–160, versus 12–40 for driven piles. A greater degree of soil displacement during installation would be expected to lead to higher stress levels and friction, a trend supported by strain path studies for closed versus open-ended piles in Chapter 3.

The double-wall suction caisson studies (Jeanjean et al. 2006) also permitted separation of axial resistance into skin friction and reverse end bearing. Figure 7.3 shows that the end resistance increases monotonically and an ultimate reverse end bearing factor $N_c = 12.2$ mobilizes at about 0.1 D. This is well above the displacement value at which peak external skin friction mobilizes; thus, the full end bearing resistance is not fully usable. At an upward displacement $0.013D$ where peak external skin friction mobilizes, the reverse end bearing factor is about $N_c = 9$. Therefore, the widely used (and API recommended) $N_c = 9$ appears to be appropriate when strain compatibility is factored into the assessment.

7.1.3 Jetted piles

Driven piles and suction caissons penetrate by displacing sufficient soil to accommodate the volume of the pile. An alternative mechanism of installation involves removal of soil through jetting, which substantially reduces the frictional resistance to penetration. As shown in Figure 3.20, jetted installation dramatically reduces the zone of disturbance surrounding the pile. However, the nature of the disturbance is such that overall stress levels, and consequently frictional resistance on the pile shaft, are substantially reduced. To assess the frictional resistance for jetted piles in normally consolidated clays, Zakeri et al. (2014) conducted a series of centrifuge test simulating jetted installation in a kaolin test bed. Adhesion values α were measured during extraction of the piles. They further compared the centrifuge data (tests T7-1 through 4 in Figure 7.4) to data from three GOM sites (A, B, and C in the figure). To facilitate comparisons between the kaolin test bed and GOM clays, they adjusted the setup times for the kaolin upward in proportion to the differences in the coefficient of consolidation c_v between the two soils. This study shows adhesion increasing from $\alpha = 0.05$ immediately following installation to $\alpha = 0.3–0.4$ after setup has substantially completed. These values are in contrast to driven pile and caisson postinstallation α values on the order of $\alpha = 0.2–0.4$ and postsetup α values in excess of 0.7 for caissons and 0.9 for driven piles in normally consolidated clays.

7.2 AXIAL CAPACITY IN SANDS

The estimates of skin friction for piles in sand have traditionally been based on friction calculations derived from a presumed horizontal effective stress at the pile wall, typically termed the "β-method." The API RP 2A-LRFD (2000) guidelines for estimating skin friction for both compression piles and anchors follow this approach. More recently, skin friction has been correlated to CPT tip resistance. At present, four CPT-based methods are in use. All methods for estimating skin friction are empirical. Axial pile capacity in sands is

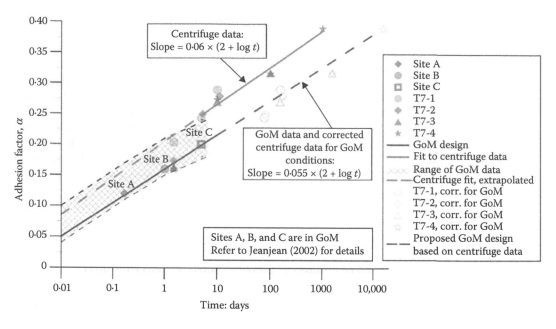

Figure 7.4 Side resistance of jetted piles.

known to be affected by the pile material, length, diameter, partial plugging, aging effects, and specific soil conditions. A description of the pile load test database and the interpretations used for developing the methods could occupy a volume in itself (Lehane et al., 2005b). Accordingly, presentation of all the nuances related to empirical estimates of skin friction is beyond the scope of this brief summary. It should nevertheless be noted that most pile load tests in the database involved pile diameters less than 0.8 m, so extrapolations to larger diameter piles should be used with caution. Most of the load tests were performed a short time after installation relative to a typical design life; therefore, the resistance may be loosely described as "intermediate term."

7.2.1 β-method

The β-method relates skin friction f to *in situ* vertical effective stress σ'_{v0} as shown in Figure 7.1. The β-factor comprises two components. The coefficient K relates σ'_{v0} to the radial stress acting on the pile boundary σ'_{rc}, inclusive of the effects of installation disturbance and equalization. The parameter δ is the interface friction angle between soil and pile. Skin friction is limited by a maximum value f_{max}. Table 7.1 shows API (2000) guidelines for δ and f_{max}, for use when other data are not available. The coefficient K is normally taken as 1.0 for solid/plugged piles and 0.8 for open-ended piles, irrespective of loading in tension or compression.

7.2.2 CPT methods

As discussed previously, four CPT-based methods are used in practice for estimating axial capacity for offshore piles. All methods consider piles in both tension and compression, and provide guidance for both skin friction and end bearing. For a focus on anchor applications, only tension is considered in this discussion. Since the pile tip is assumed to separate

Table 7.1 API Guidelines of pile skin friction in sands

Density	Soil description	δ (degrees)	f_{max} (kPa)
Very loose Loose Medium	Sand Sand–silt Silt	15	47.8
Loose Medium Dense	Sand Sand–silt Silt	20	67.0
Medium Dense	Sand–silt Silt	25	81.3
Dense Very dense	Sand–silt Silt	30	95.7
Dense Very dense	Gravel Sand	35	114.8

from the soil during uplift, the guidelines for end bearing resistance are omitted from the present discussion. Table 7.2 summarizes the relevant equations from the four methods and Figure 7.5 provides a definition sketch for the terms used in the equations. The development of these methods involved considerable exercise of judgment in dealing with scatter in the database as well as careful consideration of the details of installation, local soil conditions, loading conditions, and pile dimensions. The scope of the present discussion is limited to outlining the basic features of the relatively straightforward empirical expressions that were developed for design use. The interested reader should consult the basic references for further insights into the nuances of these methods.

The Fugro-05 (Kolk et al., 2005; Fugro, 2004) recommendations predict shaft friction directly from cone tip resistance q_c and effective overburden stress as shown in the table. The equivalent radius of an open-ended pile is simply the radius of a circle having an area equal to the annular area of the pile.

The ICP-05 design method was developed from field measurements at Imperial College in London, UK. Chow et al. (1997) presents a validation of the design method based on 65 pile tests. The method uses tip resistance q_c to predict radial effective stress following installation and equalization σ'_{rc}. The change in radial effective stress owing to dilation during axial loading, $\Delta\sigma'_{rd}$, is also explicitly calculated as a function of shear modulus G. Shear modulus G can also be estimated from CPT tip resistance. A correlation from Baldi et al. (1989) is shown in the table, but alternative correlations may be considered (Monzon, 2006). The constant volume interface friction angle δ_{cv} may be determined from interface shear tests or correlations to mean particle size.

The Norwegian Geotechnical Institute (NGI-05) design method is based on pile load tests in sand reported by Clausen et al. (2005). Friction τ_f is computed from a series of factors to account for soil relative density, loading in compression versus tension, open versus closed-ended piles, steel versus concrete and overburden stress level. The ratio z/z_{tip} in the equation reduces the skin friction at shallow depths to describe a "friction fatigue" effect, related to soil in the upper layers experiencing more disturbance from the passing pile than in the lower layers.

The UWA-05 method was developed at the University of Western Australia (UWA) (Lehane et al., 2005a,c). The general form of the shear resistance equation parallels that of the ICP-05 method, in that σ'_{rc}, $\Delta\sigma'_{rd}$ and δ_{cv} are explicitly evaluated, but the two methods differ in specific details. UWA-05 also considers partial plugging in terms of an incremental

Table 7.2 Shaft friction for tension piles in sand by CPT methods

Method	Equation for shaft friction
Fugro-05	$$\tau_f = 0.045 q_c \left(\frac{\sigma'_{v0}}{p_a}\right)^{0.15} \left[\max\left(\frac{h}{R*},8\right)\right]^{-0.85}$$
ICP-05	$$\tau_f = a(0.8\sigma'_{rc} + \Delta\sigma'_{rd})\tan\delta_{cv}$$ $a = 1.0$ for closed-ended piles $a = 0.9$ for open-ended piles $$\sigma'_{rc} = 0.029 q_c \left(\frac{\sigma'_{v0}}{p_a}\right)^{0.13} \left[\max\left(\frac{h}{R*},8\right)\right]^{-0.38}$$ $$\Delta\sigma'_{rd} = 2G\Delta r/R$$ $$G = q_c/[0.023 + 0.00125\eta - 1.21\text{x}10^{-5}\eta^2] \quad \text{where } \eta = q_c/(\sigma'_{v0}p_a)^{0.5}$$
NGI-05	$$\tau_f = z/z_{tip}p_aF_{Dr}F_{load}F_{tip}F_{mat}F_{sig} > 0.1\sigma'_{v0}$$ p_a = atmospheric pressure $F_{Dr} = 2.1(D_r - 0.1)^{1.7}$ $F_{load} = 1.0$ for tension, 1.3 for compression $F_{tip} = 1.0$ for driven open-ended, 1.6 for closed ended $F_{mat} = 1.0$ for steel, 1.2 for concrete $F_{sig} = (\sigma'_{v0}/p_a)^{-0.25}$ $z_{tip} = 1.0$ for pile tip depth $$D_r = 0.4\log_e\left[\frac{q_c}{22(\sigma'_{v0}p_a)^{0.5}}\right]$$
UWA-05 (simplified for offshore piles)	$$\tau_f = 0.75 \cdot 0.03\, q_c A_r^{0.3} \left[\max\left(\frac{h}{D},2\right)\right]^{-0.5}\tan\delta_{cv}$$ $$A_r = 1 - (D_i/D)^2$$

filling ratio (IFR). For typical offshore piles with diameter larger than 1 m, the UWA-05 method can be simplified by assuming IFR equal to unity and zero dilation $\Delta\sigma'_{rd} = 0$. Table 7.2 shows the simplified form.

Comparisons among the four CPT methods as well as to the API criteria are a focus of ongoing study, with notable publications that include Lehane et al. (2005b) and Schneider et al. (2008). All of the CPT methods listed above apply to siliceous sands. Schneider et al. (2007) suggest that shaft friction for other sand types (micaceous, calcareous) can be characterized in terms of the general framework used for characterizing siliceous sands, namely a generalized form of the equations used in the ICP-05 and UWA-5 methods:

$$\tau_f = \left(\sigma'_{rc} + \Delta\sigma'_{rd}\right)\tan\delta_f$$

$$\sigma'_{rc} = \frac{q_c A^b_{r,eff}}{a}\left[\max\left(\frac{h}{D},\nu\right)\right]^{-c} \tag{7.4}$$

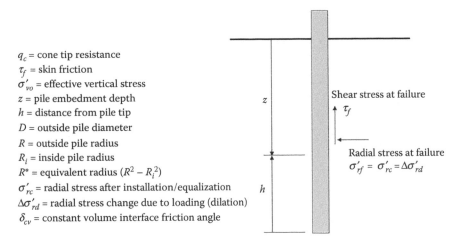

q_c = cone tip resistance
τ_f = skin friction
σ'_{vo} = effective vertical stress
z = pile embedment depth
h = distance from pile tip
D = outside pile diameter
R = outside pile radius
R_i = inside pile radius
R^* = equivalent radius $(R^2 - R_i^2)$
σ'_{rc} = radial stress after installation/equalization
$\Delta\sigma'_{rd}$ = radial stress change due to loading (dilation)
δ_{cv} = constant volume interface friction angle

Shear stress at failure
τ_f

Radial stress at failure
$\sigma'_{rf} = \sigma'_{rc} = \Delta\sigma'_{rd}$

Figure 7.5 Definition of terms for CPT methods.

Table 7.3 Shaft friction parameters for piles in sand

Sand type	Installation	a	b	c	ν	τ_{fmin}	f_t/f_c
Siliceous	Driven	33	0.3	0.5	1–2	5–25	0.7–0.8
Micaceous	Driven	33	0.3	0.5	1–2	5–25	0.7–0.8
Calcareous	Driven	33	0.4	1.0	1–2	2–15	<0.75

Source: Schneider JA, White DJ, and Lehane BM, 2007, *Proceedings of the 6th International Conference on Site Investigation and Geotechnics*, Society for Underwater Technology, SUT, London, pp 367–382.

The parameter a (Table 7.3) accounts for the reduction in radial stress during installation, b accounts for differences between open and closed-ended piles, c accounts for friction fatigue, and ν places a upper limit on the effect of $(h/D)^{-c}$. The effective area ratio is defined as $A_{r-eff} = 1 - IFR\,(D_i/D)^2$, where IFR is the incremental change in plug height in an open-ended pile divided by the incremental change in pile tip depth during installation.

7.3 HORIZONTAL CAPACITY OF PILES IN CLAY

As noted earlier, horizontal load capacity calculation lends itself to analytical treatment. In the case of clays, the primary soil input is the undrained shear strength profile $s_u(z)$. If a gap forms on the windward side of the caisson, soil unit weight γ_b also affects the calculation.

7.3.1 Horizontal-moment interaction: $L_f/D > 3$

Section 4.2.3 discusses the Murff–Hamilton analysis for lateral soil resistance on a pile from which simplified empirical equations were derived for the lateral bearing resistance factor N_p. This section describes how those equations can be used to estimate lateral load capacity H accounting for the effects of rotation as shown in Figure 7.6. A virtual work analysis assumes that the pile rotates as a rigid body about a center of rotation located at depth L_0 at angular velocity $\dot{\beta}$. The power dissipation from side resistance can be calculated in terms of (1) a linear distribution in velocity about a depth to the center of rotation L_0, (2) the Murff–Hamilton bearing resistance function $N_p(z)$, and (3) the undrained shear strength profile $s_u(z)$. The depth L_0 to the center of rotation is an optimization variable to

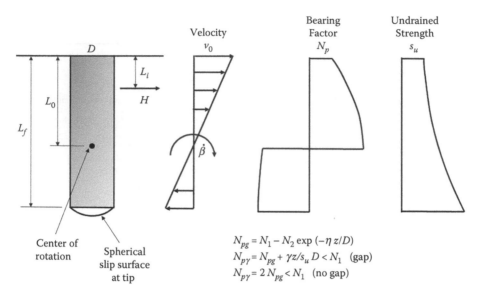

$N_{pg} = N_1 - N_2 \exp(-\eta\, z/D)$

$N_{p\gamma} = N_{pg} + \gamma z/s_u\, D < N_1$ (gap)

$N_{p\gamma} = 2\, N_{pg} < N_1$ (no gap)

Figure 7.6 Lateral loading with rotation.

be determined. The bearing resistance function is assumed to apply to both forward and reverse motions of the caisson. Although the analysis can proceed in terms of an arbitrary strength profile $s_u(z)$, evaluation of the $N_p(z)$ profile requires an approximate linearization of the profile to obtain the η factor. The total power dissipation $\dot{D}_s$ is evaluated from the following integral that is evaluated numerically:

$$\dot{D}_s = \int \dot{\beta} N_p(z)\, |\, L_0 - z\, |\, s_u(z) D\, dz \tag{7.5}$$

The base failure mechanism comprises the spherical shape assumed in the Murff–Hamilton analysis described in Section 4.2.3. Power dissipation can be calculated by numerical evaluation of the integral shown in Table 4.3; however, for desktop calculations a simpler form is desirable. Figure 7.7 shows computed base moment resistance evaluated by numerical integration for centers of rotation ranging from $R_1 = 0$ to large values that approach pure horizontal translation of the caisson. Computed moments are normalized by $M_h = \pi^2 R^2 s_{uavg}/2$, the resisting moment corresponding to a hemispherical base slip mechanism with $R_1 = 0$. An excellent fit to the numerical solution is provided by the following empirical equation:

$$\dot{D}_b = \dot{\beta} 0.5 \pi^2 R^3 s_{uavg} \left[\frac{2}{\pi} \frac{|L_f - L_0|}{R} + \exp\left(-I_4 \frac{|L_f - L_0|}{R} \right) \right] \tag{7.6}$$

With power dissipation from side and base resistance thus defined, the collapse load H is obtained by equating the time rate of external virtual work to internal power dissipation according to the following equation:

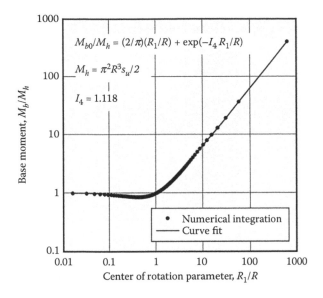

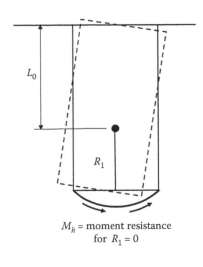

M_h = moment resistance
for $R_1 = 0$

Figure 7.7 Moment resistance versus center of rotation.

$$H = \frac{\dot{D}_s + \dot{D}_b}{\dot{\beta}(L_0 - L_i)} \tag{7.7}$$

This provides the magnitude of H for an arbitrary depth to center of rotation L_0, which is not necessarily the minimum collapse load. A minimum H is obtainable using a searching strategy for the optimal L_0 value; however, the functional form of the dissipation functions permits a more efficient approach using the methods of calculus. Differentiating Equation 7.7 and setting to zero produces the following expression for H_{min}:

$$H_{min} = \frac{1}{\dot{\beta}}\left(\frac{\partial \dot{D}_s}{\partial L_0} + \frac{\partial \dot{D}_b}{\partial L_0}\right)$$

$$\frac{\partial \dot{D}_s}{\partial L_0} = \dot{\beta} \int \mathrm{sign}(L_0 - z) N_p \, s_u(z) \, dz \tag{7.8}$$

$$\frac{\partial \dot{D}_b}{\partial L_0} = \dot{\beta} 0.5\pi^2 R^3 s_{uavg}\left[\exp\left(-I_4 \frac{L_f - L_0}{R}\right)\right]\left(\frac{I_4}{R}\right)\mathrm{sign}(L_f - L_0)$$

The location of the point of load application corresponding to this minimum collapse load is then obtained by solving for L_i in Equation 7.7.

$$L_i = L_0 - \frac{\dot{D}_s + \dot{D}_b}{\dot{\beta}H_{min}} \tag{7.9}$$

The linkage between the depth of load application L_i and the optimal center of rotation L_0 is thus established analytically, so a search is unnecessary. A sweep of calculations varying

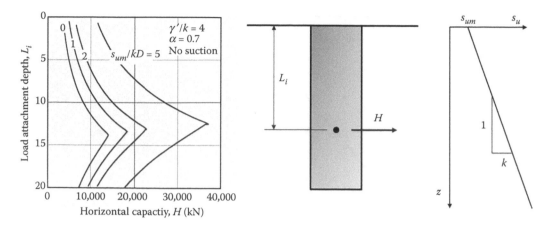

Figure 7.8 Load capacity versus attachment depth.

L_0 from minus to plus infinity using Equations 7.8 and 7.9 provides a complete definition of the relationship between L_i and H, as shown in Figure 7.8.

7.3.2 Optimal load attachment depth

Setting the depth L_0 to the center of rotation at plus or minus infinity in Equations 7.8 and 7.9 produces a condition of pure horizontal translation, which in turn produces the maximum load capacity H_{max} and the optimum depth of load application L_{iopt}. Figure 7.9 shows typical trends for various pile aspect ratios and soil strength profiles.

Noteworthy is the observation that the relatively large depth of optimal load application can lead to significant vertical load demand on the pile owing to the increased curvature in the anchor chain with increasing pad-eye depth. Figure 7.10 shows a typical example for a 5-m diameter pile with aspect ratio $L_f/D = 5$ for a catenary system having a mudline angle $\theta_0 = 0$. In this example, the anchor chain inclination angle at the pad eye is $\theta_a = 13°$ when the load is applied at the optimal depth L_{iopt}. The effects of such a load inclination are often

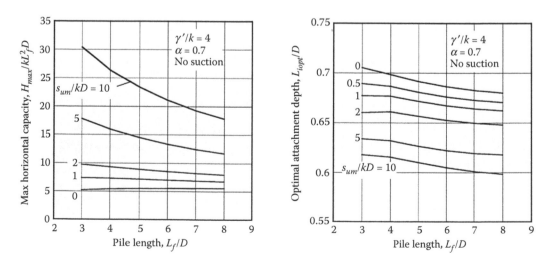

Figure 7.9 Maximum horizontal load capacity and optimal attachment depth.

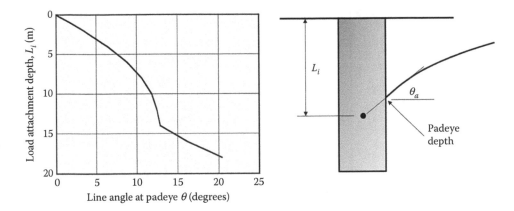

Figure 7.10 Pad-eye depth versus line angle.

relatively benign. As will be discussed subsequently, horizontal–vertical interaction effects are normally not significant for load angles less than about $\theta_a < 30°$. Nevertheless, situations do arise where the vertical load demand becomes more troublesome. For example, the low-side friction in calcareous soils may be insufficient to resist even relatively modest load inclinations, which can preclude setting the pad eye at an optimal location for maximizing horizontal load capacity. It is also noted that the depth of optimal load application is in reference to the centerline of the pile. If the load is inclined, the physical depth of the pad eye must be adjusted upward to ensure that the line of action of the pad-eye force passes through the pile centerline at depth L_{iopt}.

7.3.3 Yield locus representation

An alternative representation of the relationship between horizontal load capacity H and load application depth, L_i is the yield locus format commonly used in plasticity theory; in this case, locus takes the form of a force–moment (H–M) interaction diagram. Moment computations here will initially be taken about the depth to the optimum load application depth, $M_0 = H\,(L_{iopt} - L_i)$. Figure 7.11 shows maximum moment values M_{0max} as a function of pile aspect ratio and strength profile, while the upper left plot in Figure 7.12 shows the H–M_0 interaction for the case of zero soil strength at the mudline. The analysis shows that some moment resistance does actually develop for a load application at L_{iopt}; that is, some moment is generated for the case of pure translation when the reference point for the moment calculation is located at L_{iopt}. Nevertheless, the effect is relatively minor and calculating about the point of pure pile rotation (which may be obtained by searching) produces a similar interaction diagram. Moments M_0 calculated in this manner provide a good measure of the intrinsic moment capacity of the pile as well as an undistorted view of the H–M_0 interaction, both of which can be useful for developing empirical functions to describe the yield locus. However, in practical situations defining moments in terms of a well-defined reference point such as the mudline or top of the caisson can be advantageous, since L_{iopt} is an uncertain function of variable and uncertain soil conditions. Transforming from one reference to another is easily performed, as shown in the lower left plot in Figure 7.12. Taking the mudline as the reference point for moment calculations produces the skewed H–M yield locus shown in the figure, which is the form that sometimes appears in finite element studies investigating force–moment interactions for piles and caissons (e.g., Palix et al., 2011).

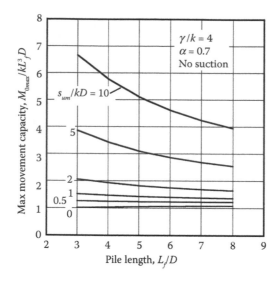

Figure 7.11 Caisson moment capacity.

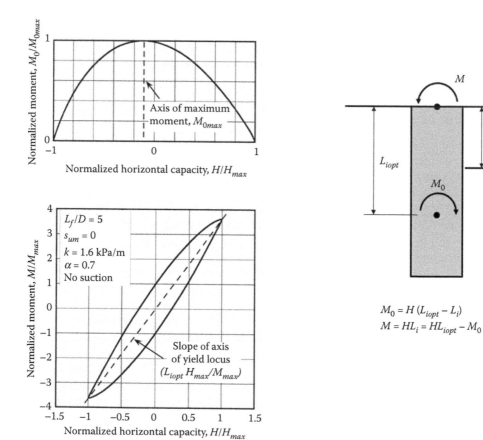

$$M_0 = H(L_{iopt} - L_i)$$
$$M = HL_i = HL_{iopt} - M_0$$

Figure 7.12 Horizontal–moment interaction.

7.3.4 Model validation

Validation of the PLA model described above is possible through comparisons to laboratory scale model tests as well as rigorous finite element analyses. Coffman et al. (2004) present the results of a program of horizontal load tests conducted at the University of Texas at Austin on 0.102 × 0.813-m (4 × 32-in.) model caissons in a normally consolidated kaolin test bed. The load attachment depth was varied for evaluating its effect on horizontal load capacity. Figure 7.13a compares PLA model predictions to their measurements. In general, the model predictions assuming full adhesion, $\alpha = 1$, compares favorably to the data. It is noted that the maximum capacity H_{max} presumes a perfectly positioned load attachment depth L_i, which is unlikely to be achievable in real situations. Thus, the physical measurements do not exhibit the sharp peak in load capacity in the vicinity of the optimum load attachment depth L_i that is seen in the predictions. The measurements also suggest a tendency for the PLA model to slightly underpredict L_{iopt}.

Figure 7.13b compares PLA predictions to finite element analyses by Palix et al. (2011) of a caisson with aspect ratio $L_f/D = 3$ in a uniform clay profile. The PLA and finite element predictions are in close agreement in regard to horizontal capacity H_{max} and intrinsic moment capacity M_0. However, the PLA prediction of M_{max} is approximately 10% less than the finite element value. As noted earlier, when moment computations are referenced to the top of the caisson, M_{max} is more a measure of the optimal load attachment depth L_{iopt} than the rotational resistance of the caisson. From this perspective, it can be concluded that the PLA load–moment $H_{max}-M_0$ capacity predictions are in agreement, but the PLA estimate of L_{iopt} is approximately 10% less than the finite element predictions in this case.

7.3.5 Effects of anisotropy

Consideration of anisotropy requires returning to the complete PLA of Murff and Hamilton (1993) for a laterally loaded pile discussed in Section 4.2.3. Recalling the collapse mechanism comprising a surface failure wedge, a flow-around zone, and a spherical slip surface at the pile tip, the same velocity and strain rate fields are applicable. However, the energy dissipation terms must be modified for an anisotropic yield locus. As discussed in Section

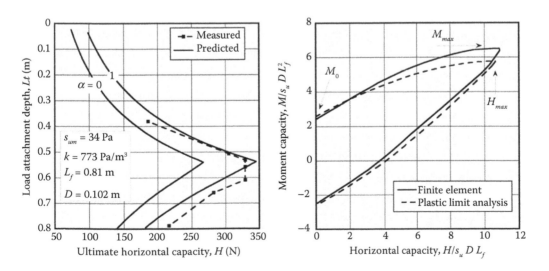

Figure 7.13 Evaluation of PLA model. (a) Tests in kaolin test bed and (b) finite element studies.

3.1.2, if a Hill model of anisotropy is adopted (Aubeny et al., 2001), the modifications are relatively straightforward. The type of anisotropy considered here is that associated with most normally consolidated clay deposits; namely, distinct strengths in both the triaxial compression/extension shearing modes, with strengths in the remaining shearing modes (cavity expansion, etc.) equal to that in the simple shear mode. The typical trend here is $s_{uTC} > s_{uDSS} > s_{uTE}$.

Referring to Table 4.3, dissipation at the surface wedge–pile interface and in the flow-around zone requires no modification for anisotropy, aside from using the simple shear strength s_{uDSS} in the calculation. Power dissipation in the continuously deforming surface wedge is easily computed by transforming conventional strain rate terms into transformed measures (Table 3.1) and employing the power dissipation rate for anisotropic materials in Equation 3.11. Computation of power dissipation across velocity jumps requires decomposition of the velocity components along the axes on anisotropy. Figure 7.14 shows the relevant velocity components. With velocity components thus defined, strain rates in the vanishingly thin zone across the velocity jump are computed from Table 3.2 and power dissipation computed from Equation 3.11.

Laboratory measurements of undrained strength ratios for K_0-normally consolidated clays (Ladd, 1991) typically show the strength in triaxial shear to differ from the direct simple shear strength by a ratio $s_{uTC}/s_{uDSS} = 1.04$–1.33 for triaxial compression and $s_{uTE}/s_{uDSS} = 0.55$–0.96 for triaxial extension. To assess the possible effects of anisotropy on ultimate load capacity calculations, Figure 7.15 presents calculated load capacities for the extreme ranges of anisotropy, $s_{uTC}/s_{uDSS} = 1.33$ and $s_{uTE}/s_{uDSS} = 0.55$. The isotropic analyses were conducted using the direct simple shear strength s_{uDSS}. The computations are presented for cases of no gapping, gapping with a weightless soil and gapping with a soil density ratio $\gamma'D/s_u = 2$. The load attachment depths considered are at the mudline, $L_i = 0$, and at

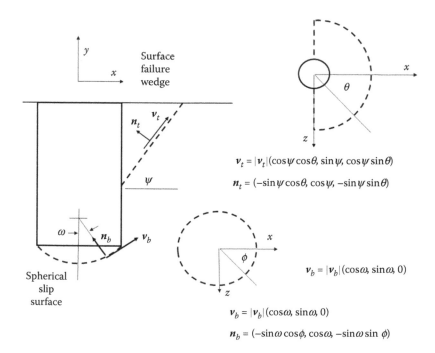

Figure 7.14 Pile in anisotropic soil.

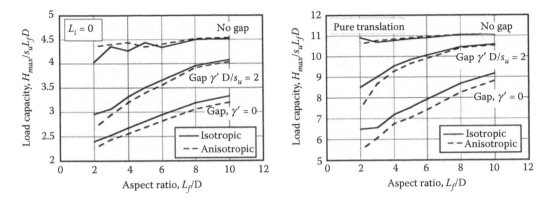

Figure 7.15 Effect of strength anisotropy on pile capacity.

the optimal load attachment depth that creates a condition of pure translation. For cases of no gapping, differences in the predictions are generally negligible, with the isotropic predictions being slightly conservative. Such a result is to be expected, since the reduced soil resistance associated with triaxial extension in the passive wedge is largely offset by increased soil resistance owing to triaxial extension in the active wedge. Thus, an isotropic analysis using a shear strength roughly matching the average of the triaxial compression and extension strength produces results that generally conform to the anisotropic analysis. By contrast, when gapping occurs this fortuitous offsetting of errors does not occur. When only the passive failure wedge develops, the isotropic analysis will tend to overestimate soil resistance. However, the effect is not as great as one might be led to expect, since soil resistance at other locations such as the flow-around region and at the pile tip are either unaffected or minimally affected by strength anisotropy. Furthermore, the contribution of soil unit weight γ' to load capacity when gapping occurs further damps out differences between the isotropic and anisotropic predictions, except for short caissons, $L_f/D = 2$. In summary, anisotropic affects are typically modest for long caissons in soils having the typical anisotropic strength characteristics of normally consolidated clays. The primary area where one might choose to investigate further into potential anisotropic effects is for short caissons where gapping is expected to occur.

7.4 COMBINED LOADING IN CLAY

Analytical description of piles subjected to combined axial–lateral loading first requires a characterization of the interaction effects on the sides of the pile and at the pile tip. Additionally, the axial component of loading generates a kinematic degree of freedom in the vertical direction that must be considered in addition to the rotational degree of freedom in a virtual work analysis. The following paragraphs describe the development and validation of the combined load analysis.

7.4.1 Side axial–lateral interaction

From three dimensional finite element analyses of a long pile subjected to pure translation, incremental forces per unit depth F_{ls} and F_{as} (the integrated resultant of stresses acting on the pile boundary) can be used for deducing dimensionless bearing factors for soil resistance

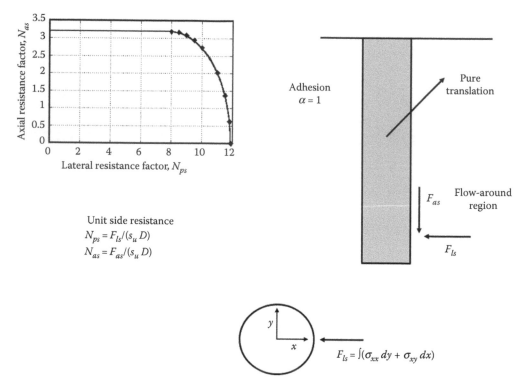

Figure 7.16 Inclined loading on a pile.

exerted by the soil on the side of the caisson, N_{ps} and N_{as} (Figure 7.16). Repeating the exercise for translation inclinations ranging from $0°$ to $90°$ permits a mapping of the N_{ps}–N_{as} interaction diagram shown in the figure. Considering the case of a rough cylinder far from any free surface effects, the bearing factor under pure horizontal loading approaches the Randolph–Houlsby plane-strain solution for a translating cylinder, $N_{ps} = 11.94$. Pure vertical translation leads to $N_{as} = \pi$. Finite element solutions for intermediate cases can be described by an empirical yield locus for side resistance f_s having the following form:

$$
\begin{aligned}
N_{ps} < 8 \quad & f_s = N_{as} - \pi = 0 \\
N_{ps} > 8 \quad & f_s = (N_{ps} - I_3)^2 + (N_{as})^2 / I_1 - I_2 = 0
\end{aligned}
\qquad (7.10)
$$

The interaction coefficients I_1, I_2, and I_3 are obtained from a curve fit to the finite element solution. This yield function takes the form of an ellipse, with I_1 controlling its aspect ratio, I_2 its size, and I_3 its origin. The region where N_{ps} is less than 8 is essentially a condition of no axial–lateral interaction. For N_{ps} greater than 8, the axial–lateral side resistance interaction relationship closely approximates an elliptical form. Adapting the yield locus to general pile roughness conditions can be achieved by assuming that the Randolph–Houlsby solution described in Section 4.2.1 varies linearly from $N_{ps} = 9.20$ to 11.94 over the interval $\alpha = 0$–1. The curve fit to the finite element solution together with an adjustment for general surface roughness conditions leads to the expressions for the interaction coefficients in Table 7.4. The interaction coefficient I_1 controls the aspect ratio of an elliptical yield locus, I_2 controls its size, and I_3 controls its origin.

Table 7.4 Caisson inclined load capacity model

Description	Equation
Axial–lateral side resistance interaction factors	$I_1 = 0.636$ $I_2 = (\pi\alpha)^2/I_1$ $I_3 = 9.20 - 1.2\alpha$
Axial side resistance	$C_1 = \|1 - z/L_o\|/\xi$ $C_2 = I_1/R_f^2 C_1$ $C_3 = I_3 I_1 s_u D/R_f C_1$ $F_{as} = C_2 F_{ls} - C_3$
Lateral side resistance	$A_1 = \dfrac{1 + I_1/R_f^2 C_1^2}{(s_u D)^2}$ $A_2 = \dfrac{R_f I_3}{(s_u D)}\left(1 + \dfrac{I_1}{R_f^2 C_1^2}\right)$ $A_3 = R_f^2\left(I_3^2 - I_2 + I_3^2 I_1/R_f^2 C_1^2\right)$ $A_4 = A_2/A_1$ $A_5 = A_3/A_1$
Vertical base resistance	$F_{ls} = A_4 + \sqrt{A_4^2 - A_5}$ $C_4 = \dfrac{1}{L_o\xi}\dfrac{M_{bmax}}{V_{bmax}}$ $V_b = \sqrt{\dfrac{1}{C_4^2 + 1}}V_{bmax}$
Moment base resistance	$C_5 = \dfrac{1}{\xi L_o}\dfrac{M_{bmax}^2}{V_{bmax}^2}$ $M_b = C_5 V_b$

Equation 7.10 applies to depths far from the free surface. To account for free surface effects, lateral resistance must be scaled down by the following reduction factors applicable to conditions of a gapping on the windward side of the pile:

$$R_f = \frac{N_1 - N_2 \exp(-\eta\, z/D)}{N_1} \qquad \text{(gap)}$$

$$R_f = \frac{2\left[N_1 - N_2 \exp(-\eta\, z/D)\right]}{N_1} \leq 1 \quad \text{(no gap)} \tag{7.11}$$

The contraction of the yield locus near the free surface is described by reducing I_2 and I_3 equally by R_f. The general form of the yield locus for side resistance, accounting for the influence of pile surface roughness and the free surface, now becomes as follows:

$$N_{ps} \leq I_3 R_f \qquad f_s = N_{as} - \pi\alpha = 0$$

$$N_{ps} > I_3 R_f \qquad f_s = (N_{ps} - I_3 R_f)^2 + \frac{(R_f N_{as})^2}{I_1} - I_2 R_f^2 = 0 \tag{7.12}$$

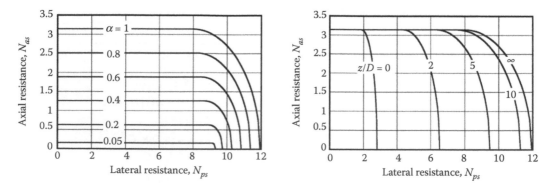

Figure 7.17 Side resistance interaction. (a) Effect of skin friction and (b) effect of free surface.

Figure 7.17 shows the effects of surface roughness and proximity to the free surface on the interaction diagram for lateral–axial side resistance

In terms of soil resistance per unit depth, $F_{ls} = N_{ps} s_u D$ and $F_{as} = N_{as} s_u D$, Equation 7.12 can be rewritten as follows:

$$N_{ps} \leq I_3 R_f \quad f_s = F_{as}/s_u D - \pi \alpha = 0$$

$$N_{ps} > I_3 R_f \quad f_s = (F_{ls}/s_u D - I_3 R_f)^2 + \frac{(R_f F_{as}/s_u D)^2}{I_1} - I_2 R_f^2 = 0 \qquad (7.13)$$

We can now consider the case of a pile experiencing the simultaneous rigid rotation and upward translation shown in Figure 7.18. Assuming an associated flow law, the ratio of axial to lateral velocity v_a/v_l at any depth must be normal to the yield locus. Enforcing this requirement (Aubeny et al., 2003b) leads to the following equation for lateral and axial soil resistance on the side of the pile:

$$F_{ls} = A_4 + \sqrt{A_4^2 - A_5}$$

$$F_{as} = C_2 F_{ls} - C_3 \qquad (7.14)$$

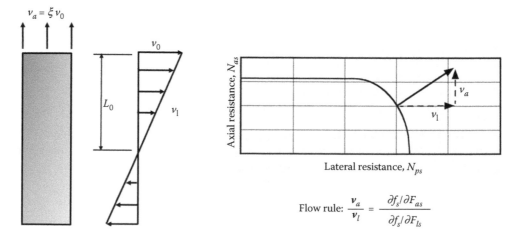

Figure 7.18 Rotating caisson with upward translation.

Table 7.4 defines the coefficients A_4, A_5, C_2, and C_3 for this equation. Note that C_1 in the table is a kinematic variable that affects all other variables. The lateral soil resistance in the interaction equations represents only that derived from soil shearing resistance for the no-suction (gapping) case on the windward side of the pile. The unit weight contribution of soil is unaffected by the axial–lateral interaction, so if gapping is assumed to occur, the total lateral resistance is as follows:

$$(F_{ls})_{total} = F_{ls} + \gamma'z < s_u N_1 D \qquad \text{(gapping)} \tag{7.15}$$

7.4.2 End axial–moment interaction

Interaction at the pile tip is characterized by Aubeny et al. (2003a,b) by the following simplified circular interaction equation:

$$f_b = \left(\frac{V_b}{V_{bmax}}\right)^2 + \left(\frac{M_b}{M_{bmax}}\right)^2 - 1 = 0 \tag{7.16}$$

The uplift capacity under pure vertical loading is $V_0 = N_{ab}\pi R^2 s_{uavg}$, where N_{ab} is the reverse end bearing factor variously estimated between 9 and 12. The moment capacity M_{b0} was defined previously in Equation 7.6. At first glance, the absence of the horizontal resisting force H_b can raise a concern as to the ability of the yield function f_b to accommodate the case of pure lateral–axial translation with no rotation. Revisiting Equation 7.6 and considering its limit as R_1 approaches infinity shows the following:

$$\lim_{R_1 \to \infty} \{M_{b0}\} = \lim_{R_1 \to \infty} \left\{0.5\pi^2 R^3 s_{uavg}\left[\frac{2}{\pi}\frac{R_1}{R} + \exp\left(-I_4\frac{R_1}{R}\right)\right]\right\} = \pi R^2 s_{uavg} R_1 = H_{bmax} R_1$$

$$\lim_{R_1 \to \infty} \left\{\frac{M_b}{M_{bmax}}\right\} = \lim_{R_1 \to \infty} \left\{\frac{H_b R_1}{H_{bmax} R_1}\right\} = \frac{H_b}{H_{bmax}} \tag{7.17}$$

Thus, Equation 7.16 implicitly accounts for H_b in the sense that it correctly characterizes the limiting conditions of pure rotation and pure horizontal translation. By invoking the flow rule using the interaction function f_b as shown in Figure 7.19 leads to the following expressions for moment resistance at the base of the pile:

$$M_b = \frac{1}{\xi L_o}\frac{M_{bmax}^2}{V_{bmax}^2}V_b = C_5 V_b \tag{7.18}$$

Back substituting M_b into the yield function f_b then produces the vertical resistance at the pile base V_b:

$$V_b = \sqrt{\frac{1}{C_4^2+1}}V_{bmax} \qquad C_4 = \frac{1}{L_0\xi}\frac{M_{bmax}}{V_{bmax}} \tag{7.19}$$

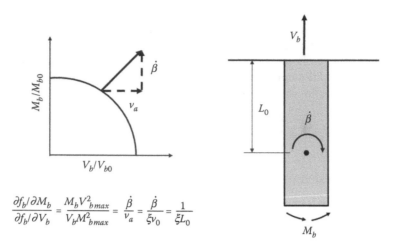

$$\frac{\partial f_b/\partial M_b}{\partial f_b/\partial V_b} = \frac{M_b V^2_{bmax}}{V_b M^2_{bmax}} = \frac{\dot{\beta}}{v_a} = \frac{\dot{\beta}}{\xi v_0} = \frac{1}{\xi L_0}$$

Figure 7.19 Vertical–moment interaction at pile base.

7.4.3 Virtual work solution

With soil resistance on the side (F_{as}, F_{ls}) and base (V_b, M_b) of the pile evaluated, the external work performed by the pad-eye force can be equated to the internal energy dissipation to obtain a solution for the collapse load. The external virtual work performed by the submerged weight W_{sub} may also be included in the calculation to obtain the following equations for horizontal and vertical load capacity, H and V:

$$H = \frac{\int \left(F_{ls}\left|1 - \frac{z}{L_o}\right| + F_{as}\xi \right) dz + \frac{M_b}{L_o} + V_b\xi}{\left(\xi\tan\psi + \left|1 - \frac{L_i}{L_o}\right| \right)}$$

(7.20)

$$V = H\tan\psi$$

The solution is a function of the kinematic optimization variables L_0 and ξ controlling the center of rotation and vertical uplift velocity of the pile. In contrast to the case of pure lateral loading, a simple calculus solution does not present itself, so a search is necessary. However, with only two optimization variables, satisfactory solutions are possible using simple search strategies.

Figure 7.20 shows an example analysis for resultant load capacity $F = (H^2 + V^2)^{0.5}$ as a function of load inclination angle and load attachment depth. The distance L_{iw} refers to the depth of load attachment on the caisson wall, as opposed to the projection of the line of action of this force to the caisson centerline. Interaction effects become significant at around $\psi > 25°$ for load attachment depths near the optimal depth and at large angles for shallower attachment depths. At large inclination angles $\psi > 60°$ no interaction occurs, leading to $F = V_{max}/\sin\psi$. The solid curves in the figure correspond to load attachment depths located above the optimum depth such that the caisson rotates forward, while the dashed curves are associated with reverse rotation. When forward rotation occurs, load capacity F shows a high sensitivity to load angle ψ. By contrast, when reverse rotation occurs load capacity F is remarkably insensitive to load angle for $\psi < 40°$, which is typically desirable from a design standpoint since it implies that predictions of F are not sensitive to uncertainty in ψ.

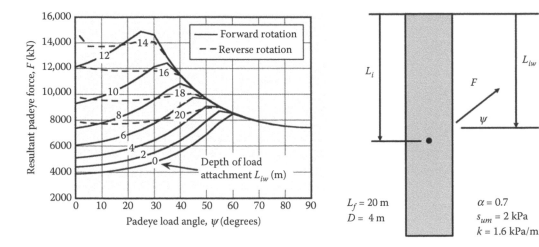

Figure 7.20 Effect of load inclination on load capacity.

7.4.4 Model validation

Model test evaluation of suction caisson load capacity under inclined loading includes a series of seven centrifuge model tests performed at the C-CORE centrifuge facility in St. Johns, Newfoundland. The tests were performed at an acceleration of 115 g with a shear strength profile representative of a normally consolidated clay soil, $s_{um} = 1.75$ kPa, $k = 1.12$ kPa/m. The model caissons had an aspect ratio $L_f/D = 5$, but the effective aspect ratio was somewhat lower than this (average $L_f/D = 4.87$) since penetration was slightly less than the full length of the caissons. All tests had a scaled diameter $D = 5.36$ m. The load attachment depth L_i was slightly under two-thirds of the caisson length. Installation was achieved through combinations of self-weight, suction, and push-in insertion. Additionally, some of the model caissons had outward facing bevels to attempt to reduce the flow of soil inside the caisson during installation. The seven tests are collectively compared to predictions from the inclined load PLA model presented in this text, recognizing that the different installation details may have generated some scatter in the measurements. For complete details on the test program, the reader is referred to Clukey et al. (2004).

Figure 7.21a shows the normalized horizontal–vertical capacity interaction diagram predicted by the PLA model for this series of tests. Since the load is attached at the caisson wall rather than the centerline, the effective depth of load application varies with load angle—the inward curvature of the interaction diagram near the horizontal axis is evidence of this effect. The applied load inclination for the centrifuge tests varied from $\psi = 24$ to 90°. The PLA predictions closely matched the measured capacity for pure vertical loading. Of particular interest in this test series was the load angle ψ at which H–V interaction effects become significant. The measured data indicate essentially no interaction effects (i.e., no reduction in vertical load capacity) for load interaction angles $\psi = 90°$ down to approximately 33°. At shallower angles a sharp reduction in vertical load capacity occurs. The PLA model predictions conform reasonably well to the measured trend.

Andersen et al. (2005) present a compilation of finite element studies using the BIFURC finite element model for caissons developed at the NGI along with parallel predictions by the UWA and the Offshore Technology Research Center (OTRC) using the finite element code ABAQUS. Among other cases considered in the study is a 5-m diameter caisson with aspect ratio $L_f/D = 5$ in a normally consolidated clay profile with strength gradient $k = 1.25$ kPa/m.

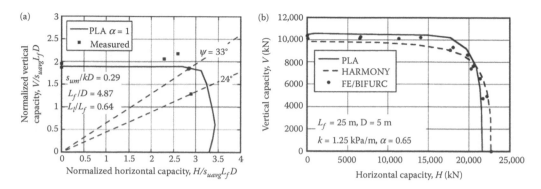

Figure 7.21 Evaluation of inclined load PLA model.

A soil–pile interface adhesion $\alpha = 0.65$ was used in the study. Kay and Palix (2011) present a completely independent analysis of this caisson in the same soil profile, using the Fourier analysis program HARMONY. Both the Andersen and the Kay–Palix studies consider pure translation of the caisson for various load inclination angles; thus, their H–V interaction diagrams are roughly representative of conditions for which the applied load is close to its optimal application depth L_f. Figure 7.21b shows the interaction diagrams from the two studies to be in general agreement. The PLA model for inclined loading presented earlier can be adapted to conditions of pure translation by suppressing the search for an optimal depth to the center of rotation and simply setting L_{0opt} to a large "infinite" value. A search is still required to determine the optimal ξ controlling the ratio of vertical to horizontal displacement. The PLA predictions compare well to the other two sets of finite element analysis, albeit the predicted vertical capacity tends to be slightly high and the horizontal capacity slightly low.

7.5 LONG TERM LOADING

Under sustained loading of sufficient duration negative excess pore pressures at the bottom of the pile will dissipate, leading to a loss of reverse end bearing resistance, $V_b = 0$. The collapse mechanism now becomes pullout of the pile, leaving behind the plug of soil in the interior. Partially compensating the loss of reverse end bearing is the mobilization skin resistance on the inner wall of the caisson. Since empirical evidence indicates that the internal adhesion factor α_{inner} is less than α_{outer}, an independently specified adhesion value would normally be used for calculating the internal skin friction. Since the soil in the interior of the pile does not resist horizontal loading, no V–H interaction reduction factor need to be applied to internal skin friction. Similarly, in the absence of reverse end bearing resistance, no V_b–M_b interaction occurs, and the full base moment resistance M_{b0} develops, irrespective of the vertical load applied to the pile. Although the loss of reverse end bearing resistance is mitigated by the factors described above, there is typically a significant net loss in load capacity when loss of suction occurs. Figure 7.22 shows an example for the effect of the loss of reverse end bearing resistance on ultimate load capacity. In this case, the effects are minor for load attachment angles ψ less than 20°. However, for ψ greater than 30° the net loss in load capacity is on the order of 25%.

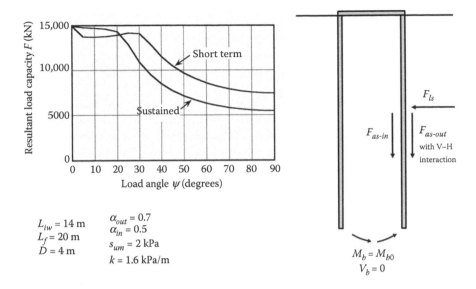

$L_{iw} = 14$ m
$L_f = 20$ m
$D = 4$ m

$\alpha_{out} = 0.7$
$\alpha_{in} = 0.5$
$s_{um} = 2$ kPa
$k = 1.6$ kPa/m

Figure 7.22 Effect of sustained loading on pile capacity.

7.6 LATERALLY LOAD PILES IN SANDS

Various methods for estimating unit lateral soil resistance P_u in sands have been developed by a number of investigators. Table 7.5 shows equations developed by Broms (1964), Petrosovitz and Award (1972), and Reese et al. (1974). Broms characterizes lateral resistance in terms of a simple multiple of passive earth pressure computed from classical Rankine theory. Petrosovitz and Award refine this by explicitly considering the active earth pressure acting on the windward side of the pile. By characterizing the resistance in terms of passive and active resistance, both methods implicitly assume that a surface wedge failure is the dominant mechanism. Conversely, neither method considers a limiting resistance associated with a flow-around zone. Accordingly, both methods would intuitively be considered appropriate for short piles or caissons. By contrast, the analysis based on Reese, Cox, and Coop is based on field tests on slender piles, 0.61 m diameter by 21 m long, for which a flow-around zone will inevitably develop. The original expressions were formulated in terms of active,

Table 7.5 P-ultimate equations for sand

Method	Broms (1964)	Petrasovitz and Award (1972)	Reese, Cox and Coop (1974)
Equation	$P_u = 3 K_p \sigma'_v D$ $K_p = \tan^2(45 + \phi'/2)$	$P_u = 3.7 (K_p - K_a) \sigma'_v D$ $K_p = \tan^2(45 + \phi'/2)$ $K_a = \tan^2(45 - \phi'/2)$	$P_u = \sigma'_v D(c_1 + c_2\, z/D) <$ $\sigma'_v D\, c_3$ for $20° \leq \phi \leq 40°$
			$c_1 = 0.124 \exp(0.091\phi)$[a]
			$c_2 = 0.58 \exp(0.051\phi)$[a]
			$c_3 = 0.73 \exp(0.123\phi)$[a]

[a] Empirical fit by Whiteside (1995).

passive, and at-rest earth pressure coefficients. The original expressions can be simplified assuming an interface friction angle $\delta = \phi/2$. Ultimate resistance is typically expressed in terms of coefficients c_1 and c_2 characterizing resistance in the shallow surface wedge region and c_3 characterizing resistance in the flow-around zone. The resulting expressions are not overly complex, but simplified empirical fits developed by Whiteside (1995) are shown in the table.

For slender, piles elastic effects are so significant that P_u is largely used as scaling factor for estimating soil stiffness, as opposed to directly computing ultimate pile capacity. By contrast, for the short aspect of suction caissons, particularly those used in sands, elastic effects are small and ultimate capacity may reasonably be estimated along the lines similar to those described above for suction caissons in clay. With the current focus being on ultimate load capacity, the following discussion centers about ultimate capacity of short caissons. Bang et al. (2011) present the results of a series of centrifuge tests on model suction piles in sands conducted by the Daewoo Institute of Construction Technology. The model pile had an aspect ratio of 2, with prototype diameter $D = 3$ m and length $L_f = 6$ m. Measured friction angle for the test bed was $\phi = 39°$ as measured in a CU triaxial test. Ultimate horizontal load capacity was measured for load attachment depths at five locations, $L_i/L_f = 0.05$, 0.25, 0.5, 0.75, and 0.95. Figure 7.23 presents the measured load capacities.

While Bang et al. (2011) interpreted the results in terms of closed-form analytical expressions, the present discussion adopts the approach discussed earlier for lateral load capacity of caissons in clay. Namely, P_u is used to generate an expression for energy dissipation for side resistance, from which a collapse load can be computed by equating it to the external work done by the anchor line. This approach also requires an internal energy dissipation term for the pile tip, which can be significant for short piles. For this purpose, the present analysis adopts the following expression for drained soil shearing resistance at the pile tip:

$$s_d = (\chi W_{net}/A_{tip} + \gamma_b L_f)\tan\phi \tag{7.21}$$

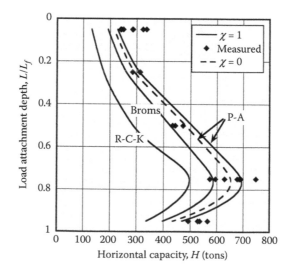

Figure 7.23 Caisson horizontal capacity in sand.

The first term in the parentheses is the contribution of caisson weight to the average normal stress at the pile tip. The net weight is the buoyant weight of the pile minus the buoyant weight of the soil displaced by the pile. The factor χ can range from 0 to 1 and denotes the portion of the caisson weight transmitted to the soil at the interior base of the pile. The present analysis uses this strength in the same manner as the undrained strength s_u was used in Equation 7.6 for computing tip energy dissipation in a cohesive soil. There are potential problems with this simplification, but it is reasonable in cases where pile rotation is small; that is, for load application near the optimal location. Comparison of prediction to measurement in Figure 7.23 indicates that the Reese, Cox, and Koop (R–C–K) equations tend to underestimate horizontal load capacity. This result is not entirely unexpected, since the R–C–K equations were developed through calibration to long, slender piles, which differ considerably from the short piles considered here. The Broms and Petrasovitz–Award (P–A) equations provide reasonably accurate predictions of capacity, particularly for load application depths L_i close to the optimal attachment depth. Including the caisson weight into the calculation ($\chi = 1$) increases predicted capacity by some 5%–10%. Discrepancies between prediction and measurement are greatest at $L_i/L_f = 0.05$ and 0.95. Pile rotation is greater at these locations, so the inaccuracy may be associated with the tip-energy dissipation calculation.

7.7 TORSION

Torsion on piles and caissons arises largely in two connections. The first relates to installation misalignment where the pad eye lies outside of the plane of the anchor line or chain. The design basis for pile anchors will typically consider a possible twist of 5–10° during installation. A second source of torsional loading can arise during partial failure of a mooring system where, owing to failure of a mooring line, the floating unit drifts off station, inducing out-of-plane loading on the remaining intact anchors. In this case, the out-of-plane load angle can be quite severe, with back-analyzed cases of damaged mooring systems by Ward et al. (2008) indicating out-of-plane load angle of 90° in some cases. In most cases, the torsional loading is unintended; so often, the main concern is that the torsional load demand on the anchor will impair its axial–lateral load capacity. In the case of extreme twist angles, a torsional failure may be self-correcting in the sense that the pile will simply spin into the

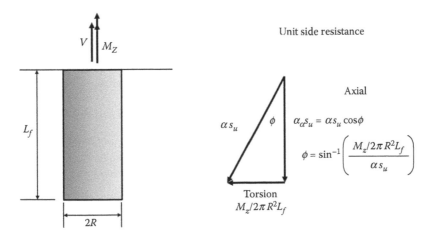

Figure 7.24 Torsion effect on pile capacity.

direction of the applied load. In this case, a pile anchor will still function, but at a reduced capacity owing to the reduction in adhesion (presumed $\alpha = 1/S_t$) accompanying the torsion failure.

In the case of large out-of-plane angles, a full finite-element analysis would normally be required for evaluating the complex combined load interactions. Taibat and Carter (2005) give an example of this type of analysis for a caisson with aspect ratio $L_f/D = 2$. However, for the more modest twist angles associated with installation misalignment, a simplified approximate analysis is often adequate.

Figure 7.24 shows the effect of torsional loading on a vertically loaded pile, neglecting any consideration of tip resistance. In this case, the side resistance αs_u will act at an angle ϕ from horizontal, owing to the horizontal component of shearing resistance arising from torsional loading. The torsion demand therefore reduces the skin resistance available for resisting vertical load by a factor $\cos\phi$. While this analysis is strictly valid only for vertical–torsion interaction, it is commonly extended to general loading by replacing the adhesion factor α used in axial–lateral load capacity calculations by α_a to account for the torsion demand. Often of interest is the effect of the out-of-plane load angle, $\theta_{op} = \tan^{-1}(M_z/RH)$, where R = caisson radius, H = horizontal load, and M_z = torque resulting from out-of-plane loading. An example calculation for a caisson with aspect ratio of 5 shows that an out-of-plane angles $\theta_{op} = 5$–$10°$ leads to approximately 0.5%–2.5% reductions in capacity, respectively.

REFERENCES

Andersen KH and Jostad HP, 2002, Shear strength along outside wall of suction anchors in clay after Iinstallation, *Proceedings of the Twelfth (2002) ISOPE Conference*, Kyushu, Japan, pp 785–794.

Andersen KH, Murff JD, Randolph MF, Clukey E, Jostad H, Hansen B, Aubeny C, Sharma P, Erbich C, and Supachawarote C, 2005, Suction anchors for deepwater applications, *Keynote lecture, International Symposium on Frontiers in Offshore Geotechnics*, Perth, Australia, pp 3–30.

API RP 2A WSD, 2000, *Recommended Practice for Designing and Constructing Fixed Offshore Platforms*, American Petroleum Institute, API Publishing Services, Washington, DC, 226p.

Aubeny CP, Han SW, and Murff, JD, 2003a, Suction caisson capacity in anisotropic soil, *Int J Geomech*, 3(4), 225–235.

Aubeny CP, Han SW, and Murff JD, 2003b, Inclined load capacity of suction caisson anchors, *Int J Numer Anal Meth Geomech*, 27, 1235–1254.

Aubeny CP, Moon SK, and Murff JD, 2001, Lateral undrained resistance of suction caisson anchors, *Int J Offshore Polar Eng*, 11(3), 211–219.

Baldi G, Bellotti R, Ghionna V, Jamiolkowski M, and Lo Presti DFC, 1989, Modulus of sands from CPTs and DMTs, *12th International Conference on Soil Mechanics and Foundation Engineering*, Rio de Janeiro, pp 165–170.

Bang S, Jones KD, Kim KO, Kim YS, and Cho Y, 2011, Inclined loading capacity of suction piles in sand, *J Ocean Eng*, 38(7), 915–924.

Broms BB, 1964, Lateral resistance of piles in cohesionless soils, *J Soil Mech Found Div ASCE*, 90(3), 123–158.

Chen W and Randolph MF, 2005, Centrifuge tests on axial capacity of suction caissons in clay, *International Symposium on Frontiers in Offshore Geotechnics*, Perth, Australia, pp 243–250.

Chow FC, Jardine FM, Nauroy J-F, and Brucy F, 1997, Time-related increases in the shaft capacities of driven piles in sand, *Geotechnique*, 47(2), 353–361.

Clausen CJF, Aas PM, and Karlsrud K, 2005, Bearing capacity of driven piles in sand, the NGI approach, *International Symposium on Frontiers in Offshore Geotechnics*, Perth, Australia, pp 677–682.

Clukey EC, Aubeny CP, and Murff JD, 2004, Comparison of analytical and centrifuge model tests for suction caissons subjected to combined loads, *ASME J Offshore Mech Arctic Eng*, 126(4), 364–367.

Coffman RA, El-Sherbiny RM, Rauch AF, and Olsen RE, 2004, Measured horizontal capacity of suction caissons, *Offshore Technology Conference*, OTC 16161, Houston, pp 1–10 (electronic format).

Fugro, 2004, Axial pile capacity design method for offshore driven piles in sand, P-1003, Issue 3, to API, August.

Jeanjean P, 2002, Innovative design method for deepwater surface casings (SPE 77357), *Proceedings of the SPE Annual Technical Conference and Exhibition*, San Antonio, TX, USA, Society of Petroleum Engineers, pp 1–14 (electronic format).

Jeanjean P, 2006, Setup characteristics of suction anchors for soft Gulf of Mexico clays: Experience from field installation and retrieval, OTC 18005, *Offshore Technology Conference*, Houston, pp 1–9 (electronic format).

Jeanjean P, Znidarcic D, Phillips R, Ko H-Y, Pfister F, Cinicioglu O, and Schroeder K, 2006, Centrifuge testing on suction anchors: Double-wall, overconsolidated clay, and layered soil profile, OTC 18007, *Offshore Technology Conference*, Houston.

Kay S and Palix E, 2011, Caisson capacity in clay: VHM resistance envelope—Part 2: VHM envelope equation and design procedures, *Frontiers in Offshore Geotechnics II*, Gouvernec S and White D, eds., Taylor & Francis Group, London, pp 741–746.

Kolk HJ, Baaijens AE, and Senders M, 2005, Design criteria for pipe piles in silica sands, *International Symposium on Frontiers in Offshore Geotechnics*, Perth, Western Australia, pp 711–716.

Ladd CC, 1991, Stability evaluation during staged construction, *ASCE J Geotech Eng*, 117(4), 537–615.

Lehane BM and Jardine RJ, 1994, Displacement–pile behavior in a soft marine clay, *Can Geotech J*, 31(2), 181–191.

Lehane BM, Schneider JA, and Xu XT, 2005a, *CPT Based Design of Driven Piles in Sand for Offshore Structures, GEO: 05345*, University of Western Australia, Perth.

Lehane BM, Schneider JA, and Xu XT, 2005b, *Evaluation of Design Methods for Displacement Piles in Sand, GEO: 05341.1*, University of Western Australia, Perth.

Lehane BM, Schneider JA, and Xu XT, 2005c, The UWA-05 method for prediction of axial capacity of driven piles in sand, *International Symposium on Frontiers in Offshore Geotechnics*, Perth, Western Australia, pp 683–690.

Monzon JC, 2006, Review of CPT based design methods for estimating axial capacity of driven piles in siliceous sand, ME Thesis, Massachusetts Institute of Technology, Cambridge, 83p.

Murff JD and Hamilton JM, 1993, P-Ultimate for undrained analysis of laterally loaded piles, *ASCE J Geotech Eng*, 119(1), 91–107.

Olsen RE and Dennis ND, 1982, *Review and Compilation of Pile test Results, Axial Pile Capacity*, PRAC Project 81-29, American Petroleum Institute, Dallas, Texas.

Palix E, Willems T, and Kay S, 2011, Caisson capacity in clay: VHM resistance envelope—Part 1: 3D FEM numerical study, *Frontiers in Offshore Geotechnics II*, Gourvenec and White, (eds.), Taylor & Francis Group, London, pp 753–758.

Petrasovitz G and Award A, 1972, Ultimate lateral resistance of a rigid pile in cohesionless soil, *Proceedings of the 5th European Conference on SMFE 3*, Madrid, pp 407–412.

Randolph MF, 2003, Science and empiricism in pile foundation design, *Geotechnique* 53(10), 847–875.

Randolph MF and Murphy BS, 1985, Shaft capacity of driven piles in clay, *Proceedings of the 17th Offshore Technology Conference*, OTC4883, Houston, pp 371–378.

Reese LC, Cox WR, and Koop RD, 1974, Analysis of laterally loaded piles in sand, *Proceedings of the 6th Offshore Technical Conference*, Houston, pp 473–483.

Schneider JA, Xu X, and Lehane BM, 2008, Database assessment of CPT-based design methods on axial capacity of driven piles in siliceous sands, *ASCE J Geotech Eng*, 134(9), 1227–1244.

Schneider JA, White DJ, and Lehane BM, 2007, Shaft friction of piles driven in siliceous, calcareous, and micaceous sands, *Proceedings of the 6th International Conference on Site Investigation and Geotechnics*, Society for Underwater Technology, SUT, London, pp 367–382.

Taibat HA and Carter JP, 2005, A failure surface for caisson foundations in undrained soils, *Frontiers in Offshore Geotechnics*, Gouvernec S and Cassidy M, eds., Taylor & Francis Group, London, pp 289–296.

Ward EG, Zhang J, Kim MH, Aubeny C, and Gilbert R, 2008, *No MODU's Adrift*, Offshore Technology Research Center, Report C188, College Station.

Whiteside WF, 1995, The behavior of laterally loaded single piles, MS Thesis Department of Civil Engineering, University of California, Berkeley.

Whittle AJ, 1992, Assessment of an effective stress analysis for predicting the performance of driven piles in clays, *Proceedings of the Conference on Offshore Site Investigation and Foundation Behaviour*, London, Vol. 28, pp 607–643.

Zakeri A, Liedke E, Clukey EC, and Jeanjean P, 2014, Long-term axial capacity of deepwater jetted piles, *Geotechnique*, 64(12), ICE Publishing, Thomas Telford, Ltd., London, pp 966–980.

Chapter 8

Elastic effects and soil–pile interaction

Elastic response of anchors can affect three aspects of pile and caisson anchor design. First, anchor displacement is often a criterion for performance. Owing to high degree of compliance of most mooring systems relative to that of anchors, displacement of anchors normally does not take on a degree of significance comparable to that provided to displacements of foundations for fixed structures. Nevertheless, the design basis for anchors often places limits on displacements, which typically requires recourse to an analysis of the elastic response of the anchor–soil system under applied loads. Second, even under circumstances where high displacements are tolerable, elastic response of the pile–soil system raises issues of strain compatibility that can effectively limit the pile/caisson capacity in situations where strain softening occurs, which is typically the case for axial side resistance. Third, the structural analysis of piles and caissons require realistic predictions of the distribution of stresses acting on the anchor. Collapse loads from PLAs or limit equilibrium analyses are often not adequate for this purpose, so elastic analysis methods should be employed.

Anchor design tends to be dominated by considerations for ultimate load capacity, with considerations for displacements and other elastic behavior imposing a reduction in available load capacity. The major exception to this observation is the case of laterally loaded flexible piles, where the displacements required to mobilize the theoretical collapse load are well beyond any reasonable tolerance and will typically induce a structural failure of the pile. Thus, the design of laterally loaded piles is often largely or completely based on the elastic analyses discussed in this chapter. In this connection, it is noted that meaningful PLAs are still possible for slender piles, provided that plastic hinges are introduced in the pile (Murff and Hamilton, 1993; Chen et al., 2017) in recognition of the fact that structural failure occurs prior to full mobilization of plastic resistance in the soil.

This chapter discusses the three facets of the elastic response of piles and caissons in Figure 8.1. The first is axial deformation. For anchor applications, the axial load is often in tension, but compressive axial loads can still occur if the pad eye is located at a depth other than the mudline. Second is the flexural bending occurring in laterally loaded slender piles, which, if combined with axial loading, induces the beam-column response characteristic of structural columns. Laterally loaded caissons and short piles tend to rotate as rigid bodies in a vertical plane, such that flexural deformation of the caisson itself is negligible. However, this should not be taken to imply that elastic deformations in caissons are negligible. Noting that caissons have relatively thin walls, with diameter-to-thickness ratios sometimes as high as $D/t = 160$, the deformation of caissons in a horizontal plane, "ovalization," can significantly impact displacement predictions.

Analysis of elastic effects requires a framework for describing the elastic behavior of the continuum in which the pile is embedded. The current prevailing practice does this by characterizing the soil as a series of soil springs, "Winkler springs" supporting the pile. Although

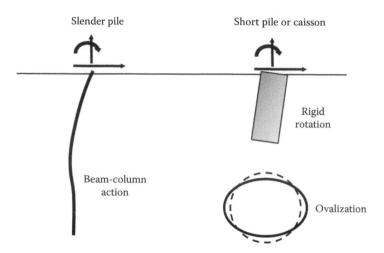

Figure 8.1 Elastic response of piles and caissons.

some analyses utilize continuum solutions to develop the springs, the current state of practice largely relies on empirically based methods derived from field or laboratory model tests. This chapter first discusses the development of axial and lateral soil springs required for supporting a general beam-column analysis for slender piles. Then, finite difference formulations for the analysis of soil–pile interaction are presented. Finally, a framework for analysis of ovalization of caissons and short piles is presented.

8.1 AXIAL LOAD TRANSFER

The elastic response of piles to axial loading is typically analyzed through separate treatment of side and tip resistance. With z denoting the vertical displacement of a point on the pile, t–z springs describe the mobilized side friction t as a function of z, and Q–z springs describe the mobilized tip resistance Q versus z.

8.1.1 Side resistance

Randolph and Wroth (1978) developed a theoretical t–z relationship for linearly elastic soil assuming that shear stress in a vertical plane τ_{rz} varies only with radial coordinate r. In the following discussion, invoking equilibrium, strain compatibility, and Hooke's law (Figure 8.2) produces a logarithmic variation in vertical displacement z. The relative displacement between a point on the pile and the one located at an infinite distance from the pile is therefore unbounded. An infinite displacement is a clearly unreasonable consequence of the assumption that shear stress decays only in the radial direction. To overcome this difficulty and account for the fact that shear stress decays with vertical coordinate also, Randolph and Wroth introduce a "magical radius" r_m at which shear stress is assumed to decay to a negligible value. The value of r_m is established by calibrations to finite element studies and varies with depth. Randolph and Wroth show that for a pile with slenderness ratio $L_f/r_0 = 40$ the average magical radius can be taken as $r_m = 2.5\,L_f\,(1 - \mu)$. Thus, vertical displacement z can be related to shear stress at the pile shaft τ_0 by the following logarithmic relationship shown in dimensional and dimensionless format (in terms of rigidity index I_r):

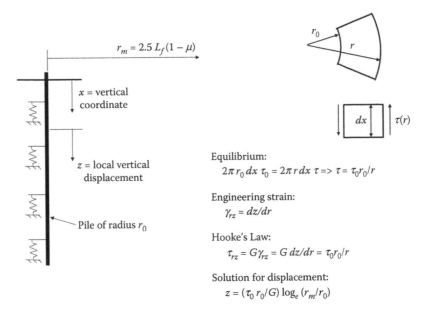

Figure 8.2 Theoretical development of axial side spring relationship.

$$z = \frac{\tau_0 r_0}{G} \log_e \left(\frac{r_m}{r_0} \right)$$

(8.1)

$$z/D = \frac{\alpha \tau_0 / \tau_{max}}{2 I_r} \log_e \left(\frac{r_m}{r_0} \right)$$

Soil nonlinearity may be introduced into the analysis, for example in the form of a second-order stress–strain law, $G = G_i [1 - (R_f \tau / \tau_f)]$, where G_i is the small-strain shear modulus, and τ_f is the maximum shear stress on the pile shaft (Kraft et al., 1981). In this case, the t–z model becomes as follows:

$$z = \frac{\tau_0 r_0}{G_i} \log_e \left[\frac{(r_m/r_0) - \psi}{1 - \psi} \right] \qquad \psi = R_f \tau_0 / \tau_{max}$$

(8.2)

An alternative empirical approach for establishing the axial side resistance versus displacement relationships involves back-calculations from pile load test measurements. Load–displacement measurements at the pile head together with a known axial pile stiffness and an assumed variation in soil resistance versus depth permits approximate back-calculation of t–z relationships. More rigorous back-calculation of t–z curves is possible with instrumented pile load tests. In this case, measured axial strains ε are integrated for obtaining a distribution of vertical displacement $z(x)$ along the pile shaft, while mobilized skin friction is calculated from measured changes in axial force ($\Delta Q = AE\Delta\varepsilon$ between any two strain gauges) in the pile. Measurement of the total settlement at the pile head is necessary to accomplish the integration to compute $z(x)$. The mobilized resistance t along any interval is computed as $t = \Delta Q/A_s$, where A_s is the surface area of the interval under consideration. Mobilized side friction is frequently normalized by peak side resistance t_{max}. The application

Table 8.1 API recommended *t–z* curves

Clays		Sands	
Deflection, z/D	Resistance, t/t_{max}	Deflection, z (m)	Resistance, t/t_{max}
0.0016	0.3	0	0
0.0031	0.5	0.0025	I
0.0057	0.75	∞	I
0.0080	0.9		
0.01	I		
0.02	0.7–0.9		
∞	0.7–0.9		

of this approach is described by Coyle and Reese (1966) for clays and Coyle and Sulaiman (1967) for sands. In the absence of site-specific criteria, API (1993) recommends the *t–z* relationships for noncarbonate soils shown in Table 8.1.

Figure 8.3 compares *t–z* curves computed by various methods for a pile embedded using rigidity indices $I_r = G/s_u = 100–300$ and $\alpha = 1$; that is, conditions roughly representative of normally consolidated clay. Up to the point of maximum mobilized side resistance, $t = t_{max}$, the linear model is seen to bracket the API curve for the range I_r under consideration. The nonlinear model with $I_r = 300$ is fairly consistent with the API curve up to the point at which peak resistance occurs. It is emphasized that this comparison is simply a snapshot for a single case, so the reader should not generalize these observations as being representative of the performance of the three models for all conditions.

8.1.2 Tip resistance

Randolph and Wroth (1978) provide the following relationship between pile tip force Q and tip displacement z in a linearly elastic medium as follows:

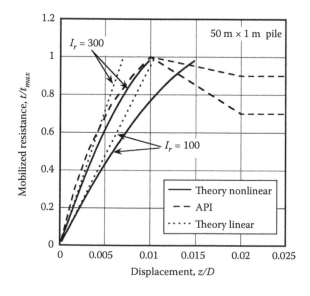

Figure 8.3 Comparisons of *t–z* curves.

Table 8.2 API recommended tip spring

Deflection, z/D	Resistance, Q/Q_p
0.002	0.25
0.013	0.5
0.042	0.75
0.073	0.9
0.1	1
∞	1

$$z = \frac{Q(1-\mu)}{4r_0 G}\eta \tag{8.3}$$

The factor η accounts for the depth of the pile tip below the surface. For a detailed discussion of the selection of η, the reader is referred to Randolph and Wroth (1978). However, they generally conclude that for a straight pile η is between 0.85 and unity. When considering incremental loads after side friction has been fully mobilized, they conclude that η is approximately 0.85 for a straight shaft. If a nonlinear shear modulus is adopted in a manner similar to that described above for side resistance (Chow, 1986), the pile tip Q–z takes the following form:

$$z = \frac{Q(1-\mu)}{4r_0 G_i[1-(R_f Q/Q_p)]^2} \tag{8.4}$$

where Q_p is the ultimate tip resistance, R_f is the hyperbolic fitting parameter, and G_i is the small-strain shear modulus. Finally, API recommends the Q–z relationship shown in Table 8.2.

Figure 8.4 compares the model predictions for axial tip resistance, assuming a circular pile with tip bearing factor $N_c = 9$ and $R_f = 1$ for rigidity indices $I_r = 100$ and 300. In general, the theoretical predictions using $I_r = 300$ are in better agreement with the API

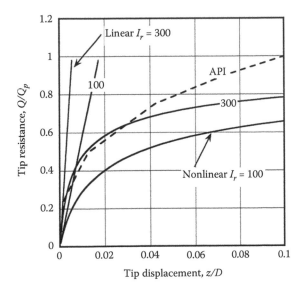

Figure 8.4 Comparisons of Q–z curves.

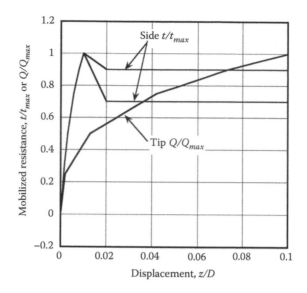

Figure 8.5 Comparisons of side to tip resistance.

recommendation, although at larger displacements, $z/D > 0.03$, even the nonlinear curve deviates significantly from the API guideline.

Figure 8.5 compares API recommended curves for describing mobilized soil resistance on the sides and at the pile tip. Since side resistance peaks at much smaller displacements than tip resistance, the ultimate axial load capacity of the pile cannot be considered as a simple sum of the respective ultimate side and tip resistance values.

8.2 LATERAL SOIL RESISTANCE

Rigorous analysis of a laterally loaded pile treats the soil in which the pile is embedded as a continuum for which a variety of solution methods are possible, such as boundary integrals and finite elements. While this approach has considerable potential as a research tool, it requires a fairly high degree of sophistication in regard to the soil constitutive model used in the analysis, as well as an analytical framework that can account for the effects of pile installation disturbance. Thus, an alternative approach involving empirically derived Winkler springs is widely adopted for design of piles. By this approach (Figure 8.6) a line

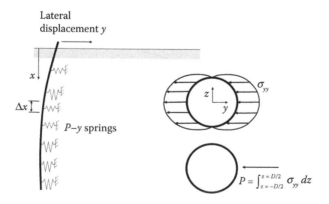

Figure 8.6 Lateral Winkler springs.

load having intensity P is the integrated resultant of the reaction stress σ_{yy} acting around the circumference of the pile on an elemental length of pile Δx. Implicit in this analysis is an assumption that in-plane deformation of pile (ovalization) is negligible. When flexure is in fact the dominant mode of deformation, the distributed stresses around the circumference can appropriately be considered as a statically equivalent line load. In conventional analyses of this type, the spring connects the pile to a "ground," that is, to a point outside the pile experiencing zero displacement. Thus, the springs along the pile are completely uncoupled. If one chooses to model interaction effects, a number of finite element programs have capabilities for more sophisticated user elements, where the springs can be linked to other degrees of freedom along the pile. The discussion throughout the remainder of this chapter pertains to the simplest case of springs linked to a ground.

8.2.1 Experimental measurement

Following Matlock (1970), P–y curves can be back-calculated from measured bending moment data collected from instrumented pile load tests (Figure 8.7). The starting point of the analysis is to recognize that, for small deflections y, the bending moment M in a pile having modulus E and moment of inertial I relates to curvature as follows:

$$\frac{d^2 y}{dx^2} = \frac{M}{EI} \tag{8.5}$$

The same database of moment (curvature) measurements are used to compute both y and P. Displacement y is computed by double integration of the curvature:

$$\frac{dy}{dx} = \int \frac{M}{EI} dx + C_1$$
$$y = \int \frac{dy}{dx} dx + C_1 x + C_2 \tag{8.6}$$

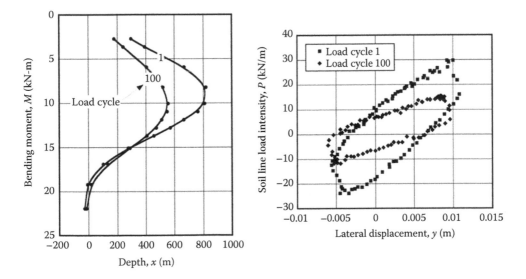

Figure 8.7 Interpretation of laterally loaded pile test data.

Evaluating the integrals requires two constants of integration, which must be provided by auxiliary measurements. For example, an inclinometer measurement of slope $(dy/dx)_0$ at the pile head provides the information required for evaluation of C_1, and a displacement measurement y_0 at the pile head permits evaluation of C_2. Numerical integration of measured data is relatively insensitive to noise, so displacements computed by this approach are generally accurate.

Successive differentiation of bending moment M produces profiles of shear V and line load intensity P from beam theory:

$$V = \frac{dM}{dx}$$
$$P = \frac{d^2M}{dx^2}$$

(8.7)

In contrast to integration, numerical differentiation of discrete data measurements is typically highly sensitive to noise, particularly when dealing with higher order derivatives; therefore, data smoothing of some type is typically necessary. A number of least squares and cubic spline procedures are available, with no particular means of judging in advance which procedure performs best. For this reason, curve fitting using several procedures is advisable to assure a reasonable degree of agreement among the methods. The procedure outlined below involves a simple local least squares curve fit based on measured data points in the vicinity of the point of interest. Since second derivatives are required, the fit must be at least a second-order polynomial. Higher order polynomials may also be tried, but they will not necessarily lead to improved local curve fit. If a second-order polynomial with fitting coefficients c_i is selected, the local curve fit takes the following form:

$$M = c_1 x^2 + c_2 x + c_3$$

(8.8)

We can now construct matrices comprising the polynomial terms of the measurement depths x_i, measured moments M_i, and weighting coefficients w_i:

$$[X] = \begin{bmatrix} x_1^2 & x_1 & 1 \\ x_2^2 & x_2 & 1 \\ . & . & . \\ x_n^2 & x_n & 1 \end{bmatrix} \quad [M] = \begin{bmatrix} M_1 \\ M_2 \\ . \\ M_n \end{bmatrix} \quad [W] = \begin{bmatrix} w_1 & & & \\ & w_2 & & \\ & & . & \\ & & & w_n \end{bmatrix}$$

(8.9)

The matrix $[W]$ simply provides greater weight to data points in the vicinity of the local curve fit. It can be a simple step function for which $w_i = 1$ within the window of the curve fit and $w_i = 0$ outside the window. Selection of the size of the window depends on data quality and the spacing of strain gauge locations; therefore, different forms of $[W]$ may be tried to ascertain its influence on the quality of the final results. Finally, the system of equations is "squared" by premultiplying all terms by $[X]^T[W]^T$ as shown below:

$$[X]^T [W]^T [W][X][C] = [X]^T [W]^T [M] \quad [C] = \begin{bmatrix} c_1 \\ c_2 \\ c_3 \end{bmatrix}$$

(8.10)

The fitting coefficients c_i are determined by solution of the resulting 3×3 system of equations. The soil resistance line load intensity at the point under consideration is then evaluated from the second derivative of the curve fit, $P = 2c_1$. As a localized curve-fitting procedure requires no assumption as to the form of the global moment distribution, it is fairly robust. Nevertheless, curve fitting is more of an art than a science, so, as noted above, trial of more than one curve-fitting procedure is advisable. Jeanjean (2009) describes various alternative curve-fitting procedures for interpreting pile lateral load test measurements.

Figure 8.7 illustrates the data interpretation for a centrifuge test of a laterally loaded pile in normally consolidated clay. The left side of the figure shows moment profiles corresponding to peak positive applied lateral displacement. For the case of cyclic loading, similar profiles are produced for each measured time step of loading. Resistance degrades with cycling, as indicated by the comparison of the bending moment profiles for the first and 100th load cycles. Computation of y and P from the bending moment profiles at each time step of a load cycle permits construction of the P–y loops shown in the right side of the figure. The P–y loops show peak displacements y to be slightly out of phase with peak soil resistance P, which could be owing to a viscous component of soil resistance. However, it is difficult to draw definitive conclusions given the extensive numerical processing of the measured data.

The methodology outlined above requires a pile that is sufficiently flexible to permit measurable bending moments. This condition does not occur in the case of short piles. In this case, some alternative methods are required for deducing the P–y behavior from experimental data. Gilbert (2016) describes an experimental approach where contact stress is measured around the circumference of the pile. At a given pile displacement y at the depth x in question, the stresses can be integrated (Figure 8.6) to obtain a resultant P, from which a P–y curve is deduced.

8.2.2 Application to design

Figure 8.8 shows a typical P–y history at a selected depth on the pile. Only selected loading cycles are included for clarity. The following three features of this history are relevant to pile design:

1. The P–y relationship during the initial lateral push of the pile into the soil, the "backbone" curve, characterizes the soil response under monotonic loading conditions. Use of the backbone curve is appropriate for analyses directed toward predicting the ultimate load capacity of the pile. It is noted that cycling at low to intermediate load levels has negligible influence on the ultimate mobilized soil resistance P_{ult}. Thus, in situations where monotonic loading to failure is preceded by episodes of cyclic loading, the use of the backbone curve in the ultimate pile load capacity evaluation is still appropriate (Gilbert, 2016).
2. Fatigue analysis of piles involves large numbers (10^4–10^5) of load cycles at relatively small displacement magnitudes. As is apparent from Figure 8.7, degradation of soil stiffness to a steady-state P–y loop occurs after some 50–100 cycles. Thus, the transient phase of degradation in soil resistance from the backbone curve to a steady-state condition has negligible effect on fatigue life assessments, and the steady-state P–y loop adequately describes the soil response. The use of equivalent secant stiffness (Jeanjean, 2009) provides an effective method for characterizing soil response for this type of study.
3. Loading of short duration and high intensity, as may occur during earthquake or impact loading, involve cyclic loading of the pile from the backbone curve to various

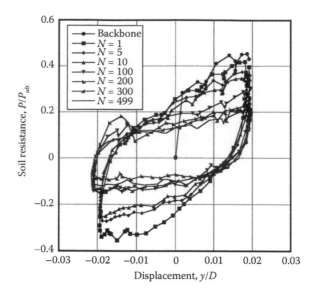

Figure 8.8 Example P–y data from cyclic load test.

stages of degraded stiffness. Additionally, the loading involves variable amplitudes of loading as the pile motions decay over time. Soil response for this type of loading is not well described by either the backbone curve or a steady-state secant stiffness, which requires recourse to a true nonlinear description of soil P–y behavior.

8.3 P–Y CURVES FOR MONOTONIC LOADING

A number of simplified descriptions of P–y behavior have been developed to make the analysis amenable to routine design calculations. The methods are generally classified according to the soil profiles for which they are considered applicable. Some significant contributions on the topic include seminal contributions on the topic that includes the work of Matlock (1970) for soft clays; Reese et al. (1975), Gazioglu and O'Neill (1984), and Dunnavant and O'Neill (1989) for stiff clays; and Reese et al. (1974), Parker and Reese (1979), and Murchison and O'Neill (1984) for sands. Experimental determination of lateral resistance is a widely researched topic, with a number of studies focused toward specific methods of pile installation and soil condition. The author is aware of studies that are currently in progress that are likely to be of relevance to anchor application. Full coverage of this topic could occupy a text of its own, and this is beyond the scope of this book. The following discussion is limited to a presentation of widely used methods for driven piles in soft clays, stiff clays, and sands.

8.3.1 Soft clay

Based on lateral load tests on a 0.328-m (12.75 in.) driven pipe pile in 12.8 m of clay, Matlock (1970) developed the model shown in Table 8.3 for describing P–y behavior in soft clays. The model first describes ultimate resistance P_{ult} as shown in the table. Since a soil unit weight term occurs in the expression for P_{ult}, gapping near the free surface is implicitly assumed. The limiting bearing factor is $N_p = 9$, which approximates the theoretical

Table 8.3 P–y curves for laterally loaded piles in soft clay

Model Component	Equation
Ultimate resistance	$P_{ult} = \left[3 + \dfrac{\gamma_b}{s_u}x + \dfrac{J}{D}x\right]s_u D \leq 9s_u D$ $J = \begin{cases} 0.5 & \text{Soft clay} \\ 0.25 & \text{Medium clay} \end{cases}$
Monotonic loading	$P = 0.5P_{ult}(y/y_{50})^{1/3} \quad y/y_{50} < 8$ $P = P_{ult} \qquad\qquad\quad y/y_{50} \geq 8$ $y_{50} = 2.5\varepsilon_{50}D$ $\varepsilon_{50} = \begin{cases} \text{Soft-0.020} \\ \text{Medium-0.010} \\ \text{Stiff-0.005} \end{cases}$
Cyclic loading	$P = 0.5P_{ult}(y/y_{50})^{1/3}$ $\qquad\qquad\qquad y/y_{50} < 3$ $P = 0.72P_{ult}$ $\qquad\qquad\qquad\qquad y/y_{50} \geq 3 \text{ and } x \geq x_r$ $P = 0.72P_{ult}\left[1 - (1 - x/x_r)\dfrac{y/y_{50} - 3}{12}\right] \quad 3 \leq y/y_{50} \leq 15 \text{ and } x < x_r$ $P = 0.72P_{ult}(x/x_r)$ $\qquad\qquad\qquad y/y_{50} \geq 15 \text{ and } x < x_r$ $x_r = \dfrac{6s_u D}{\gamma' D + Js_u}$

Source: Matlock H, 1970, *Proceedings of the 2nd Annual Offshore Technology Conference*, Houston, TX, Vol. 1, pp 577–594.

flow-around solution for a smooth pile. The parameter J controls the depth from the free surface at which a full flow-around condition develops. The plot on the left side of Figure 8.9 compares Matlock's equation to the simplified Murff–Hamilton N_p profile discussed in earlier chapters. The Murff–Hamilton analysis was conducted with and without suction but, as is apparent from the figure, the differences are relatively minor. To facilitate a comparison,

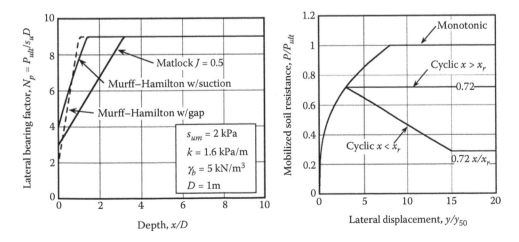

Figure 8.9 P–y curve for soft clay.

a smooth pile condition ($N_1 = 9$) was used in the Murff–Hamilton analysis. This particular comparison shows the Matlock equation to be somewhat more conservative than Murff–Hamilton from the standpoint of the depth at which a flow-around condition develops, with the Murff–Hamilton flow-around depth being about one-third of that predicted from the Matlock equation. The estimation of P_{ult} using the Matlock model appears to have two elements of conservatism: the limiting bearing resistance corresponds to a perfectly smooth pile, and the depth at which a flow-around condition develops exceeds that predicted from a PLA. The monotonic loading component of the model comprises a power law relationship between P and y. Displacement y is normalized by y_{50}, which is related to the strain level ε_{50} in a triaxial compression test at which 50% of full resistance is mobilized.

For cyclic loading, the Matlock model degrades the soil resistance according to three ranges of lateral displacement. At small displacements, $y/y_{50} < 3$, there is no reduction in resistance. At large displacements, $y/y_{50} > 15$, soil resistance is reduced to its fully degraded level. In the flow-around region at depths x greater than x_r (Table 8.3) resistance is capped at $0.72\,P_{ult}$. At depths shallower than x_r, the soil resistance is further degraded directly proportional to depth such that zero resistance mobilizes at the mudline, $0.72\,P_{ult}\,x_r/x_r$. At intermediate levels of displacement, soil resistance is computed by linear interpolation of P computed at $y/y_{50} = 3$ and $y/y_{50} = 15$. The description of cyclic degradation of soil resistance in the Matlock model assumes that a large number of load cycles has occurred such that a steady-state condition exists.

As a practical matter of implementation of the model, it should be noted that a power law relationship predicts an infinite secant stiffness as deflection y approaches zero, which can cause numerical difficulties for the soil–pile interaction analyses discussed subsequently in this chapter. Use of the API (1993) tabular form of this model can avoid this problem.

Although the Matlock model has seen extensive usage, subsequent investigations observed that it tended to underestimate soil stiffness. Stevens and Audibert (1979) postulated that the low soil stiffness could be attributed to the relatively small diameter test pile ($D_M = 0.328$ m) used in the development of the Matlock model. To mitigate this problem they proposed the following square root scaling law for pile diameters larger than those used in the Matlock study:

$$y_{50} = 2.5\varepsilon_{50}(D/D_M)^{0.5}D_M \tag{8.11}$$

The Stevens–Audibert adjustment certainly moves toward a more realistic characterization of lateral soil resistance in soft clays. However, re-evaluation of the original data on which Matlock based his model (Gilbert, 2016) indicates that the model underestimates soil stiffness irrespective of pile diameter.

Disparities in prediction of P–y curves by various approaches raises the question as to which predictions can be considered conservative. For predicting displacements, softer soil springs are unambiguously more conservative. However, the issue is more nuanced when considering bending moments. First, the degree of conservatism will depend on whether one is considering a point on the pile within the soil column or in a portion of the pile projecting above the seabed. High estimates of soil stiffness tend to underestimate curvature of the pile in the soil column; thus, computed bending moments will be too low in this region. However, overestimating soil stiffness can also lead to overestimates of bending moments in the portion of the pile projecting above the seabed. One can convince oneself of this by considering the case of a pile embedded in a rigid soil. Zero curvature implies no bending moments in the embedded portion of the pile. However, without the cushioning effect of a compliant soil, a severe bending moment occurs in the pile just above the seabed.

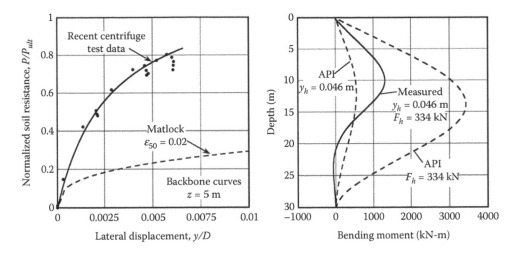

Figure 8.10 Effect of P–y spring stiffness on pile response.

Second, one should keep in mind whether a boundary condition of force control versus displacement control exists at the pile head. The left-hand plot in Figure 8.10 compares the P–y backbone curve from a centrifuge test in a normally to lightly over-consolidated kaolin to the soft clay API curve using the common assumption of $\varepsilon_{50} = 0.02$ in the y_{50} calculation. Consistent with a number of observations, the measured curve is significantly stiffer than the API curve. The right-hand plot compares the measured bending moment profile to that computed from the measured API curve. The test in question had an imposed displacement $y_h = 0.046$ m at the pile head, which also corresponded to a pile head force of 334 kN. When the analysis is conducted using the API curves imposing a $F_h = 334$-kN force at the pile head, the predicted bending moment profile considerably exceeds the measured profile; hence, the API curves give a conservative prediction. By contrast, when the equivalent pile head displacement $y_h = 0.046$ is imposed at the pile head, use of the softer API springs produce a clearly unconservative bending moment profile. Many applications exist, which can be approximated by force-controlled loading; for example, the pile foundation supports for steel jacket structures. However, when the pile is coupled to a more complex system, a simple assumption of force control may not be warranted. For example, the analysis of a pile embedded in a soil mass subjected to seismic shaking is more closely approximated by displacement control. In such situations, the most prudent approach is to base the analysis on the best estimate of the true soil resistance and make decisions on the appropriate level of conservatism based on the best estimate of pile response.

8.3.2 Stiff clays

The method developed by Dunnavant and O'Neill (1989) for stiff clays was based on three pile load tests in a submerged over-consolidated clay profile with undrained shear strengths ranging from 50 to 150 kPa. All piles had a length of approximately 11.5 m. Two of the tests were conducted on open-ended driven piles with respective diameters 0.272 and 1.22 m. The third test was on a bored reinforced concrete pile with diameter 1.83 m. The model considers the effect of soil–pile interaction; thus, the ratio of pile flexural stiffness EI to soil stiffness E_s appears in the equation for y_{50}. Table 8.4 summarizes the relevant equations. The equation for P_{ult} has the same form as the Matlock equation for soft clays, but with different coefficients. The form of the equation for P_{ult} implicitly assumes a smooth pile with

Table 8.4 P–y curve for laterally loaded pile in stiff clay

Loading case	Equation
Ultimate resistance	$P_{ult} = \left(2 + \dfrac{\gamma_b x}{s_{u-avg}} + 0.4\dfrac{x}{D}\right) s_u D \leq 9 s_u D$
	$N_p = \left(2 + \dfrac{\gamma_b x}{s_{u-avg}} + 0.4\dfrac{x}{D}\right) \leq 9$
Monotonic loading	$P = 1.02 P_{ult} \tanh\left[0.537\left(y/y_{50}\right)^{0.70}\right] \leq P_{ult}$
	$y_{50} = 0.0063\,\varepsilon_{50}\,D\left(\dfrac{L^4}{EI/E_s}\right)^{0.875}$
	$L \leq L_c = 3D\left(\dfrac{EI/E_s}{D^4}\right)^{0.286}$
Cyclic loading	$\left(\dfrac{N_{cm}}{N_p}\right)_N = 1 - \left(0.45 - 0.18\dfrac{x}{x_0}\right)\log_{10} N \leq (1 - 0.12\log_{10} N)$
	$P_{cm} = N_{cm} s_u D \quad x_0 = 1\text{m}$
	$\left(\dfrac{P_r}{P_{cm}}\right)_N = 1 - \left(0.025 - 0.07\dfrac{x}{x_0}\right)\log_{10} N \leq 1$
	$y_{cm} = 2.43 y_{50}\left[\tanh^{-1}\left(\dfrac{0.98 P_{cm}}{P_{ult}}\right)\right]^{1.428}$

gapping in the active region behind the pile. The plot on the left side of Figure 8.11 illustrates the variation of P_{ult} versus depth. A PLA using the procedure of Murff and Hamilton (1993) for a smooth pile with gapping shows the depth to the flow-around zone predicted from the Dunnavant–O'Neill equations significantly greater than the PLA.

The method can account for stiffness degradation under cyclic loading, with degradation expressed as a logarithmic function of the number of load cycles. The number of load cycles

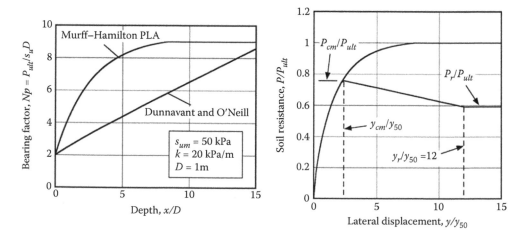

Figure 8.11 P–y curves in stiff clay.

Table 8.5 P–y curve in sand

Loading case	Equation
Ultimate resistance	$P_{ult}(x) = \gamma_b \times D \ (c_1 + c_2 \ x/D) < \gamma_b \times D \ c_3$ c_1, c_2, c_3 from Figure 8.12
Monotonic loading	$P = AP_{ult} \ \tanh\left[\dfrac{kx}{AP_u} y\right]$ $A = 3 - 0.8 \ x/D > 0.9 \ k$ from Figure 8.12
Cyclic loading	$P = AP_{ult} \ \tanh\left[\dfrac{kx}{AP_u} y\right]$ $A = 0.9 \ k$ from Figure 8.12

N for which stiffness will continue to degrade is unclear, but Dunnavant and O'Neill note that, for the load test data they interpreted, stabilization to a steady state still did not occur after 200 cycles. Two additional parameters P_{cm} and P_r are introduced to describe cyclic degradation, as shown in the plot on the right side of Figure 8.11. No degradation in stiffness is assumed to occur for load levels $P < P_{cm}$ and $y < y_{cm}$. For displacements $y > 12 \ y_{50}$, a fully degraded resistance P_r occurs. At intermediate displacement levels between y_{cm} and 12, resistance P is obtained by linear interpolation between P_{cm} and P_r.

8.3.3 Sand

The method described here appears in the API guidelines, which follow Murchison and O'Neill (1984). Ultimate resistance P_{ult} is predicted from the equation shown in Table 8.5. Two empirical coefficients, c_2 and c_2, control P_{ult} at shallow depths, while a third coefficient c_3 controls P_{ult} in the flow-around zone. Figure 8.12 depicts plots of these coefficients along with curve fits to assist implementation in spreadsheet-type calculations. The P–y curves for both monotonic and cyclic loading are described by a hyperbolic tangent function. An empirical parameter A distinguishes monotonic from cyclic behavior. A horizontal coefficient of subgrade reaction k, dependent on relative density D_r as shown in Figure 8.12, controls stiffness. Figure 8.13 illustrates the hyperbolic tangent equation used for characterizing the P–y relationship.

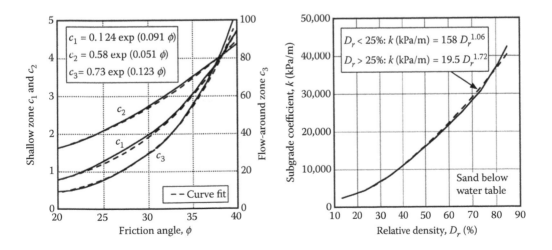

Figure 8.12 Ultimate capacity subgrade reaction coefficients for piles in sand.

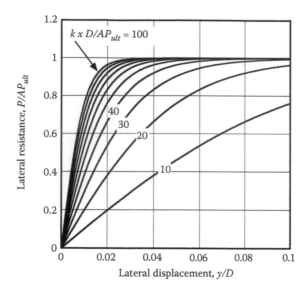

Figure 8.13 P–y curve in sand.

8.4 SECANT STIFFNESS

The P–y behavior in a given load cycle can be described in terms of increments of displacement Δy and soil resistance ΔP relative to the most recent load reversal as defined in Figure 8.14. In clays, the relationship between ΔP and Δy is typically linear on a log–log scale; therefore, the relationship can be described by a power law having the following form:

$$\Delta P/P_{ult} = K_0(\Delta y/D)^n \tag{8.12}$$

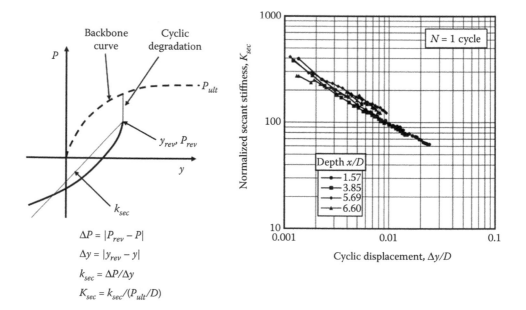

Figure 8.14 Normalized secant stiffness.

The coefficients K_0 and n can be obtained from simple linear regression of data measurements. For illustration, power law curve fits are superimposed on to the basic P–y loops in Figure 8.7. To unify data measurements obtained at various depths along the pile, it is useful to normalize soil resistance ΔP by the ultimate soil resistance $P_{ult} = N_p\, s_u\, D$ described in 0. Equivalent linear analysis of a laterally loaded pile (described subsequently in this chapter) can proceed in terms of a secant stiffness K_{sec}, defined as the slope of the line connecting the current P–y point to the point of the last reversal. When the power law relationship holds, the normalized secant stiffness can also be described in terms of the coefficients K_0 and n:

$$K_{sec} = (\Delta P/\Delta y)/(P_{ult}/D) = K_0(\Delta y/D)^{n-1} \tag{8.13}$$

The plot on the right side of Figure 8.14 shows normalized secant stiffness K_{sec} versus displacement Δy for the first cycle of loading at various depths along the pile shaft.

As is apparent from comparison of the P–y loops for load cycles 1 and 100 in Figure 8.7, significant stiffness degradation occurs during cyclic loading. Figure 8.15 illustrates the stiffness degradation in more detail for cycles $N = 1, 10, 30, 50,$ and 100. In this instance, a steady-state condition develops after approximately 30 cycles of loading. Figure 8.16 expresses degradation in terms of the ratio of the secant stiffness at cycle N to that of cycle 1, K_{secN}/K_{sec1} for various depths along the pile shaft. Since secant stiffness is displacement dependent, such comparisons should be linked to a specified displacement amplitude $\Delta y/D$. The curves in Figure 8.16 are generated for secant stiffness at displacements equal to the cyclic displacement amplitude $\Delta y_{cyc}/D$; the distance between the reversals in the P–y loop is illustrated in Figure 8.15. In a pile load test, displacement is imposed at the pile head and decays with depth. In general, as the amplitude of cyclic displacement increases, the rate of degradation also increases. Also shown in Figure 8.16 is a conservative estimate of stiffness degradation as a function of load cycle N proposed by Jeanjean (2009) for normally consolidated clay, which takes the following form:

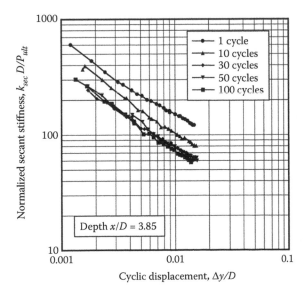

Figure 8.15 Secant stiffness during cyclic loading.

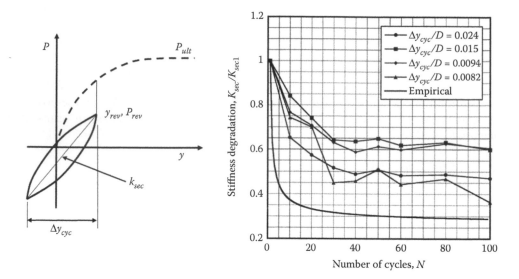

Figure 8.16 Cyclic degradation of secant stiffness.

$$\frac{K_{secN}}{K_{sec1}} = \frac{0.9}{0.9 + 2.5 \tanh(0.7 \log_{10} N)} \tag{8.14}$$

The secant stiffness framework employs an equivalent linear method of analysis, with K_{sec} selected to represent the computed displacement magnitude by a process of trial and error. Lost in the process of linearization is the energy absorbed through hysteresis. Following a similar approach commonly used in soil dynamics, this energy dissipation can be approximately modeled by introducing viscous damping to the soil springs. If ΔW is the energy dissipated in the hysteresis loop, and W_e is the maximum stored elastic energy, the corresponding damping ratio β for a spring obeying a power law load–displacement relationship (Equation 8.12) is as follows:

$$\beta = \frac{1}{4\pi} \frac{\Delta W}{W_e} = \frac{1}{\pi} \left(\frac{1}{n+1} - \frac{1}{2} \right) \tag{8.15}$$

An exponent $n = 1$ implies a perfectly linear spring, which Equation 8.15 shows to have zero damping. Considering a range of n between 0.2 and 0.5 implies damping ratios in the range $\beta = 0.1$–0.05. Studies on well conductors (Zakeri et al., 2015) show that fatigue life can be significantly underestimated when damping effects are neglected; therefore, incorporating damping into the soil springs is advisable when employing an equivalent linear secant stiffness approach.

8.5 TRUE NONLINEAR BEHAVIOR

The simplified P–y and secant stiffness relationships described above largely apply to monotonic or steady-state cyclic loading. However, situations such as earthquake loading or extreme storm events, which involve transient or random loading of the pile, can arise. In such cases, the evolution of soil resistance during transient loading from the initial intact state may have to be more carefully modeled. This section presents two approaches for modeling the $P-y$ behavior: a phenomenological description and incremental plasticity models.

8.5.1 Phenomenological description

One approach to characterize lateral soil resistance under transient cyclic loading is to use empirical equations derived from field or model test data. Figure 8.17 illustrates the general nature of the backbone curve and the unload–reload loops during cycling under conditions of displacement control with a uniform amplitude of cyclic displacements. Mobilized soil resistance P is normalized by ultimate resistance P_{ult}, which is evaluated using the methods outlined in 1 and 2. As shown in the figure, the backbone curve is well represented by a hyperbolic equation, which utilizes two fitting parameters: K_{max} defining the initial tangent stiffness and f defining curvature. Unload–reload behavior during cyclic loading is well represented by a power law function, which requires two additional parameters, K_0 and n, which vary with depth and evolve over repeated cyclic loading. The power law function shown is formulated in terms of the changes in soil resistance ΔP and Δy relative to the last load reversal, as depicted in Figure 8.17. Under conditions of nonuniform cyclic loading, P_{ref} is described by a stiffness degradation formulation, which is discussed subsequently.

As depicted schematically in Figure 8.18 the $P-y$ loops ratchet downward during cycling as stiffness degrades. Following an approach developed by Idriss et al. (1976) for describing cyclic triaxial data, the change in soil resistance occurring during any load cycle can be described in terms of the power law function shown in the figure, where the parameter t controls the rate of degradation in soil resistance. Experimental evaluation of t is by defining a soil resistance reduction factor $R_f = P_N/P_1$. The left-hand plot in Figure 8.19 illustrates the variation in R_f as cycling progresses. The parameter t controlling the rate of soil resistance

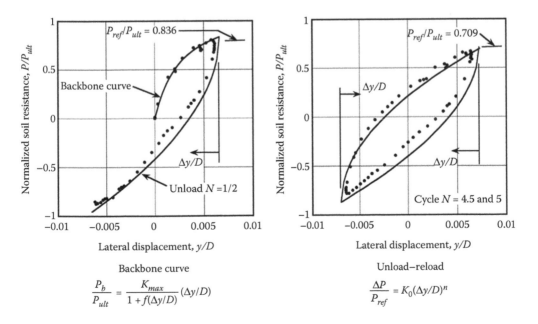

Backbone curve

$$\frac{P_b}{P_{ult}} = \frac{K_{max}}{1 + f(\Delta y/D)}(\Delta y/D)$$

Unload–reload

$$\frac{\Delta P}{P_{ref}} = K_0(\Delta y/D)^n$$

Figure 8.17 Nonlinear backbone and unload–reload behavior.

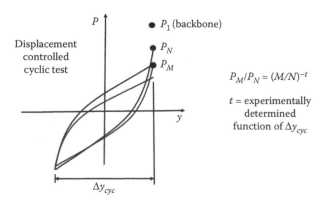

$$P_M/P_N = (M/N)^{-t}$$

t = experimentally determined function of Δy_{cyc}

Figure 8.18 Definition sketch for soil resistance degradation during cycling.

degradation increases with increasing magnitude of cyclic displacement Δy_{cyc}, as illustrated by the example plot on the right side of Figure 8.19.

The framework for describing stiffness degradation presented above suffices for uniform cyclic displacements. However, under general conditions of cyclic loading, the amplitudes of Δy_{cyc} are not necessarily uniform. Additionally, under random loading conditions, the amplitude of the reload displacement does not necessarily match that of the unload displacement (Figure 8.20). Thus, some framework is needed to update the level of stiffness degradation (through the parameter t) at the level of each half-cycle of loading. At this point, we can define a reference soil resistance P_{ref}, which is analogous to the soil resistance at the load reversal point in Figure 8.18. With some algebra, Idriss et al. (1976) show that the power law for stiffness degradation can be reformulated to permit updating after each load reversal; that is, at the level of each half-cycle of loading. Equation 8.16 displays the relevant relationships. After each reversal, the parameter t is updated to reflect the appropriate magnitude of cyclic displacement (Figure 8.19), while P_{ref} is updated for use in the power law unload–reload P–y equation in Figure 8.16.

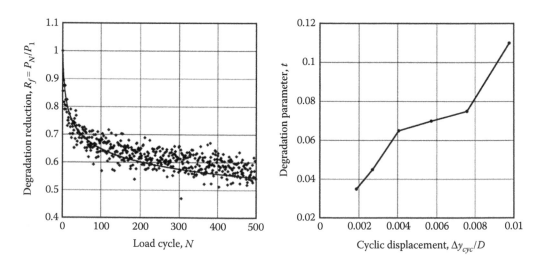

Figure 8.19 Determination of the soil resistance degradation parameter t.

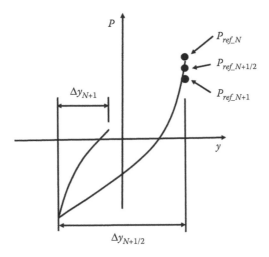

Figure 8.20 Nonuniform cyclic displacement amplitudes.

$$P_{ref} = R_f P_1$$
$$R_{f\_N+1/2} = R_{f\_N} \left(\frac{N+1/2}{N} \right)^{-t}$$

(8.16)

An illustration of the approach can be made by considering a pile subjected to an exponentially decaying motion at the pile head shown in the upper left plot in Figure 8.21. The plots on the right-hand side in Figure 8.21 illustrates a typical computed P–y response to this type of loading, with the degradation in soil resistance updated at each load reversal as per Equation 8.16. The plot in the lower left shows the bending moment profiles corresponding to the decaying pile head motions.

8.5.2 Incremental plasticity models

The model described above is a purely phenomenological description of pile response to lateral cyclic loading. Bounding surface plasticity formulations have also been used for describing cyclic P–y springs, examples include McCarron (2015) for clay soils and Choi et al. (2015) in sands. As an example of the methodology, the model of Choi et al. is described below. As summarized in Table 8.6, the model employs four parameters at a given depth along the pile: an elastic modulus K^e, a plastic deformation parameter C, an ultimate soil resistance P_{ult} defining the bounding surface, and a parameter P_y defining the size of the elastic region. Figure 8.22 graphically illustrates the latter two parameters. Kinematic hardening occurs during plastic loading, such that the center of the elastic region—defined by P_α—migrates in accordance with the hardening law listed in Table 8.6. Within the yield surface ($f < 0$), the spring behavior is linear. Outside the yield surface, the spring is both nonlinear and includes inelastic deformations. The second state variable is P_{in}, which is simply the value of P at which inelastic behavior is initiated in the current load cycle. The formulation of the plastic modulus is designed for K^p to decrease with increasing proximity to the bounding surface. Examination of the tangent elasto-plastic modulus shows that $K = K^e$ at the onset of plastic loading and $K = 0$ as the bounding surface is approached.

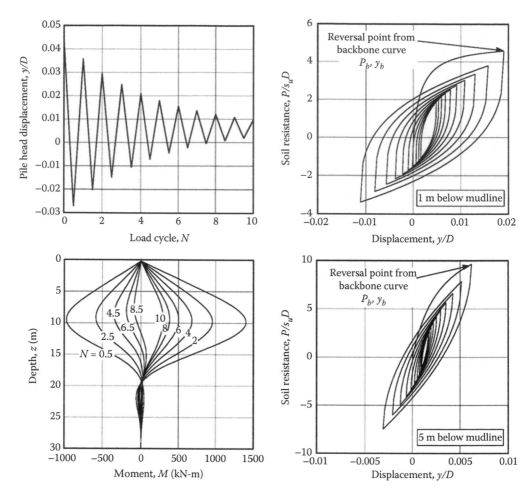

Figure 8.21 Pile response to exponentially decaying lateral motions.

Table 8.6 Bounding surface P–y model

Model parameters	K^e = elastic modulus C = parameter controlling plastic modulus P_{ult} = ultimate soil resistance (bounding surface) P_y = size of elastic region
State variables	P_α = center of elastic region P_{in} = P at onset of plastic loading during current cycle
Yield criterion	$f = |P - P_\alpha| - P_y$
Load–displacement equations	Elastic : $\dot{P} = K^e\dot{y} = K^e(\dot{y} - \dot{y}^p)$ Elasto-plastic : $\dot{P} = K\dot{y} = \dfrac{K^eK^p}{K^p + K^e}\dot{y}$ $K^p = CK^e\dfrac{|P_{ult}\ \text{sign}(\dot{y}) - P|}{|P - P_{in}|}$
Hardening law	$\dot{P}_\alpha = K^p\dot{y}^p$

Source:　Choi JI, Kim MM, and Brandenberg SJ, 2015, ASCE J Geotech Geoenviron Eng, 141(5): 04015013.

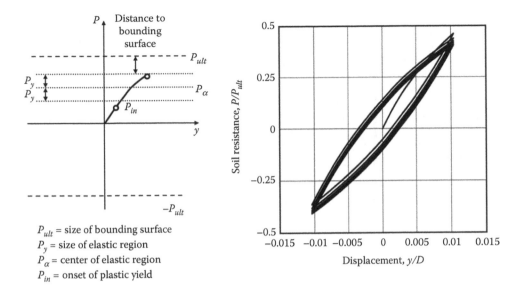

P_{ult} = size of bounding surface
P_y = size of elastic region
P_α = center of elastic region
P_{in} = onset of plastic yield

Figure 8.22 Incremental plasticity model.

The model can be calibrated to experimental test measurements. However, as discussed by Choi et al., several of the model parameters can be estimated from basic soil properties. For example, the elastic modulus K^e is related to the Young's modulus of the soil E_s, which, in turn, is related to the shear wave velocity of the soil. Similarly, the displacement at yield can be taken as proportional to the strain at which inelastic behavior initiates, $y_{yield} = 2.5\,D$ $\gamma_{yield}/(1 + \mu)$, where μ is Poisson's ratio (Kagawa and Kraft, 1980) and γ_{yield} is on the order of 0.001% (Darendeli, 2001). Based on an estimated yield displacement the size of the elastic region can be estimated as $P_y = K^e\, y_{yield}$.

Figure 8.22 shows a typical output from the model for 10 cycles of displacement-controlled loading. Evident from the figure is the ability of this type of formulation to simulate the downward ratcheting of the $P–y$ loop with a concomitant overall reduction in the soil spring stiffness.

8.6 SOIL–PILE INTERACTION

Analysis of a pile supported by Winkler springs is possible using analytical methods, finite differences, and finite elements. Analytical solutions are generally restricted to linear springs. In spite of this restriction, the closed-form solutions can provide insights into behavior that can be obscured by the relative complexity of numerical solutions. Examples of analytical solutions can be found in Poulos and Davis (1980). Nevertheless, as discussed in previous sections, the nonlinear response of the soil is clearly significant. Thus, numerical analyses capable of supporting nonlinear soil springs are widely employed. Either finite difference or finite element methods can be applied to the problem. The development of finite difference solutions is relatively straightforward and is presented in the following sections.

8.6.1 Axial

Figure 8.23 shows a schematic representation of an axially loaded pile with cross-sectional area A and elastic modulus E supported on idealized soil springs. When loaded, points along

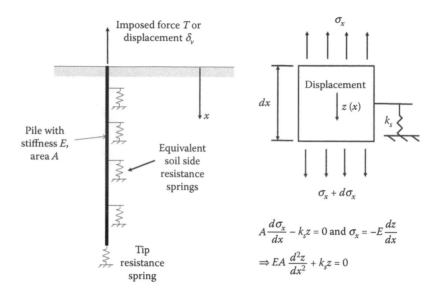

Figure 8.23 Axially loaded pile.

the pile undergo a vertical displacement $z(x)$ relative to their initial position. Mobilized soil skin resistance varies with depth and displacement, and can be described in terms of a secant spring stiffness $k_s(x, z)$. If compressive stresses and strains are taken as positive, tip resistance can similarly be represented by a displacement-dependent spring at the pile tip. The governing differential equation is derived by invoking static equilibrium together with Hooke's law to produce the following second-order ordinary differential equation:

$$EA\frac{d^2z}{dx^2} + k_sz = 0 \tag{8.17}$$

This differential equation can be solved numerically using finite differences by formulating n equations at discrete points, as illustrated in Figure 8.24. The nonlinear soil springs require iterative solution for unknown displacements. The notation used here uses the subscript i to denote the calculation point under consideration and the superscript j to denote the iterative trial solution. Noting that the spring stiffness used in the calculation for the current (jth) iteration is based on displacements from the previous ($j-1$) iteration, a central difference expression for Equation 8.17 is as follows:

$$z_{i-1}^j - \left(2 + k_{si}^{j-1}\Delta x^2 / EA\right)z_i^j + z_{i+1}^j = 0 \tag{8.18}$$

At least two boundary constraints should be specified. If an upward displacement δ_v is specified at the pile head, the first equation is simply $z_1 = -\delta_v$. Although tip resistance is modeled as a spring, the mathematical implementation actually imposes a force boundary constraint here. The force Q^{j-1} is computed from the $Q-z$ relationship selected for the analysis together with the tip displacement computed from the previous iteration. Specifying a tip force essentially imposes a strain constraint at this location, so the nth equation becomes as follows:

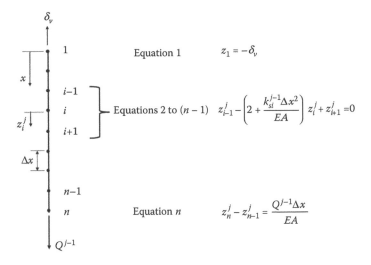

Figure 8.24 Finite difference discretization of axially loaded pile.

$$z_n^j - z_{n-1}^j = \frac{Q^{j-1}\Delta x}{EA} \qquad (8.19)$$

Other variants are possible for specific situations. For example, for an embedded pad eye, displacement is specified at the pad-eye depth and a zero stress condition is specified at $x = 0$. The assembled matrix equation takes the tridiagonal form, as shown below.

$$\begin{bmatrix} 1 & & & & & & \\ 1 & d_2^j & 1 & & & & \\ & 1 & d_3^j & 1 & & & \\ & & 1 & d_4^j & 1 & & \\ & & & & \ddots & & \\ & & & & 1 & d_{n-1}^j & 1 \\ & & & & & -1 & 1 \end{bmatrix} \begin{bmatrix} z_1^j \\ z_2^j \\ z_3^j \\ z_4^j \\ \vdots \\ z_{n-1}^j \\ z_n^j \end{bmatrix} = \begin{bmatrix} \delta_v \\ \\ \\ \\ \\ \\ Q^{j-1}\Delta x / EA \end{bmatrix} \qquad (8.20)$$

$$d_i^j = -2 - \frac{k_{si}^{j-1}\Delta x^2}{EA}$$

Solution requires an initial estimate of the distribution of vertical displacements along the pile shaft. Iteration continues until the maximum difference in displacement lies below a specified tolerance, say 10^{-4} pile diameters. Fortunately, rapid convergence is possible even for simplistic initial estimates, such as a uniform distribution with displacement set to δ_v at all points.

Figure 8.25 shows an example analysis of a 50-m long, 1.2 diameter pile with 0.03-m wall thickness embedded in a normally consolidated clay (mudline strength $s_{um} = 2$ kPa, strength gradient $k = 1.6$ kPa/m). An adhesion factor $\alpha = 0.7$ is used in the analysis. The

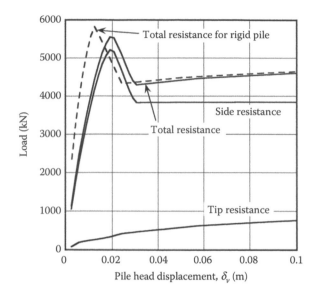

Figure 8.25 Example load–displacement curves for axially loaded pile.

figure shows the individual side and end bearing contributions to total capacity. Also shown in dashes is the load–displacement curve corresponding to a completely rigid pile. This example illustrates two aspects of elastic effects on pullout capacity. First, skin resistance mobilizes much more rapidly than end bearing. Thus, peak capacity in this case mobilizes at relatively small displacement, such that less than half of the available reverse end bearing contributes to the peak capacity. At larger displacements, additional reverse end bearing mobilizes; however, the resulting total capacity in this case never reaches the peak achieved at small displacements. Second, comparison of total capacity estimates for an elastic pile versus a rigid pile illustrates the effect of nonuniform mobilization of skin resistance along the pile shaft owing to elastic deformation of the pile. This is further illustrated in Figure 8.26 comparing the mobilization of side resistance at the top, middle, and bottom of the pile as a function of pile head displacement. The different rates of mobilization associated preclude the simultaneous mobilization of peak shearing resistance at all locations. In this case, the effect is somewhat mitigated by the fact that most resistance arises near the bottom of the pile, owing to the higher soil strength occurring at greater depth. Thus, the net reduction in pile capacity owing to elastic effects in this case is on the order of 4%. Noting that ring shear tests indicate the softening factor along the pile shaft may be in the range of 0.5–0.8—in contrast to the API guideline of 0.7–0.9—Randolph (2003) indicates that the reduction in axial capacity owing to this effect can be as high as 35% in some instances.

8.6.2 Lateral

A laterally loaded flexible pile is analyzed within the framework of classical beam bending theory (Figure 8.27), which neglects shear deformations and relates the fourth derivative of lateral displacement y to distributed line load $w(x)$ in Equation 8.21.

$$EI \frac{d^4y}{dx^4} = w(x) \tag{8.21}$$

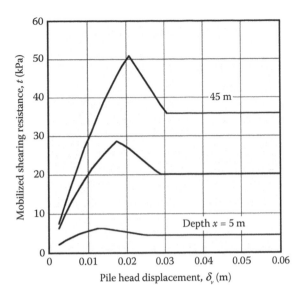

Figure 8.26 Mobilized shearing resistance versus depth along pile shaft.

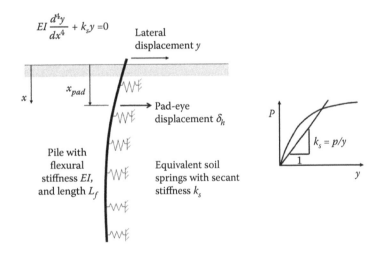

Figure 8.27 Equivalent spring model for laterally loaded pile.

If supported by the springs of secant stiffness $k_s(x)$, the line load acts in a direction opposite to displacement, $w = -k_s y$, generating the following fourth-order differential equation:

$$EI \frac{d^4 y}{dx^4} + k_s y = 0 \qquad (8.22)$$

As for the case of an axially loaded pile, finite difference solution for y is possible. Also, following the approach for the axial load case, iterative solution is required owing to the nonlinear springs. With the subscript i denoting the point about which the finite difference calculation is made and the superscript j denoting the trial solution in the iteration, a central difference approximation to Equation 8.22 produces the following equation:

$$y^j_{i-2} - 4y^j_{i-1} + \left(6 + \frac{k^{j-1}_{si}\Delta x^4}{EI}\right)y^j_i - 4y^j_{i+1} + y^j_{i+2} = 0 \tag{8.23}$$

At least four boundary constraints should be imposed. For a floating pile, that is one not embedded in a stiff layer, the shear and bending moment at the pile tip must be zero: shear $V(L_f) = EI\, y'''(L_f) = 0$ and moment $M(L_f) = EI\, y''(L_f) = 0$. Unless connected to a rigid pile cap, the moment at the pile head is also zero, $M = EI\, y''(0) = 0$. If the pad eye is at the pile head, the final boundary constraint may simply be specified as $y(0) = \delta_h$. Alternatively, if the pad eye is embedded at a depth x_{pad}, then the displacement is specified at this depth and a zero-shear boundary constraint is imposed at the pile head: $y(x_{pad}) = \delta_h$ and $V(0) = EI$ $y'''(0) = 0$. Figure 8.28 shows the finite difference expressions associated with the boundary constraints for an embedded pad eye. Assemblage of Equation 8.23 and the additional boundary constraint equations into matrix form parallels that are shown previously for the axially loaded pile case, except that the resulting matrix has a penta-diagonal form.

Figure 8.29 shows the results of an analysis pure lateral loading on a 30-m long, 1.2-m diameter, 0.03-wall thickness pile embedded in the normally consolidated clay profile used in the earlier example of an axially loaded pile: mudline strength $s_{um} = 2$ kPa and strength gradient $k = 1.6$ kPa/m for a displacement, resistance w mobilized for a displacement at the pad eye of $\delta_h = 0.2$ m or approximately 16% of the pile diameter. Also shown in the figure is the ultimate soil resistance P_{ult}. Even with relatively large deformations, only a fraction (some 40%) of the ultimate soil resistance is utilized. Continued displacement of the pad eye to $\delta_h = 2$m begins to approach the ultimate geotechnical capacity of the pile. Such a large pad-eye displacement would actually induce structural yielding in the pile itself and—even without yielding—would not be acceptable in most design situations. Nevertheless, the example does illustrate the dominance of elastic effects in long pile anchors. This behavior is in contrast to the other anchors considered in this book, where capacity is largely controlled by ultimate soil resistance with some modest downward adjustment being necessary to account for elastic effects.

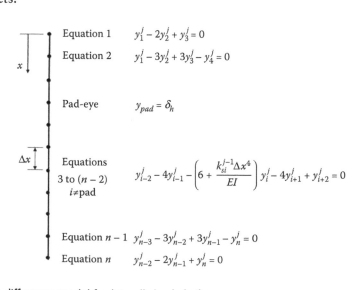

Figure 8.28 Finite difference model for laterally loaded pile.

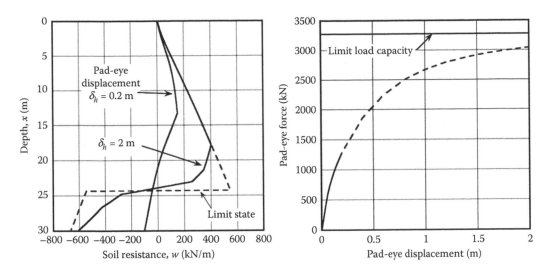

Figure 8.29 Response of a laterally loaded pile with pad-eye depth of 5 m.

In spite of the low geotechnical efficiency of laterally loaded piles used as anchors, increased capacity can be realized through a more judicious choice of load attachment depth. For example, Figure 8.30 shows predicted load–displacement curves for the pile considered in the previous example, but with the pad-attachment depth varied from 5 to 25 m. As is the case for suction caissons, an optimized load attachment depth provides a substantial (in this example fivefold) increase in load capacity. It is noted that, although the deeper attachment depths provide substantial additional load resistance, they still have relatively low geotechnical efficiency in terms of their usable load capacity relative to their limit state capacity. The profiles of computed lateral displacement in Figure 8.31 show the expected trend of rotation with kick-back at the base of the pile for the case of a shallow pad-eye connection depth, $x_{pad} = 5$ m. By contrast, the deformed configuration for the case of a 20-m deep pad-eye connection shows more of a translational or plowing motion. Examination of the bending moment profiles in Figure

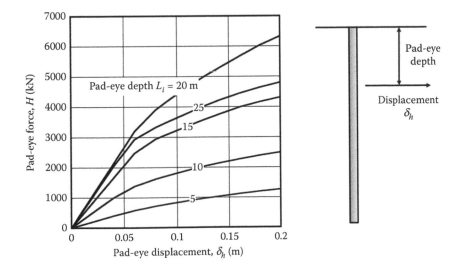

Figure 8.30 Load–displacement response for laterally loaded pile.

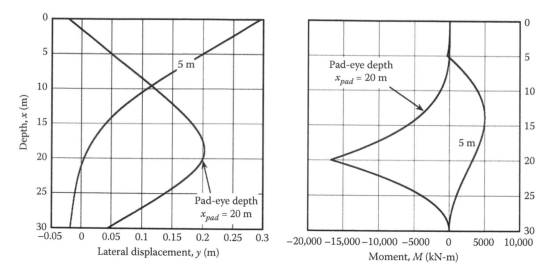

Figure 8.31 Displacement and bending moment response in laterally loaded pile.

8.31 shows that, as the pad-eye position is varied, the structural load demand on the pile does not increase in proportion to the increase in lateral load capacity of the pile. For example, moving the pad-eye downward from 5 to 20 m produces a fivefold increase in capacity, but only a roughly threefold increase in the maximum bending moment in the pile.

8.6.3 Beam-column effects

Figure 8.32 now considers pile behavior under combined axial and lateral loading. The analysis discussed herein assumes no interaction between the respective axial and lateral soil springs. The variation in axial force $T(x)$ is assumed to be governed completely by the load transfer analysis for a pile computed for purely axial loading. Likewise, the $P-y$ curves defining the equivalent lateral soil spring response are assumed to be independent of axial loading. However, a variable distribution of lateral displacement $y(x)$ along the pile shaft generates moments $T \cdot dy$ as shown in Figure 8.32. Neglecting high-order differential terms produce a modification of Equation 8.23 having the following form:

$$EI \frac{d^4 y}{dx^4} - T \frac{d^2 y}{dx^2} - \frac{dT}{dx}\frac{dy}{dx} + k_s y = 0 \qquad (8.24)$$

A central difference approximation to Equation 8.24 parallels that of Equation 8.22 with additional terms for axial force T_i and its derivative as follows:

$$y_{i-2}^j + (-4 - a_1 + a_2)y_{i-1}^j + (6 + 2a_1 + a_3)y_i^j + y_{i+1}^j(-4 - a_1 - a_2) + y_{i+2}^j = 0$$

$$a_1 = \frac{T_i \Delta x^2}{EI}$$

$$a_2 = \frac{(T_{i+1} - T_{i-1})\Delta x^2}{4EI} \qquad (8.25)$$

$$a_3 = \frac{k_{si}^{j-1}\Delta x^4}{EI}$$

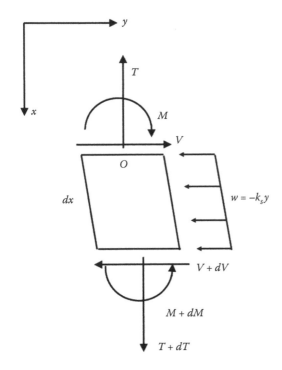

A. Force equilibrium $\Sigma F_y = 0$:

$$\frac{dV}{dx} = -k_s y$$

B. Moment equilibrium $\Sigma M_0 = 0$:

$$dM + Tdy - Vdx = 0 \text{ or } \frac{dM}{dx} + T\frac{dy}{dx} - dV = 0$$

C. Substitute A into B and differentiate:

$$\frac{d^2M}{dx^2} + T\frac{d^2y}{dx^2} + \frac{dT}{dx}\frac{dy}{dx} + k_s y = 0$$

D. Relationship between moment and curvature:

$$M = EI\frac{d^2y}{dx^2}$$

E. Governing differential equation:

$$EI\frac{d^4y}{dx^4} + T\frac{d^2y}{dx^2} + \frac{dT}{dx}\frac{dy}{dx} + k_s y = 0$$

Figure 8.32 Formulation for combined axial-lateral loading of pile.

The boundary constraint equations that should be imposed with Equation 8.25 are identical to those described earlier for pure lateral loading. Assuming no interaction between axial and lateral springs leads to a sequentially coupled formulation for which the axial forces T_i at each finite difference point are calculated from Equations 8.18 through 8.20 in advance of solving Equation 8.24.

A compressive (negative) axial force T leads to larger bending moments, while tension has the opposite effect. Figure 8.33 shows the bending moment distribution in a 40-m long by 1-m diameter pile with wall thickness 0.025 m embedded in a normally consolidated clay. The load is attached at the top ($x_{pad} = 1$ m) of the pile and inclined at 25°. Bending moment profiles for a 0.2-m pad-eye displacement are shown based on calculation with and without the beam-column effect. In this case, considering the beam-column effect makes a relatively modest 2% difference in the calculation. While tension is generally beneficial in anchor piles from the standpoint of reducing bending moments, it is not the case that an anchor pile will always be in tension. For example, if the pad eye is placed at the two-thirds depth for the pile considered in Figure 8.33, the portion of the pile above the pad eye will actually be in compression.

8.7 SHORT PILES AND CAISSONS

To this point, the discussion of elastic effects has centered on slender piles that behave as beam-columns for which flexure dominates response. Short piles and caissons actually do not experience significant bending under lateral loads, which introduces several issues. First, since bending is negligible, the approach for interpreting pile load tests—based on double integration to compute y and double differentiation to compute P—is no longer

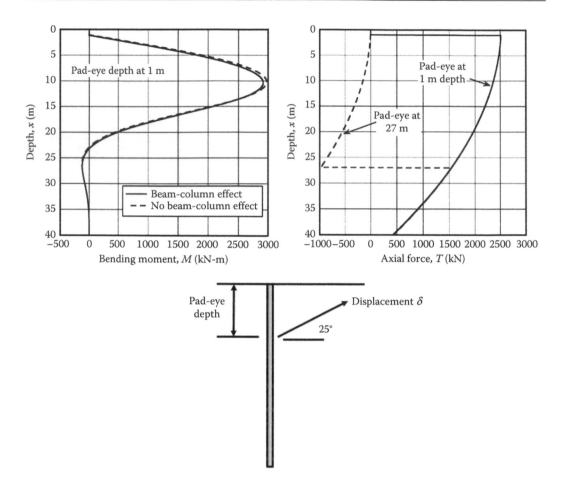

Figure 8.33 Example beam-column analysis.

viable. Xue et al. (2016) describe an optimization approach for interpreting pile load tests on short rigid piles that avoids the pitfalls associated differentiating measured moment profiles. Second, the line loads obtained from a conventional P–y analysis inadequately characterize the soil stresses governing the stress distribution in the pile shell–the difference being comparable to that of a point load versus a distributed load acting on a beam. Third, pile displacements computed from a conventional P–y analysis on a short pile will be comprised entirely from deformations of the soil, since flexural deformations of the pile are negligible. Such an analysis neglects the ovalization that occurs when loading a slender ring (Figure 8.1).

Compared to the extensive database of measurements on flexible piles, there is a relative paucity of information on deformation of flexible shells in a horizontal plane. This section seeks to shed some light on the topic through finite element predictions of stress distributions along the pile circumference for rigid and flexible piles. These stress distributions are then related to the net resultant line load P, which currently provides the basis for the design of laterally loaded piles. Finally, a three-dimensional finite element predictions of pile load versus pad-eye deflection are presented to illustrate the effects of pile elasticity on the behavior of short piles.

8.7.1 Circumferential distribution of soil resistance

Conventional analysis of a laterally loaded pile proceeds in terms of a line load of intensity P per unit height of pile, with P representing the resultant of radial and tangential stresses, σ_{rr} and $\sigma_{r\theta}$, integrated around the circumference of the pile. This framework suffices for slender piles where bending about a horizontal axis is the dominant mode of deformation. However, for short piles where ovalization is the dominant mode of deformation, details of the distribution of stresses around the pile circumference are required. In this connection, deformation of the pile shell is driven by net stresses acting on the pile, that is, the difference between stresses acting on the outer and inner faces of the pile. Additionally, the stresses contributing to the net force P should be resolved into radial and tangential components. In general, the stiffness of the pile itself will affect the distribution of stresses. This stiffness should not be confused with the flexural stiffness from classical beam bending theory. Rather, it is the stiffness of a ring section, with the governing moment of inertia defined in Figure 8.34. For the case of no soil inside or outside of a pile of radius R with concentrated loads P acting at $\theta = 0°$ and 180°, Housner and Vreeland (1966) gives the analytical solution for deflection δ shown in Equation 8.26. This equation, of course, does reflect the actual stress distribution for piles—but it does provide a convenient reference solution for validating and calibrating numerical models.

$$\delta = 0.149 \frac{PR^3}{EI_o} \tag{8.26}$$

We can now proceed to a two-dimensional finite element analysis of a segment of a 5-m diameter pile located at the pad eye, with a rigid plate stiffener connecting the nodes at $\theta = 0°$ and 180°. The analysis is conducted for a soil with rigidity index $I_r = G/s_u = 200$, with no slippage permitted at the soil–pile interface (approximately $\alpha = 1$). The analyses are conducted for a rigid pile and a flexible steel pile with a wall thickness $t_w = 0.04$ m. Figure 8.35 shows the circumferential distribution of radial and tangential stresses acting on the pile. In the case of the flexible pile, reported stresses are the net stress acting on the shell, that is the differences in stresses acting on the outer and inner surfaces of the shell. In the rigid case, the pile is modeled as a hollow cylinder. Tangential stresses are seen to mobilize much more rapidly than radial stresses, providing another reason for developing

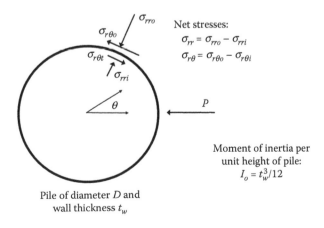

Net stresses:
$$\sigma_{rr} = \sigma_{rro} - \sigma_{rri}$$
$$\sigma_{r\theta} = \sigma_{r\theta o} - \sigma_{r\theta i}$$

Moment of inertia per
unit height of pile:
$$I_o = t_w^3 / 12$$

Pile of diameter D and
wall thickness t_w

Figure 8.34 Ring stiffness of a pile.

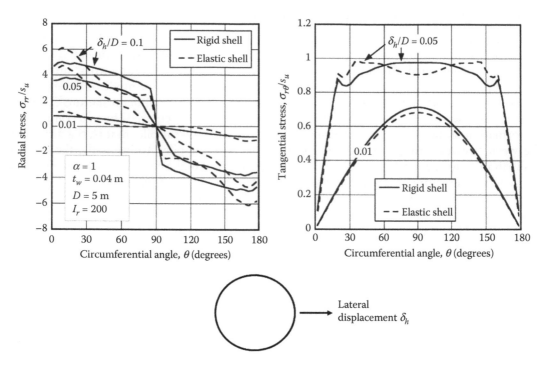

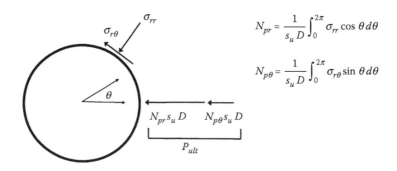

Figure 8.35 Circumferential stress distribution around rigid and flexible piles.

independent springs for characterizing soil resistance in the radial and tangential directions. In the direction of pull ($\theta = 0°$ and $180°$) radial stresses around a flexible pile noticeably exceed those for a rigid pile (by approximately 20%), while the opposite occurs at directions greater than $30°$ from the direction of pull. Pile flexibility exerts only a minor influence on the distribution of tangential stresses.

Since the ultimate line load $P_{ult} = s_u D$ can often be established with a reasonably high degree of confidence on a given project, it is desirable to adopt a normalization that facilitates estimating the distribution of radial and tangential stresses from a known P_{ult}. This requires distinguishing between the radial and tangential contributions to ultimate resistance (Figure 8.36). Defining N_{pr} as the bearing resistance derived from radial stresses acting on the pile, the radial resistance bearing resistance for the case of full adhesion ($\alpha = 1$) turns out to be $N_{pr1} = 10.02$. Proceeding in like fashion for tangential resistance leads to

$$N_{pr} = \frac{1}{s_u D} \int_0^{2\pi} \sigma_{rr} \cos\theta \, d\theta$$

$$N_{p\theta} = \frac{1}{s_u D} \int_0^{2\pi} \sigma_{r\theta} \sin\theta \, d\theta$$

Figure 8.36 Bearing resistance from radial and tangential stresses.

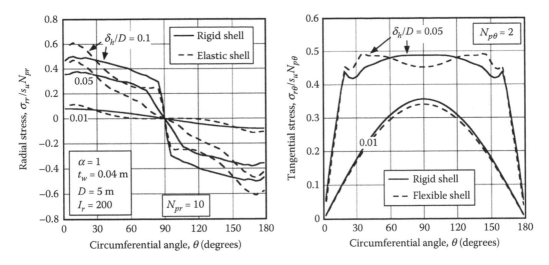

Figure 8.37 Normalized circumferential stress distribution.

$N_{p\theta 1} = 1.92$. For the case of zero adhesion, the radial resistance bearing factor is known from the Randolph–Houlsby solution for a smooth cylinder, $N_{pr0} = 9.14$, while by definition $N_{p\theta 0} = 0$. Thus, as adhesion factor α increases from 0 to 1, the radial bearing factor increases from approximately 9 to 10, or $N_{pr} \approx 9 + \alpha$. Similarly, the tangential bearing factor increases with adhesion according to the equation $N_{p\theta} \approx 2\alpha$. Figure 8.37 shows the distribution of radial and tangential stresses around the circumference of the pile, with stresses normalized N_{pr} and $N_{p\theta}$. It is noted that integrated resultants of stresses are essentially identical for the elastic and rigid cases, which is consistent with the notion that collapse loads are independent of elastic deformations (Chen, 1975).

8.7.2 Elastic effects and ovalization

The circumferential distribution of radial and tangential stresses can be reformulated to develop equivalent soil springs as shown in Figure 8.38. The finite element analyses can now be extended to three dimensions by attaching the springs to structural shell elements. Since the springs in Figure 8.38 apply to mobilized soil resistance at depth in the flow-around region, the procedure adopted herein scales the spring stiffness down in proportion to be Murff–Hamilton $N_p(z)$ profile described in Chapter 4 to account for the reduction in soil resistance near the ground surface. The example presented here considers the case of pure horizontal translation of a caisson in a typical normally consolidated clay profile. Figure 8.39 from Zhang (2016) shows predicted relationships between pad-eye displacement and pad-eye force for (1) a rigid caisson, (2) an elastic caisson with wall thickness $t_w = 6.25$ cm ($D/t_w = 80$), and (3) an elastic caisson with wall thickness $t_w = 3.125$ cm ($D/t_w = 160$). The rigid case essentially corresponds to a conventional beam-column analysis of a short pile, such that all displacements are associated with the soil springs, while the latter cases consider the interaction effect owing to ovalization of the caisson. In all three cases considered, the caisson has an aspect ratio ($(L_i/D = 5)$, a pad-eye depth L_i at two-thirds of the caisson length, a perfectly rigid plate stiffener at the pad-eye depth, and a linearly varying strength profile. The predictions show displacements for the "thick-wall" ($D/t_w = 80$) elastic caisson to be comparable to those for a rigid caisson. By contrast, displacements for the "thin-wall" ($D/t_w = 160$) elastic caisson are approximately three times those of a rigid caisson. Such a

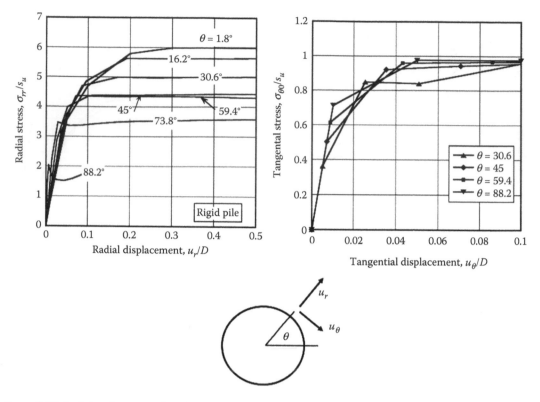

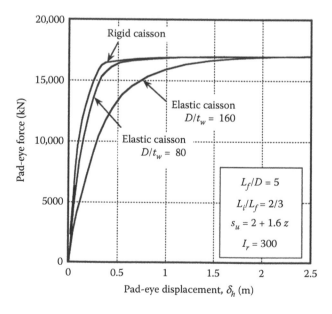

Figure 8.38 Radial and tangential soil springs.

Figure 8.39 Predicted load versus displacement for rigid and flexible caissons.

difference can be significant when the design basis has criteria for limiting displacements. The example shown here suggests that the flexural rigidity of short piles and caisson—that is negligible beam bending in a vertical plane—does not necessarily imply the absence of elastic effects, since ovalization in a horizontal plane can have a significant effect on overall stiffness.

REFERENCES

API RP 2A LRFD, 1993, *Recommended Practice for Designing and Constructing Fixed Offshore Platforms*, American Petroleum Institute, API Publishing Services, Washington, DC, 224p.

Chen WF, 1975, *Limit Analysis and Soil Plasticity*, Elsevier Publishing Co., Amsterdam, The Netherlands, 636 p.

Chen J, Gilbert R, Choo YS, Marshall P, and Murff J, 2017, Two-dimensional lower bound analysis of offshore pile foundation systems, *Int J Numer Anal Meth Geomech*, 40, 1321–1338.

Choi JI, Kim MM, and Brandenberg SJ, 2015, Cyclic *p–y* plasticity model applied to pile foundations in sand, *ASCE J Geotech Geoenviron Eng*, 141(5): 04015013.

Chow YK, 1986, Analysis of vertically loaded pile groups, *Int J Numer Anal Meth Geomech*, 10(1), 59–72.

Coyle HM and Reese LC, 1966, Load transfer for axially loaded piles in clay, *ASCE J Soil Mech Found Eng*, 92(SM2), 1–26.

Coyle HM and Sulaiman IH, 1967, Skin friction of steel piles in sand, *ASCE J Soil Mech Found Div*, 93(6), 261–278.

Darendeli M, 2001, *Development of a New Family of Normalized Modulus Reduction and Material Damping Curves*, Ph.D. thesis, Univ. of Texas, Austin, TX.

Dunnavant TW and O'Neill MW, 1989, Experimental *p–y* model for submerged stiff clay, *ASCE J Geotech Eng*, 115(1), 95–114.

Gazioglu SM and O'Neill M, 1984, Evaluation of P–Y relationships in cohesive soils, *Proceedings of the Symposium on Analysis and Design of Pile Foundations*, ASCE, pp 192–213.

Gilbert R, 2016, P–y knowledge gaps, *Geotechnical Input to Well Integrity Assessment, Workshop organized by Norwegian Geotechical Institute, Texas A&M University and University of Texas at Austin*, April 29, 2016, Houston.

Housner GW and Vreeland T, 1966, *The Analysis of Stress and Deformation*, Macmillan, New York, 440p.

Idriss IM, Dobry R, Doyle EH, and Singh RD, 1976, Behavior of soft clays under earthquake loading conditions, *8th Offshore Technology Conference*, OTC 2671, Houston, pp 605–617.

Jeanjean P, 2009, Reassesment of *P–Y* curves for soft clays from centrifuge testing and finite element modeling, *2009 Offshore Technology Conference*, OTC 20158, Houston, TX.

Kagawa T and Kraft LM, 1980, Seismic p-y response of flexible piles, *J Geotech Eng Div*, 106(GT8), 899–918.

Kraft LM, Ray RP, and Kagawa T, 1981, Theoretical *t–z* curves, *ASCE J Geotech Eng*, 107(GT11), 1543–1561.

Matlock H, 1970, Correlations for design of laterally loaded piles in soft clay, *Proceedings of the 2nd Annual Offshore Technology Conference*, Houston, TX, Vol. 1, pp 577–594.

McCarron WO, 2015, Bounding surface model for soil resistance to cyclic lateral pile displacements, *Comput Geotech*, 65, 285–290. Elsevier Ltd.

Murchison JM and O'Neill MW, 1984, Evaluation of P–Y relationships in cohesionless soils, *Proceedings of the Symposium on Analysis and Design of Pile Foundations*, ASCE, pp 174–191.

Murff JD and Hamilton JM, 1993, P-Ultimate for undrained analysis of laterally loaded piles, *ASCE J Geotech Eng*, 119(1), 91–107.

Parker F and Reese LC, 1979, *Experimental and Analytical Studies of Behavior of Single Piles in Sand under Lateral and Axial Loading*, Research Report 117-2, Center for Highway Research, University of Texas at Austin, 251p.

Poulos HG and Davis EH, 1980, *Pile Foundations Analysis and Design*, John Wiley and Sons, New York, 397p.

Randolph MF, 2003, Science and empiricism in pile foundation design, *Geotechnique*, 53(10), 847–875.

Randolph MF and Wroth CP, 1978, Analysis of deformation of vertically loaded piles, *ASCE J Geotech Eng*, 104(GT12), 1465–1488.

Reese LC, Cox WR, and Koop FD, 1974, Analysis of laterally loaded piles in sand, *Proceedings 6th Annual Offshore Technology Conference*, Houston, TX, Vol 2, pp 473–484.

Reese LC, Cox WR, and Koop FD, 1975, Field testing and analysis of laterally loaded piles in stiff clay, *Proceedings of the 7th Annual Offshore Technology Conference*, Houston, TX, Vol 2, pp 671–690.

Stevens JB and Audibert JME, 1979, Re-examination of P–y curve formulations, *11th Annual OTC, Offshore Technology Conference*, Houston, OTC 3402, pp 397–403.

Xue J, Gavin K, Murphy G, Doherty P, and Igoe D, 2016, Optimization technique to determine the p-y curves of laterally loaded stiff piles in dense sand, *Geotech Test J*, 39(5), 842–854.

Zakeri A, Clukey E, Kabadze B, Jeanjean P, Piercy G, Templeton J, Connelly L, and Aubeny C, 2015, Recent advances in soil response modeling for well conductor fatigue analysis and development of new approaches, *Offshore Technology Conference*, Houston, TX, OTC-25795-MS, pp 1–30 (electronic format).

Zhang Y, 2016, Finite element analysis of elastic behavior of a suction caisson, Doctoral Dissertation, Texas A&M University, College Station, TX, 160p.

Chapter 9

Drag embedded anchors

Drag embedded anchors (DEAs) are bearing plates inserted into the seabed by dragging horizontally with a chain or wire rope. They are an attractive anchorage alternative owing to their relatively low installation cost and the relative high geotechnical efficiency of plate anchors in general. They can also be easily retrieved for reuse on other projects. A potential drawback of DEAs is the inability to achieve precise positioning. Uncertainty in vertical positioning translates into uncertainty in holding capacity. For this reason, they have traditionally been used largely for temporary moorings rather than for permanent facilities. Nevertheless, uncertainty in holding capacity can be mitigated to a great extent by proof testing to verify holding capacity, so DEAs can certainly be considered as a viable option for permanent facilities.

Historically, DEA designs have been developed by trial and error based on field tests and experience with actual deployments. More recently, the principles of geomechanics have been applied for understanding the DEA behavior. Such analyses include limit equilibrium analyses (Stewart, 1992; Neubecker and Randolph, 1996; Dahlberg, 1998), finite element studies, and PLAs of soil–anchor interactions (Rowe and Davis, 1982; O'Neill et al., 2003; Murff et al., 2005). Understanding the interaction between the mooring line and the anchor is crucial for predicting DEA performance, and contributions on this front include studies by Vivatrat et al. (1982), Neubecker and Randolph (1995a), and Aubeny and Chi (2010).

9.1 BASIC DESCRIPTION OF DEAs

A large number of anchor design have been developed, as shown in Figure 9.1. Early models featured a stock, a crosspiece at the top of the anchor. The stockless anchor, invented in 1821, removed this crosspiece, enabling the fluke to fully penetrate into the soil. The stockless anchor is still in use today. DEA designs typically feature stabilizing fins to restrict roll and yaw; hence, the complex anchor geometries shown in Figure 9.1. This is in contrast to the relatively simple rectangular or circular configurations characteristic of the directly embedded plate anchors considered in Chapter 10. Few analyses attempt to model the actual complex geometry of DEAs. Therefore, analytical models of anchor capacity should be regarded as providing useful insights into anchor performance but they require calibration to field or model test measurements.

Figure 9.2 illustrates the basic elements of a DEA. The fluke is a bearing plate that provides most of the anchor's load capacity. The shank controls the angle and location at which the line of action of the mooring line force passes through the fluke. A freely rotating shank is possible, but most DEAs have a fixed shank. Most DEAs also feature multiple settings for the angle between the shank and fluke. Small fluke–shank angle settings—on the order of 30°—are typically used in sand or stiff clay seabeds. Larger fluke–shank angles—on the

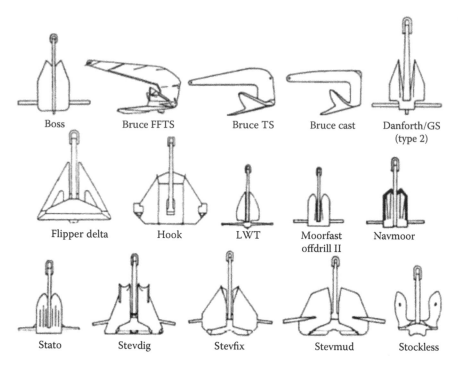

Boss   Bruce FFTS   Bruce TS   Bruce cast   Danforth/GS (type 2)

Flipper delta   Hook   LWT   Moorfast offdrill II   Navmoor

Stato   Stevdig   Stevfix   Stevmud   Stockless

Figure 9.1 Drag embedment anchors. (From NAVFAC, 2011, *SP-2209-OCN Handbook for Marine Geotechnical Engineering, Naval Facilities Engineering Command*, Engineering Service Center, Port Hueneme, USA.)

order of 50°—are used in soft clays. The projected area of the shank of many DEAs is relatively large normal to the direction of drag embedment (Figure 9.2), which is beneficial from the standpoint of reducing roll during drag embedment as well as providing resistance to out-plane loading (OOP). Figure 9.3 shows two widely used modern anchors. These anchors have twin shanks, which reduce the projected area of the shank in the direction of drag embedment. The anchor is attached to a chain or wire line at the pad eye.

Vertically load anchors (VLAs) are a variant of DEAs that have been developed for taut mooring systems. They are designed to develop high vertical load capacity to resist high angle loading. Figure 9.4 shows two commonly used VLAs. During installation, they function essentially in the same manner as a DEA. However, after installation, the anchor configuration is modified such that the mooring line load acts essentially normal to the fluke,

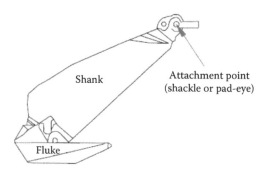

Shank   Attachment point (shackle or pad-eye)

Fluke

Figure 9.2 Components of DEA system.

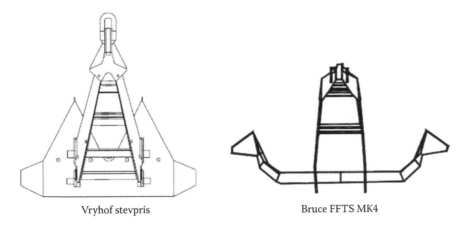

Vryhof stevpris   Bruce FFTS MK4

Figure 9.3 Widely used DEAs.

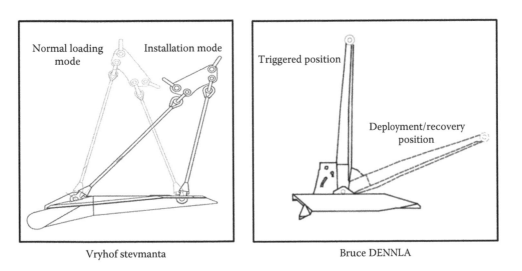

Vryhof stevmanta   Bruce DENNLA

Figure 9.4 Vertically loaded anchors (VLAs).

which can enhance the load capacity of the anchor. A common mechanism for releasing the shank to its postinstallation orientation is a shear pin, which ruptures when the load exceeds a predetermined value. Some shear pin designs require an increased mooring line angle—achieved by shortening the mooring line—to trigger the shear pin. While maximum load capacity is achieved by setting the postinstallation fluke–shank angle at 90° to the fluke, a "near-normal" VLA design sets this angle at a slightly lower value, say 75°, to reduce the potential for a brittle pullout failure. Thus, an overloaded anchor will tend to dive in a downward motion parallel to the fluke, rather than an upward motion normal to the fluke.

DEA embedment in stiff clays and sands is relatively shallow, typically less than 1–2 fluke lengths (NCEL, 1987). By contrast, DEA embedment in soft clay in soft clays can be on the order of tens of meters, with a correspondingly long drag distance. Since soil strength increases with depth, deep embedment is essential for mobilizing the full load capacity of the anchor. Figure 9.5 shows a conceptual sketch of the trajectory of a DEA installation in a soft clay seabed. Accurate prediction of anchor load capacity requires (1) accurate prediction of

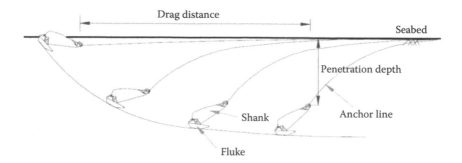

Figure 9.5 DEA installation.

the anchor trajectory during installation and (2) prediction of the load capacity at the actual depth at its installed depth. This is in contrast to most other anchor types, where vertical position is fairly well defined. The complex geometric configurations of most DEAs and VLAs together with their complex trajectories during installation pose a significant challenge from the standpoint of formulating analytical models of the DEA performance, thus, the heavy reliance until recently on an empirical design approach.

9.2 EMPIRICAL METHODS

Methods for estimating anchor load capacity and other aspects of anchor performance, such as embedment depth and required drag distance, have been developed by the US Navy as well as anchor manufacturers. The various correlations that have been developed are described below.

9.2.1 Design guide of NCEL

Based on prototype scale tests under various seabed soil conditions on diverse Navy and commercial drag anchors, the Naval Civil Engineering Laboratory (NCEL, 1987) developed a convenient empirical prediction method to predict anchor capacity. Their method correlates anchor weight to ultimate holding capacity. Holding capacity is more directly related to fluke areas; however, since fluke area correlates to anchor weight, anchor-specific correlations of ultimate holding capacity to anchor weight have validity. The studies suggest a power law relationship between ultimate holding capacity F and anchor weight W_a described by Equation 9.1. Since the original NCEL correlations were formulated in terms of English unit of kilopounds (1 kip $= 0.4536$ metric tons-force), those units are retained here.

$$F = a\left(\frac{W_a}{C}\right)^b \tag{9.1}$$

The term a is a capacity parameter having units of kips, b is a dimensionless exponent, and $C = 10$ kips. Figure 9.6 shows NCEL estimates of ultimate load capacity for various anchor types in mud (soft clay). The NCEL correlations were based on anchor installation using chain anchor lines. As will be discussed subsequently, embedment depth and capacity can be considerably greater when wire line is used.

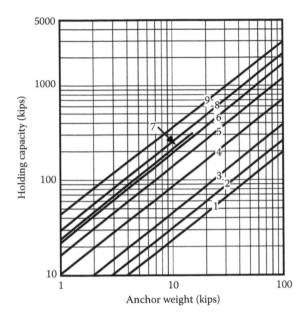

1. Stockless (movable fluke)
2. Bruce cast
3. Stockless (fixed fluke)
4. Danforth, GS, LWT*, Moorfast Offdrill II
5. Flipper delta, Slevin, Stevdig
6. Bruce TS, Hook, Stevfix*, Stevpris MK3
7. Navmoor stato boss
8. Bruce FFTS, Stevmud
9. Bruce FFTS MK4**, Stevpris MK5**

*Anchors require special handling to ensure fluke tripping (possibly fixed open flukes)

**Up to 20% increase when used with wire forerunner.

Figure 9.6 NCEL design chart for anchor capacity in soft clay.

The capacity estimates from Equation 9.1 and Figure 9.6 presume that the drag distance is sufficient for ultimate anchor capacity to fully mobilize. Various scenarios can occur where this is not the case. Therefore, the NCEL guidance includes estimates on mobilized anchor load capacity as a function of drag distance (Figure 9.7).

Occasions arise where anchor embedment depth can be of interest. For example, a stiff clay or sand stratum may occur in a predominantly soft clay soil profile, which would effectively limit the embedment depth and, therefore, the achievable load capacity of the anchor under consideration. Figure 9.8 provides estimates of fluke tip depth d_t normalized by fluke length L as a function of drag distance D for various anchor types in soft clay.

Drag embedment penetration depth and the overall performance of anchors in sands differ substantially from that in soft clays; therefore, NCEL developed a completely different set of correlations for sandy seabed soils, shown in Figure 9.9. Although anchor penetration depth is much lower in sands than soft clays, holding capacity in nonetheless relatively high, since drained loading conditions prevail. Despite the relatively high load capacity of DEAs in sand, it should be recalled that their shallow embedment depth precludes the possibility of significant vertical load capacity. Thus, their usage is largely restricted to catenary mooring systems.

NCEL (1987) indicates that 8 fluke lengths of drag distance is required for fixed fluke anchors and 10 fluke lengths are required for movable fluke varieties. Estimates of anchor embedment depth may also be required owing to constraints imposed by buried utility lines or details of the soil site stratigraphy. Table 9.1 provides guidance on embedment depth for various anchor types and soil conditions.

9.2.2 Manufacturer guidelines

Vryhof (2015), a major manufacturer of DEAs, follows the approach of NCEL in adopting a power law function to characterize the relationship between anchor weight and ultimate holding capacity. Figure 9.10 shows an example design chart for a Vryhof Mk6 Stevpris

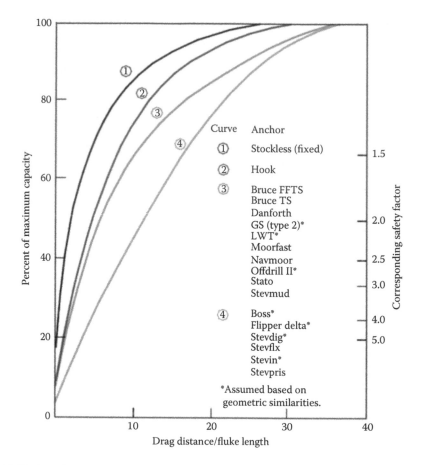

Figure 9.7 NCEL chart for development of anchor capacity with drag distance in soft clay.

anchor. The chart includes capacity estimates for anchors in sand, stiff clay, and soft clay seabed soil conditions. Information on the drag distance required to achieve ultimate holding capacity and anchor penetration depth is conveniently included on the same graph.

As discussed earlier, mobilized holding capacity (UHC) as a function of drag distance is often of interest. Vryhof (2015) provides the estimates shown in Table 9.2 of percentage of UHC mobilized as a function drag distance expressed as a percentage of the drag distance required to mobilize UHC.

Manufacturer data may also account for the influence of chain versus wire line forerunners on anchor performance, which can be very significant for DEA installations in soft clay. Figure 9.11 shows the anchor capacity curves for a Bruce FFTS MK4 anchor in sand and soft clay. For soft clay, separate curves are provided for chain and wire line systems. This particular correlation shows UHC to be approximately 25% greater when a wire forerunner is used.

9.3 UNDRAINED BEHAVIOR OF AN EMBEDDED PLATE

Empirical correlations undoubtedly have great practical utility in predicting anchor performance. However, a purely empirical approach has inherent limitations in that the expense of field tests typically precludes full evaluation of all of the variables that can affect anchor

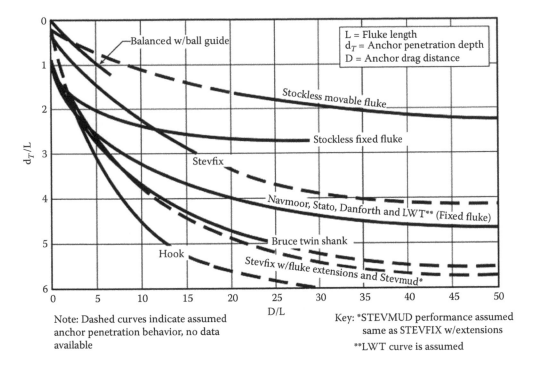

Figure 9.8 NCEL chart for anchor penetration in clay.

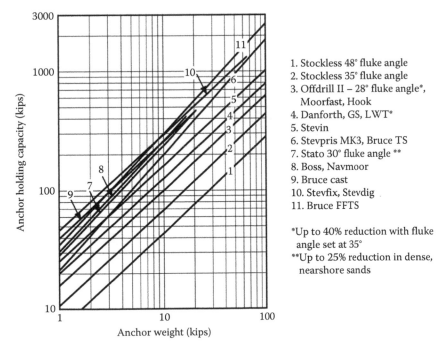

Figure 9.9 NCEL design chart for anchor capacity in sand.

Table 9.1 Anchor penetration in various soil conditions

| | Normalized fluke tip penetration (fluke lengths) | |
Anchor type	Sands/stiff clays	Mud (e.g., soft silts and clays)
Stockless	1	3
Moorfast Offdrill II	1	4
Stato Stevfix[a] Flipper delta Boss Danforth LWT[a] GS (type 2)	1	4.5
Bruce twin shank Stevmud	1	5.5
Hook	1	6

Source: NCEL, 1987, *Drag Embedment Anchors for Navy Moorings*, Techdata Sheet
83-08R, Port Hueneme, California: Naval Civil Engineering Laboratory.

[a] In mud, anchor flukes fixed fully open or held open initially.

performance—mooring line diameter, chain versus wire forerunners, anchor fluke and shank geometry, and details of the soil strength profile. Furthermore, field tests often exhibit scatter and uncertainty, which is difficult to resolve without a reliable theoretical framework. Analytical models can thus play an important role for interpreting and understanding field test measurements, as well as providing a rigorous basis for extrapolating predictions of anchor performance to conditions outside of the range for which the field tests were

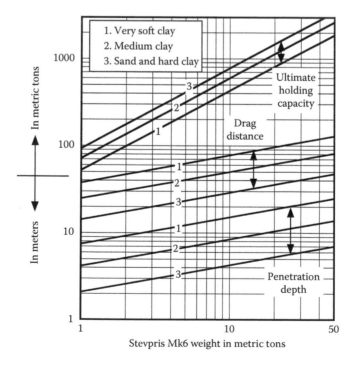

Figure 9.10 Load capacity chart for Vryhof Stevpris MK5 anchor.

Table 9.2 Mobilized capacity versus normalized drag distance

Capacity/UHC (%)	Drag/max drag distance (%)	Penetration/max penetration (%)
70	48	80
60	37	68
50	27	55
40	18	42
30	9	23

Source: Vryhof, 2015, *Anchor Manual 2015 The Guide to Anchoring*, Vryhof Anchors. Capelle a/d Yssel, The Netherlands, 168p.

conducted. This section discusses the application of plasticity theory to characterize the load resistance and kinematic behavior of DEAs. The focus will be on the undrained response of deeply embedded anchors in clay.

9.3.1 Load capacity

DEAs typically have complex geometric configurations to maintain stability during embedment. Such complexity is generally beyond the reach of the plastic limit solutions for strip and

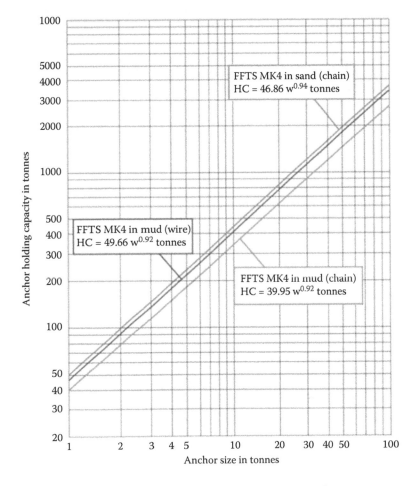

Figure 9.11 Load capacity chart for Bruce FFTS MK4 anchor (Bruce, undated).

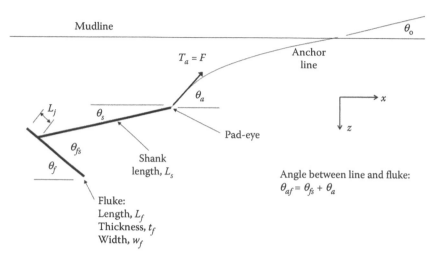

Figure 9.12 Definition sketch of idealized drag embedment anchor.

rectangular plates presented in Section 4.1. Nevertheless, simplified analyses based on ideal-ized plate anchors comprising rectangular plates of uniform thickness are useful for gaining insights into the fundamental mechanics of drag anchors without the distractions associated with numerical solutions for complex geometric configurations. With this in mind, we will consider the case of the fluke comprising a simple strip anchor in Figure 9.12 of length L_f and thickness t_f. A shank of length L_s is rigidly attached to the fluke at a fluke–shank angle θ_{fs}. At a given instant in the anchor trajectory, the fluke and shank are oriented at θ_f and θ_s, respectively. The anchor is attached to an anchor line which, at an incipient state of collapse, has a line tension T_a at the pad eye uplifted at an angle θ_a. The analysis here is performed for the limit state such that the applied tension T_a equals the ultimate load capacity F of the anchor.

The analysis is simplified by describing the applied load at the pad eye in terms of its posi-tion and orientation relative to the center of the plate, as shown in Figure 9.13. Specifically,

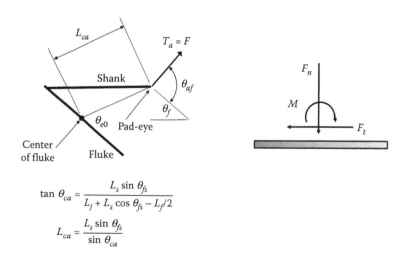

Figure 9.13 System of forces on anchor fluke.

the angle θ_{ca} and the length L_{ca} define the direction and length of a line segment connecting the center of the fluke to the pad eye. Also of particular significance to the capacity calculation is the direction of the applied load relative to the orientation of the fluke, $\theta_{af} = \theta_a + \theta_f$. The normal force F_n, tangential force F_t, and moment M acting on the fluke can defined by the following equations.

$$F_n = F\sin\theta_{af} = Fc_1 \tag{9.2}$$

$$F_t = F\cos\theta_{af} = Fc_2 \tag{9.3}$$

$$M = FL_f[\cos\theta_{af}\sin\theta_{ca} - \sin\theta_{af}\cos\theta_{ca}](L_{ca}/L_f) = FL_fc_3 \tag{9.4}$$

Normalizing by soil strength and fluke length leads to the following non-dimensional expressions for anchor capacity N_e, and normal (N_n), tangential (N_t), and moment (N_m) resistance:

$$N_e = F/s_u A_f \tag{9.5}$$

$$N_n = F_n/s_u A_f = N_e c_1 \tag{9.6}$$

$$N_t = F_t/s_u A_f = N_e c_2 \tag{9.7}$$

$$N_m = M/S_u A_f L_f = N_e c_3 \tag{9.8}$$

We may now first consider the case in which no interaction effects occur under combined loading; that is, load capacity in one direction of loading is unaffected by load demand from another direction. In this case, the load capacity for the anchor will be the least of the following three collapse loads:

Normal Mechanism : $N_{en} = N_{nmax}/|c_1|$ $\hspace{3cm}$ (9.9)

Tangential Mechanism : $N_{et} = N_{tmax}/|c_2|$ $\hspace{2.8cm}$ (9.10)

Rotational Mechanism : $N_{em} = N_{mmax}/|c_3|$ $\hspace{2.8cm}$ (9.11)

Figure 9.14 presents an example computation of the bearing factor N_e as a function of load angle θ_{af} for the case of a strip anchor with pad-eye distance $L_{ca} = L_f$ and a pad-eye angle $\theta_{ca} = 60°$. Bearing factors in this example are computed for a soil sensitivity $S_t = 3$. The resistance associated with a normal collapse mechanism is well in excess of that associated with tangential or rotational collapse. This is typical of drag anchors; in fact, the basic concept of the DEA is for it to dive downward in the tangential direction, as opposed to plowing upward in a normal collapse mode. The rotational collapse mechanism is seen to control when the load angle θ_{af} deviates significantly from the pad-eye angle θ_{0a}, in this case, for θ_{af} less than approximately 30° and greater than 70°. As θ_{af} approaches θ_{ca}, the moment resistance approaches infinity, in which case translation parallel to the fluke becomes the operative collapse mechanism. It is noted that if a rotational collapse actually occurs, the

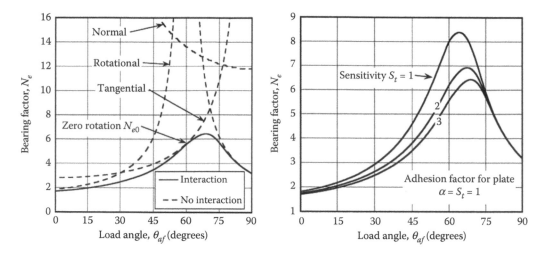

Figure 9.14 Bearing factor versus load angle.

anchor will reorient itself such that the difference between θ_{af} and θ_{ca} is small, in which case the tangential mechanism becomes the controlling collapse mode.

A more realistic characterization of plate anchor behavior must consider the effects of interaction under combined loading. Section 4.1.3 earlier addressed this topic, and invoking yield locus equation for combined loading on a plate Equation 4.5 in conjunction with Equations 9.5 through 9.8 yields the following equation defining the combined loading bearing factor N_e for a plate considering interaction effects:

$$f = \left[\left(\frac{|c_1| N_e}{N_{n\,max}} \right)^q + \left[\left(\frac{|c_3| N_e}{N_{m\,max}} \right)^m + \left(\frac{|c_2| N_e}{N_{t\,max}} \right)^n \right]^{\frac{1}{p}} \right] - 1 = 0 \tag{9.12}$$

The effective bearing factor for the anchor is then taken as the root of the equation f $(N_e) = 0$. Figure 9.14 shows the load capacity curve for the case considered in the previous paragraph. The interaction curve is predictably a subdued version of the controlling no-interaction curves. A loading condition of great relevance to drag anchor behavior is when zero moment loading occurs; that is, when the line of action of the applied load passes through the center of the plate, in this case when $\theta_{af} = \theta_{ca} = 60°$. The bearing factor corresponding to zero rotation, N_{e0}, computed from Equation 9.12 is nearly identical to the no-interaction bearing factor N_{et} for tangential loading Equation 9.10. In effect, the interaction effects are sufficiently small for this loading case to be safely neglected, in which case the operative bearing factor for the anchor can be simply estimated from Equation 9.10. This simplification often proves to be quite useful; however, it is emphasized that this observation applies only to the pad-eye angles typical of DEAs, say $\theta_{0a} < 60°$. At larger angles, the normal load demand becomes substantially greater; therefore, the interaction effect becomes more significant and N_{e0} computed from Equation 9.12 is significantly less than N_{et}. The right-hand plot in Figure 9.14 shows the effect of soil sensitivity on anchor load capacity. Depending on the direction of applied loading, when considering increasing soil sensitivity over a range $S_t = 1$–3 the bearing factor N_e can decline by a factor of up to one-third.

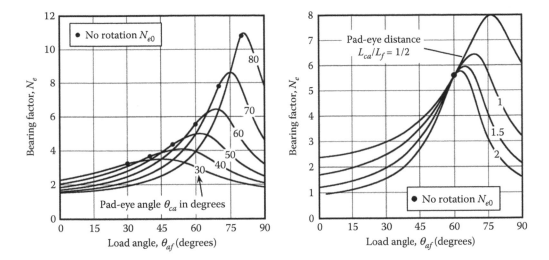

Figure 9.15 Effect of shank geometry on anchor bearing resistance.

Figure 9.15 shows the influence of basic drag anchor geometric variables on load capacity characteristics. The plot on the left shows the influence of pad-eye angle θ_{ca}, which is closely linked to the fluke–shank angle setting θ_{fs} for the anchor. Since plate anchors tend to rotate to an orientation where the line of action of the applied load passes through the center of the plate, $\theta_{af} = \theta_{ca}$, the bearing resistance N_{e0} often becomes the operative bearing factor for the anchor. The left-hand plot in Figure 9.15 shows N_{e0} to be highly sensitive to pad-eye angle, increasing by a factor of 3–4 as θ_{0a} increases from 30° to 80°. However, it will be seen that this increased load capacity comes at the price of an increased tendency for upward movement normal to the fluke, with a concomitant loss of embedment. The concept of the VLA is to install the anchor using a smaller fluke–shank angle to promote deep embedment, and then open the shank following installation to maximize load capacity. The right-hand plot in Figure 9.15 shows the effect of the distance between the pad eye and the center of the plate, L_{ca} (closely related to the shank length L_s). The no-rotation bearing factor N_{e0} is completely unaffected by this distance. Larger shank length imposes greater moment demand on the anchor, thereby reducing its overall capacity. The complex geometry of actual DEAs do not permit a precise determination of the pad-eye distance; however, L_{ca} for DEAs used in practice is roughly equal to the fluke length.

9.3.2 Kinematic response

During drag embedment, the relative magnitudes of translational and rotational motions can be of comparable importance to load capacity. Assuming an associated flow law, the angular and tangential velocity of the fluke, $\dot{\beta}$ and v_t, can be computed by taking appropriate partial derivatives of f:

$$\dot{\beta} = \lambda\, \partial f / \partial M \tag{9.13}$$

$$v_t = \lambda\, \partial f / \partial F_t \tag{9.14}$$

$$v_n = \lambda\, \partial f / \partial F_n \tag{9.15}$$

where λ is a scalar multiplier. The value of the scalar multiplier is indeterminate, but it turns out to be unnecessary for evaluating relative motions. The ratio of normal to tangential translation R_{nt} is then evaluated by the following equation:

$$R_{nt} = v_n/v_t = \frac{\left(N_{tmax}/N_{nmax}\right)(pq/n)}{\left[\left(|N_m|/N_{mmax}\right)^m + \left(|N_t|/N_{tmax}\right)^n\right]^{(1/p)-1}} \frac{\left(|N_n|/N_{nmax}\right)^{q-1}}{\left(|N_t|/N_{tmax}\right)^{n-1}} \tag{9.16}$$

Plotting a range of R_{nt} spanning from zero to infinity is facilitated by introducing a normal motion parameter ψ_n defined as follows:

$$\psi_n = \frac{2}{\pi}\tan^{-1}(v_n/v_t) \tag{9.17}$$

A motion parameter $\psi_n = 0$ denotes pure translation parallel to the fluke, while $\psi_n = 1$ denotes pure normal motion. The ratio of rotation to tangential translation R_{rt} may similarly be defined as follows:

$$R_{rt} = \dot{\beta}L_f/v_t = \text{sign}(c_3)\frac{m}{n}\frac{N_{tmax}}{N_{mmax}}\frac{\left(|N_m|/N_{mmax}\right)^{m-1}}{\left(|N_t|/N_{tmax}\right)^{n-1}} \tag{9.18}$$

Again, to map a range of values from zero to infinity, we can define rotation parameter ψ_r as follows:

$$\psi_r = \frac{2}{\pi}\tan^{-1}(\dot{\beta}L_f/v_t) \tag{9.19}$$

where ψ_r is positive for positive N_m and vice versa. For pure tangential translation $\psi_r = 0$, while for pure counter-clockwise rotation about the center of the plate $\psi_r = 1$.

Figure 9.16 plots ψ_n as a function of θ_{af} for the case considered in the previous section. Soil sensitivity S_t has some influence on the behavior. This is predictable, since the greater tangential resistance associated with lower sensitivity leads to a greater proportion of the motion being normal to the fluke. In general, for load angles less than about $\theta_{af} = 50°$, motions normal to the fluke are relatively small. For load angles greater than about $\theta_{af} = 75$–$80°$, the magnitude of normal motion begins to exceed that of tangential motion. As noted earlier, greater load angles generate greater bearing resistance, but for angles greater than 75–80°, this is seen to create an undesirable tendency for upward movement of the anchor, with a consequent loss of embedment and load capacity. The right-hand side of Figure 9.16 shows the rotational characteristics of the anchor over a sweep of load angles $\theta_{af} = 0$–$90°$ for the cases of interactive and noninteractive bearing resistance behavior. Naturally, zero rotation occurs when the line of action of the applied load passes through the center of the fluke; that is, $\theta_{af} = \theta_{0a} = 60°$ in this example. Accurate modeling of the interaction effect is seen to be important for predicting rotational behavior of the anchor. For example, the interactive model predicts significant rotations to occur when θ_{af} deviates from θ_{0a} by more than 5–10°, behavior which the noninteractive model fails to capture.

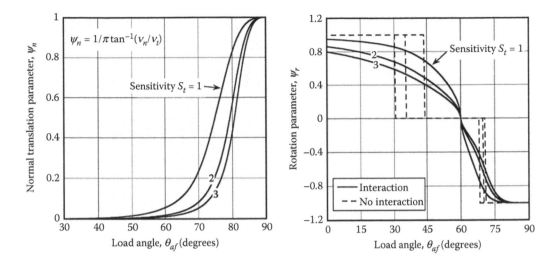

Figure 9.16 Kinematic behavior of a drag embedded anchor.

9.3.3 Shank resistance

The preceding discussion applies to the performance of anchors having shanks of negligible thickness; that is, the positioning of the shank affects the system of forces acting on the fluke, but otherwise offers no soil resistance itself. Some anchor designs actually conform to this idealization, particularly some of the VLAs having a bridle shank comprised of chains. On the other hand, many DEA designs feature shanks having a large cross-sectional area that should be considered in a model of anchor behavior. The analysis presented here will be for a shank comprised of two plate sections similar to that shown in Figure 9.17. In a manner similar to the idealization of the fluke as a simple rectangular plate, the geometry of the shank is analyzed as two parallel rectangular plates of length L_s, width w_s, and thickness t_s. The focus of attention will be on a particularly important aspect of anchor behavior, load capacity under conditions of zero moment loading. Toward this end, the analysis utilizes the equations described earlier for fluke load capacity Equation 9.12 and motions Equation 9.16, with zero inserted for all moment terms. The shank resistance yield locus will have a similar form, defined by Equation 9.20.

$$f_s = \left(\frac{N_{ns}}{N_{n\,max\,s}}\right)^r + \left(\frac{N_{ts}}{N_{t\,max\,s}}\right)^s - 1 = 0 \tag{9.20}$$

Information on interaction coefficients r and s for translating plates is relatively scant, but finite element studies for low aspect ratio plates, $L_s/w_s = 2$, indicate $r = s = 2.5$. Maximum resistance for uniaxial loading follow the form of Equation 4.2 in Section 4.1.1 for loading normal and parallel to the long axis of the shank:

$$N_{n\,max\,s} = \frac{F_{n\,max\,s}}{L_s w_s} = 2\alpha + 15\frac{t_s}{w_s}$$
$$N_{t\,max\,s} = \frac{F_{t\,max\,s}}{L_s w_s} = 2\alpha + 7.5\frac{t_s}{L_s} \tag{9.21}$$

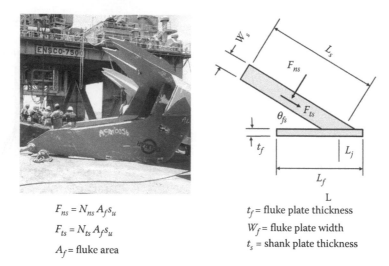

$$F_{ns} = N_{ns} A_f s_u$$

$$F_{ts} = N_{ts} A_f s_u$$

$$A_f = \text{fluke area}$$

t_f = fluke plate thickness

W_f = fluke plate width

t_s = shank plate thickness

Figure 9.17 DEA with thick shank.

Only one end resistance term is assigned to the parallel load resistance term, since soil end bearing resistance cannot mobilize where the shank joins to the fluke. A question arises as to whether the shank resistance can fully mobilize, since interference with the fluke yield mechanism inevitably occurs. Nevertheless, for the purpose of gaining insights into the effect of shank resistance interference effects are not considered here. Finally, the analysis requires the definition of shank motions, which following the approach described earlier for the fluke, has the following form:

$$R_{nts} = \frac{v_{ns}}{v_{ts}} = \frac{r}{s} \frac{N_{tmaxs}}{N_{nmaxs}} \frac{(N_{ns}/N_{nmaxs})^{r-1}}{(N_{ts}/N_{tmaxs})^{s-1}} \tag{9.22}$$

In addition to the equations governing the yield loci and the conditions of plastic flow, the analysis requires the definition of the kinematic and equilibrium constraints on the individual anchor elements—fluke and shank—of the anchor. These are illustrated, respectively, in Figures 9.18 and 9.19.

The analysis proceeds in the following iterative sequence:

1. Select a trial load angle for the soil resistance on the fluke $\theta_i = \tan^{-1}(F_{nf}/F_{tf})$.
2. Compute the fluke bearing factor N_{e0} for this load angle using Equation 9.12, and compute the fluke soil resistance forces $F_{nf} = N_{e0} s_u A_f \sin \theta_i$ and $F_{tf} = N_{e0} s_u A_f \cos \theta_i$.
3. Compute the fluke ratio of normal to tangential motion R_{ntf} using Equation 9.16.
4. Compute the ratio of normal to tangential motion R_{nts} for the shank using the kinematic relationship shown in Figure 9.18.
5. Compute the direction of the soil resistance bearing factors N_{ns} and N_{ts} that satisfy the kinematic constraint imposed by Equation 9.22 and the yield locus constraint imposed by Equation 9.20.
6. From equilibrium the pad-eye force T_a and the moment at the junction of the shank and fluke M_j can be computed as shown in Figure 9.19.
7. The value of the joint moment required for zero net moment on the fluke, M_{j0}, is also shown in Figure 9.19. If this requirement is not satisfied, the fluke load angle θ_i in Step 1 is increased and the calculation sequence is repeated.

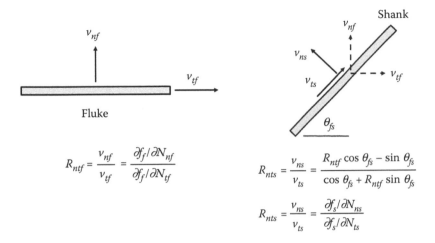

Figure 9.18 Kinematics of fluke and shank.

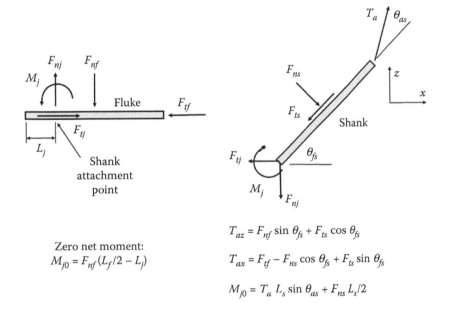

Figure 9.19 Equilibrium of fluke and shank.

Figure 9.20 shows the effect of shank resistance on anchor properties for various shank plate thickness t_s. The analysis considers a shank comprising two parallel plates having aspect ratio $L_s/w_f = 6$. The zero shank thickness case shown corresponds to a condition of no soil resistance acting on the shank. The analysis shows shank resistance to have two significant effects on anchor behavior. Firstly, the effective bearing factor N_{e0} increases with increasing soil resistance on the shank. This obviously results in a direct increase in anchor capacity. It will also be shown that a greater N_{e0} leads to increased anchor penetration during drag embedment. Shank resistance also increases the angle θ_{e0} between the pad-eye load and the fluke. This has the effect of tilting the fluke downward, which also promotes greater embedment.

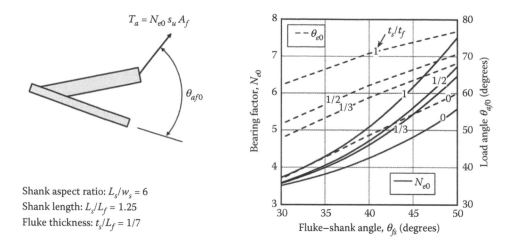

Figure 9.20 Effect of shank resistance on anchor bearing factor.

9.4 EMBEDMENT TRAJECTORY

To this point anchor, the discussion of DEA behavior has centered solely on the load capacity and kinematic behavior of the anchor, where it was shown that the yield locus equation combined with equilibrium equations produces a load capacity curve relating load angle relative to the fluke to ultimate load capacity. Determination of a unique pad-eye tension at a given instant of drag installation requires a second relationship between load angle and direction; the Neubecker and Randolph (1995a) anchor line equation discussed in Section 5.2 provides the required second equation. For the typical case of zero mudline angle ($\theta_0 = 0$) that prevails during drag embedment and for a linearly varying strength profile, their equation is as follows:

$$T_a\theta_a^2/2 = z_a E_n N_c b(s_{um} + kz_a/2) \tag{9.23}$$

z_a = depth of pad eye
s_{um} = soil strength at mudline
k = soil strength gradient
E_n = area multiplier for chains
N_c = bearing factor for anchor line
b = anchor line diameter
T_a = anchor line tension at pad eye

For a given fluke orientation θ_f, the use of Equation 9.23 together with the anchor load capacity–direction curves described in the previous section define a unique T_a–θ_a combination, as illustrated in Figure 9.21. The interaction with the anchor line forms the basis for the analysis of two stages of anchor embedment: initial setting (or keying) of the anchor and the post-set trajectory.

9.4.1 Anchor set

Depending on the initial orientation of the fluke, the anchor will be subjected to some level of moment loading. Anchor set in this discussion will refer to the process during which the

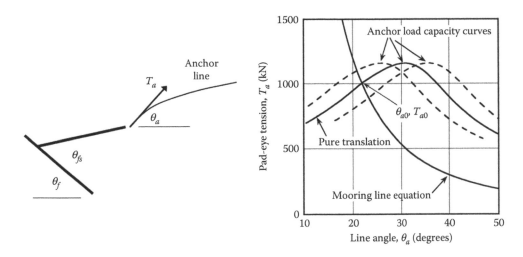

Figure 9.21 Anchor interaction with mooring line.

anchor rotates in response to the applied moment until it aligns itself into an orientation at which zero moment acts on the anchor. For the case of a rectangular fluke with an infinitely thin shank, this orientation corresponds to a condition where the line of action of the pad-eye force T_a passes through the center of the fluke. For a given initial fluke orientation θ_{fi}, analysis of anchor set can proceed according to the following steps:

1. Compute the load capacity curve from Equation 9.12, the normal motion R_{nt} versus θ_{af} relationship from Equation 9.16, and the rotational motion versus θ_{af} relationship R_{rt} from Equation 9.18. Figures 9.15 and 9.16 show example results from these calculations.
2. For a trial pad-eye load angle θ_a, compute the pad-eye tension T_a from Equation 9.23 and the load–fluke angle $\theta_{af} = \theta_a + \theta_f$.
3. Compare the anchor load capacity $F = N_e(\theta_{af}) \, s_{ua} \, A_f$ to T_a. Iterate on θ_a until F and T_a agree to be within a prescribed tolerance.
4. Calculate the motion parameters $R_{nt}(\theta_{af})$ and $R_{rt}(\theta_{af})$.
5. Advance the anchor a distance Δt parallel to the fluke, and compute the vertical and horizontal components of motion $\Delta z_a = \Delta t \, \sin\theta_f - \Delta t \, R_{nt} \, \cos\theta_f$ and $\Delta x = \Delta t \, \cos\theta_f + \Delta t \, R_{nt} \, \sin\theta_f$.
6. Compute the fluke rotation $\Delta\theta_f = \Delta t \, L_f \, R_{rt}$, and update the fluke angle $\theta_f = \theta_f + \Delta\theta_f$. Repeat Steps 2 through 6 until an anchor orientation corresponding to a condition of zero moment loading occurs.

Figure 9.22 shows an example anchor set calculation for an anchor with fluke length $L_f = 3$ m and fluke–shank angle $\theta_{fs} = 50°$ for two initial fluke orientations. In the first case the shank rests horizontally such that the initial fluke orientation is $\theta_{fi} = 50°$. The second case considers the other extreme, with the fluke oriented horizontally $\theta_{fi} = 0°$. The analyses are initiated assuming an initial self-weight penetration of the anchor into the seabed, $z_a = 1$ m. In the case of $\theta_{fi} = 50°$, the anchor is nearly in its keyed orientation at the outset; therefore, keying is seen to occur almost immediately. By contrast, $\theta_{fi} = 0°$ considerable anchor rotation must occur for keying to occur. In this case, the model predicts a required drag distance of about one fluke length for the necessary fluke rotation to occur.

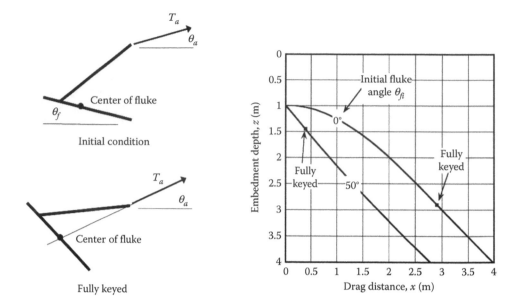

Figure 9.22 Anchor trajectory during setting stage.

9.4.2 Post-set trajectory

After keying occurs, the analysis simplifies somewhat since the bearing factor is now a constant $N_e = N_{e0}$, which avoids the requirement for an iterative analysis to identify the intersection between the load capacity and anchor line tension curves. Additionally, the relative magnitude of normal to tangential motion is also constant, $R_{nt} = R_{nt0}$. While anchor continues to occur, as pointed out by Aubeny and Chi (2010), rotation is entirely controlled by changes in the pad-eye line angle θ_a, as illustrated in Figure 9.23. From implicit differentiation of Equation 9.23 of θ_a with respect to z_a, Aubeny and Chi (2010) reformulated this equation to the following equation for computing the rate of anchor rotation that drives the trajectory calculation:

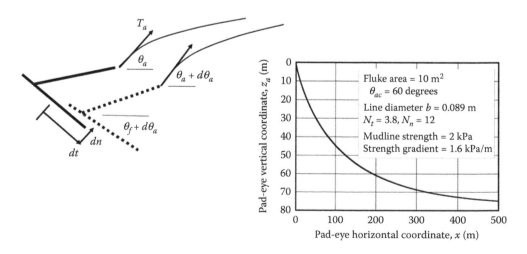

Figure 9.23 Post-setup anchor trajectory.

$$\frac{d\theta_a}{dz} = \frac{1}{\theta_a}\left(\frac{E_n N_c b}{N_{e0} A_f} - \frac{k\theta_a^2}{2s_{ua}}\right)$$

(9.24)

where z is the shackle depth, N_{e0} is the anchor bearing factor for zero rotation state, s_{ua} is the soil shear strength at the shackle, and A_f is the fluke area.

Two constant properties of the anchor itself, N_{e0} and A_f, influence the trajectory calculation, which is otherwise dominated by mooring line behavior. This equation provides the basis for a recursive algorithm for predicting anchor trajectory comprising the following steps:

1. Advance the anchor a distance Δt parallel to the fluke and resolve the motion into a vertical and horizontal components $\Delta z_a = \Delta t \sin\theta_f - \Delta t\, R_{nt0} \cos\theta_f$ and $\Delta x = \Delta t \cos\theta_f + \Delta t\, R_{nt0} \sin\theta_f$.
2. Compute the change in the anchor line angle $\Delta\theta_a$ over the depth increment Δz using Equation 9.24.
3. Update the anchor line angle, $\theta_a = \theta_a + \Delta\theta_a$, and fluke angle, $\theta_f = \theta_f - \Delta\theta_a$.
4. Steps 1 through 3 are repeated until the drag distance of interest has been reached or to the ultimate state where the fluke reaches a horizontal orientation $\theta_f = 0$.

Figure 9.23 shows an example trajectory prediction. As with many trajectory calculations, the anchor approaches its ultimate embedment depth ($\theta_f = 0$) at a drag distance that considerably exceeds a typical drag distance in practice. Additionally, the large embedment depth shown here will well lead to an anchor load capacity that exceeds the bollard pull of many AHVs. Nevertheless, computation of the ultimate embedment depth provides a useful picture of the full potential of the anchor.

A question that naturally arises is in regard to the effect of the initial anchor set process on the subsequent post-set behavior. Figure 9.24 shows two simulations of anchor trajectory for the case considered in Figure 9.22. The left-hand plot shows anchor embedment versus drag distance, while the right-hand plot shows pad-eye tension T_a. The analyses indicate that, depending on the initial orientation of the anchor θ_{fi}, the setting process will produce

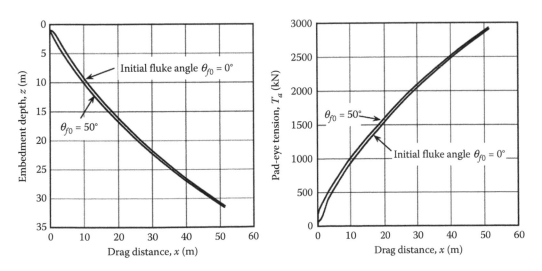

Figure 9.24 Effect of keying on subsequent trajectory.

an offset in the trajectory. Otherwise, details of what occurs during setting exert little influence on the subsequent anchor trajectory calculation.

9.5 FACTORS AFFECTING DEA PERFORMANCE

Based on the models described above for estimating anchor load resistance and kinematic behavior, the relative significance of basic anchor, soil, and mooring line characteristics can be explored. Toward this end, we will consider a DEA with area $A_f = 10$ m² with an aspect ratio $w_f/L_f = 1.25$ and fluke thickness $t_f/L_f = 7$ (Figure 9.12). In regard to the selected fluke thickness, it is noted that this represents a typical *projected* thickness of a DEA fluke, which is not typically perfectly planar. The physical thickness of the actual plate is often less than the projected thickness. When shank resistance is considered, the shank plate will have a thickness $t_s/t_f = 1/3$, with a shank plate aspect ratio $L_s/W_s = 6$ (Figure 9.17). In all cases, the shank is joined to the fluke at $L_f/L_f = 1/4$. The base studies consider an anchor wire line diameter $b = 0.1$ m, although this value will be varied in that portion of the parametric study considering mooring line properties.

Anchor parameters varied in the parametric study are the fluke–shank angle θ_{fs} and the shank thickness (Table 9.3). Anchor installation is in a normally consolidated clay profile, $k = 1.6$ kPa/m, with soil sensitivity varied over a range $S_t = 1-3$. Basic anchor and soil characteristics are used in conjunction with the plastic limit model formulation described in this chapter to produce three operative parameters governing the post-set behavior of the anchor: the bearing factor N_{e0}, the pad-eye load angle relative to the fluke θ_{e0}, and the ratio of normal to tangential motion $R_{nt0} = v_n/v_t$. All three operative parameters correspond to a condition of zero net moment load on the anchor, which represents post-set conditions. The two aspects of anchor performance of ultimate interest are embedment depth z and mudline mooring line tension T_0 as a function of drag distance x.

9.5.1 Anchor properties

Figure 9.25 compares computed anchor trajectory and line tension history for three different fluke–shank angles, with and without shank resistance. Increasing the fluke–shank angle increases the operative bearing factor N_{e0}, which provides the dual benefit of decreasing the

Table 9.3 Factors considered in parametric study

	Basic anchor properties		Operative anchor parameters		
Sensitivity S_t	Fluke–shank angle θ_{fs} (degrees)	Shank resistance t_s/t_f	Bearing factor N_{e0}	Load–fluke angle θ_{e0} (degrees)	Normal motion R_{nt0}
3	30	0	3.51	36.9	0.0010
3	40	0	4.25	48.6	0.0043
3	50	0	5.58	60.0	0.020
3	30	1/3	3.57	47.8	0.00085
3	40	1/3	4.60	58.8	0.0041
3	50	1/3	6.47	68.2	0.021
1	30	0	5.17	36.9	0.0053
1	50	0	8.01	60.0	0.089
2	30	0	3.93	36.9	0.0017
2	50	0	6.21	60.0	0.0297

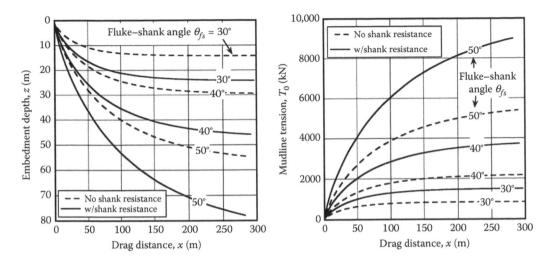

Figure 9.25 Effect of fluke–shank angle and shank resistance.

rate of anchor rotation during drag installation and increasing the load capacity. A larger fluke–shank angle also increases the load angle θ_{e0} for the anchor, which is crucial for tilting the fluke downward at the onset of drag installation, thereby promoting more diving action. A larger fluke–shank angle will finally increase the ratio of normal to tangential motion of the anchor during installation, which tends to reduce overall embedment depth. However, Figure 9.25 indicates that the benefits of increased bearing factor and load angle far outweigh the potential down side of increased normal motion, with resultant anchor load capacity for a 50° fluke angle being at least six times greater than that for a 30° fluke angle. Shank resistance tends to increase N_{e0} and θ_{e0}, thereby promoting deeper embedment and greater load capacity.

9.5.2 Soil properties

Figure 9.26 shows the effect of soil sensitivity on anchor performance. Reduced soil sensitivity has some of the same effects as increased fluke–shank angle, namely, an increase in both the operative bearing factor N_{e0} and the ratio of normal to tangential motion R_{nt0}. Since variations in sensitivity have no influence on load angle (for the case of an infinitely thin shank), the sole effect of reducing sensitivity is to reduce the rate of anchor rotation, thereby leading to increases in embedment depth on the order of 15%–30% when comparing trajectories for sensitivity $S_t = 1$ and 3. The effect of sensitivity on load capacity is more marked, since the increase in bearing factor owing to reduced sensitivity translates directly into increased load capacity. Variation in sensitivity over a range of 1–3 can lead to variation in load capacity on the order of 60%–100%, depending on fluke–shank angle.

9.5.3 Mooring line properties

Examination of Equation 9.24 shows that as anchor line resistance increases, the rate of anchor rotation during drag embedment also increases. The three terms controlling line resistance are the line diameter b, the line bearing factor N_c, and, for the case of a chain, the multiplier E_n. Recognizing that the N_c parameter can vary, particularly owing to variations

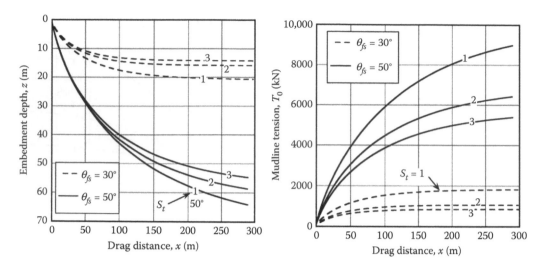

Figure 9.26 Effect of soil sensitivity.

in soil sensitivity, this parametric study adopts a single value $N_c = 10$ for all cases considered. The line diameter b will be considered in terms of its size relative to the anchor fluke. Relative fluke areas $A_f/b^2 = 800$–1200 represents the range of mooring system designs observed in practice, and this range will be considered in the parametric study. Finally, the use of wire rope versus chain is considered, which corresponds to E_n values of 1.0 and 2.5, respectively. Since the evaluations of anchor capacity are in terms of tension at the mudline, compute values of tension at the pad eye must be adjusted upward according to the Neubecker and Randolph (1995a,b) equation, which for a horizontal mudline angle is $T_0 = T_a \exp(\mu\theta_a)$. From Section 5.2, the parameter μ is the ratio of tangential to normal soil resistance acting on the line. This parametric study adopts $\mu = 0.1$ for wire rope and $\mu = 0.5$ for chain.

Figure 9.27 shows mooring line dimensions to have a pronounced effect on trajectory. Varying the factor A_f/b^2 over the range 800–1200 affects embedment depth by 20%–25%.

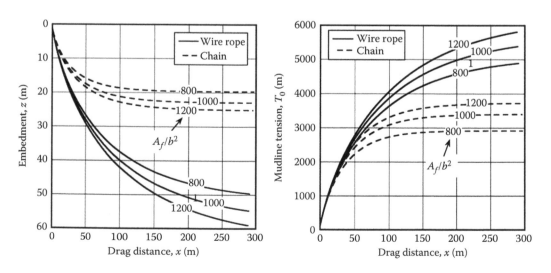

Figure 9.27 Effect of mooring line resistance.

More remarkable is the effect of using wire rope as opposed to chain, with embedment depths using wire being more than twice those for a chain installation. Although chain systems do not embed as deeply into the seabed as wire systems, chains have the benefit of contributing more to mudline tension T_0 by virtue of the additional line friction associated with a higher μ value. However, Figure 9.27 shows that even with the added contribution of chain friction, T_0 for an anchor with a wire forerunner still substantially exceeds that for the same anchor attached to a chain forerunner.

9.6 CALIBRATION TO CASE HISTORY DATA

Theoretical parametric studies provide useful insights into the relative effects of soil properties and anchor geometry on drag anchor performance. However, despite the relatively large number of input parameters required to characterize plate anchor behavior (e.g., three uniaxial bearing factors and four interaction coefficients), the analysis still falls short of accounting for the relatively complex anchor geometries occurring in practice. Lacking a database of parameters derived specifically for the complex configurations of DEAs used in practice, this section considers a modified version of the analytical model presented earlier to make it more amenable to interpreting field load test data. Basic elements of this approach are as follows:

1. The primary field measurements include anchor embedment depth z and mudline tension T_0 as a function of drag distance x.
2. The focus of the data interpretation is the operative bearing factor of the anchor N_{e0} and the load angle between the pad eye and fluke θ_{e0}, both corresponding to the zero moment loading condition for the anchor. These two factors are selected by trial and error to provide a match to both measured trajectory and measured line tension data.
3. The interpretation does not explicitly attempt to model the setting process, although it implicitly accounts for setting by shifting the origin of the analytical equations to match the data measurements.
4. Anchor and mooring line data include anchor fluke area A_f, wire line or chain stock diameter b, and the type of mooring line—wire line or chain.
5. Soil data includes the undrained shear strength at the mudline s_{um} and the strength gradient k. For the most part, these parameters are relatively crude estimates based on limited site data
6. Two equations are used in the data interpretation. Equation 9.23 is used to establish the fluke orientation (Figure 9.28) at the onset of embedment. Equation 9.24 is used to predict fluke rotation as described in the earlier discussion of trajectory prediction.
7. Parameters assumed *a priori* in the calibration exercises are the mooring line bearing factor $N_c = 10$ and the chain area multiplier $E_n = 2.5$. The mooring line friction ratio is taken as $\mu = 0.5$ for chains and $\mu = 0.1$ for wire rope. Since the ratio of normal to tangential motion R_{nt} is relatively small for DEAs, no attempt is made to deduce it through back-analysis of field measurements. Rather, a uniform assumption of $R_{nt} = 0.01$ is made throughout.

Since the field measurements at best provide two independent measurements of anchor performance—embedment depth and load resistance—the interpretation procedure here limits the number of anchor parameters to those two considered to have the greatest impact on performance. The operative bearing factor N_{e0} is chosen as being of prime importance,

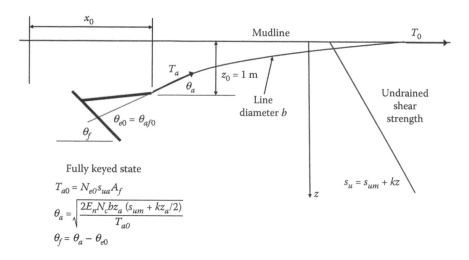

Figure 9.28 Framework for interpreting field data.

since it not only governs load capacity directly, it also controls the rate of rotation of the anchor and, consequently, the rate of anchor penetration. Second in order of importance is the angle at which the line of action of the mooring line load passes through the center of rotational resistance of the fluke θ_{e0}, since this controls the direction of the fluke during early stages of embedment, which has a large effect on anchor trajectory.

Figure 9.28 illustrates the parameters and approach used in the back-analysis. The analysis initiates at a depth $z_0 = 1$ m; trial of other initial depth values were found to not materially affect the analysis. As noted earlier, x_0 is selected to provide a match to the measured data. The initial fluke orientation is computed assuming a fully keyed condition. This is the only contribution that the parameter θ_{0e} makes to the analysis, but the analysis proves to be sensitive to this value. Line tension is not always measured at the optimal location for model calibration, either T_0 at the mudline or T_a at the pad eye. Measurement of tension from the deck of an anchor handling vehicle requires additional calculations—with concomitant additional uncertainty—to estimate mooring line tension at the anchor. In regard to mudline angle, the conditions of installation are not always well documented. The back-analyses of case histories presented below assume a horizontal mooring line angle at the mudline $\theta_0 = 0$, but some lift can occur in the mooring line during the latter stages of embedment. In regard to the fixed variables, the soil strength properties typically generate the greatest uncertainty, since the soil strength parameters s_{um} and k are crude estimates based on fragmentary data and general descriptions of soil conditions. Additionally, even ostensibly simple input variables can entail considerable uncertainty. For example, the fluke area A_f is actually the projected area of a nonplanar fluke element; computed areas can depend on how one chooses to compute the projection. Inclusion or omission of various notches and projections from the fluke can also affect the calculation.

The following discussion presents back-analysis of six instrumented field anchor installations. Table 9.4 summarizes the basic features of these case histories. Measurements ranged from the ideal case of continuous measurement of position (x–z trajectory) and tension (T_0 or T_a) to measurements at discrete locations, such as position or anchor line tension at the end of installation. The anchors used in these studies were commonly used commercial anchors produced by two leading manufactures, Vryhof Anchors and Bruce Anchors. Details of anchor dimensions, mooring line data, estimated soil strength profiles, and other

Table 9.4 Summary of case history interpretations

Location/project	Anchor	Mooring line	Soil description	Type of measurements	Interpreted parameters
Gulf of Mexico (1990)	68.6 kN Vryhof Stevpris $A_f = 10.38$ m²	Wire $b = 0.089$ m	Soft clay $s_{um} = 0$ kPa $k = 1.6$ kPa/m	Continuous position and tension	$N_{e0} = 5.3$–5.6 $\theta_{e0} = 41$–$44°$ $x_0/L_f = 3.4$
Gulf Mexico S. Timbalier Block 295 (1996)	12.74 kN Bruce MK2 DENNLA $A_f = 4.58$ m²	Wire $b = 0.073$ m	Soft clay $s_{um} = 0$ kPa $k = 1.6$ kPa/m	Continuous position and tension	$N_{e0} = 5.8$ $\theta_{e0} = 58°$ $x_0/L_f = 7.5$
Gulf Mexico S. Timbalier Block 295 (1996)	32 kN Vryhof Stevmanta $A_f = 5$ m²	Wire $b = 0.073$ m	Soft clay $s_{um} = 0$ kPa $k = 1.6$ kPa/m	Final position, continuous tension	$N_{e0} = 4.4$ $\theta_{e0} = 67°$ $x_0/L_f = 5.8$
S. China Sea Liuhua 11-1 field (1996)	392 kN Bruce FFTS MK4 $A_f = 16.38$ m²	Chain $b = 0.086$ m	Soft clay $s_{um} = 9$ kPa $k = 1.6$ kPa/m	Final position and tension for 11 lines	$N_{e0} = 9$ $\theta_{e0} = 51°$ $x_0/L_f = 1.2$
Offshore Brazil P-13 Site (1997)	63.7 kN Bruce MK3 DENNLA $A_f = 10$ m²	Wire $b = 0.086$ m	Soft clay $s_{um} = 3.6$ kPa $k = 1.6$ kPa/m	Continuous position	$N_{e0} = 5.8$ $\theta_{e0} = 57°$ $x_0/L_f = 0.6$
Brazil Voador P-27 Campos Basin (1998)	102 kN Vryhof Stevmanta $A_f = 11$ m²	Wire $b = 0.102$ m	Soft clay $s_{um} = 5$ kPa $k = 2$ kPa/m	Installation position and tension for 12 anchors	$N_{e0} = 4.7$ $\theta_{e0} = 57°$ $x_0/L_f = 0$

pertinent information relating to the case studies are compiled in a thesis by Yoon (2002) and a dissertation by Kim (2005). The last column in the table report the interpreted anchor parameters. Also listed is the shift in the horizontal coordinate to account for anchor set. The axis shift x_0 is reported in terms of equivalent fluke lengths, where fluke length is taken as the square root of fluke area.

9.6.1 Joint industry project Gulf of Mexico

As part of a Joint Industry Project (JIP), anchor tests were conducted in a soft clay by Omega Marine Services International (1990) using a catenary mooring system. Water depth at the test site was 90 m. Anchor line tension was measured with an instrumented link 122 m from the pad eye. The estimated depth of this link never exceeded 0.2 m; hence, the measured tension reasonably approximates the mudline tension T_0. Estimated soil strength parameters are $s_{u0} = 0$ and $k = 1.57$ kPa/m. The anchor line had a diameter $b = 0.089$ m. Four tests (7-1 to 7-4) were conducted with a Stevpris 68.6 kN anchor (Vryhof, 2015). Figure 9.29 shows details of the anchor geometry. The fluke–shank angle was set at the 50° setting suitable for soft clay.

The first test showed erratic results. Figure 9.30 shows plots of embedment and mudline tension T_0 for the remaining tests, 7-2, 7-3, and 7-4. As is evident from the plots, several meters of drag—up to 15 m—occurs before diving commences. The back-analysis procedure does not attempt to characterize this early stage of embedment and instead focuses on that portion of the trajectory where diving behavior is fully established. Back-calculated parameters lie in a relatively narrow band for the three tests. Specifically, for tests 7-2, 7-3, and 7-4, the respective operative bearing factors are $N_{e0} = 5.4, 5.3$, and 5.6; and the respective operative load angles are $\theta_{e0} = 42°, 44°$, and $41°$.

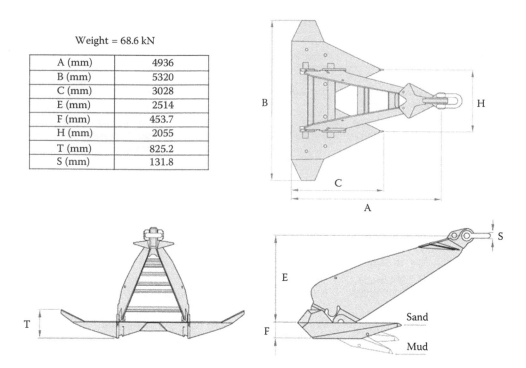

Weight = 68.6 kN

A (mm)	4936
B (mm)	5320
C (mm)	3028
E (mm)	2514
F (mm)	453.7
H (mm)	2055
T (mm)	825.2
S (mm)	131.8

Figure 9.29 Vryhof Stevpris anchor data for 1990 Gulf of Mexico JIP.

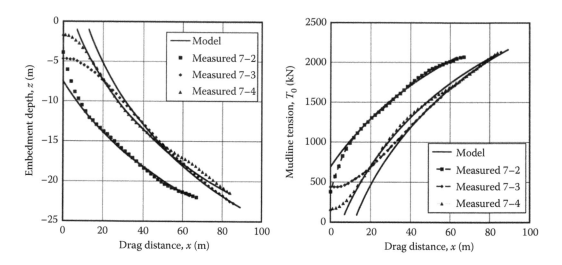

Figure 9.30 Model analysis for 1990 Gulf of Mexico JIP.

9.6.2 South Timbalier Block 295 Gulf of Mexico

Large-scale anchor tests were carried on in South Timbalier Block 295 in the Gulf of Mexico in soft clay by Aker Maritime Contractors in 1996 as part of a JIP. The water depth at this site is 91 m. Two anchor installation tests are considered here: The first is a 12.74 kN Bruce DENNLA MK2 (Figure 9.31). Test results provide deck tension versus

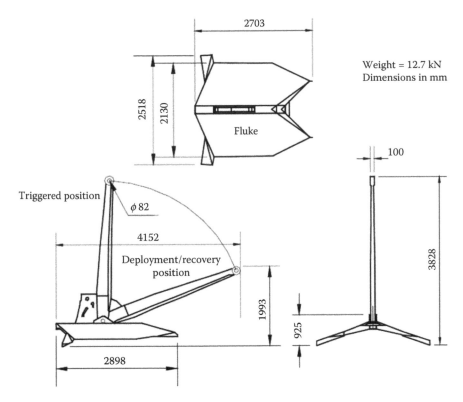

Figure 9.31 Bruce DENNLA MK2 anchor data for South Timbalier Block 295 test.

drag distance. Anchor line tension at the mudline is estimated to be 15 kN less than the deck tension.

Figure 9.32 shows the trajectory and mudline tension history for the DENNLA MK2 installation. Diving of the anchor commences after about 20 m of drag. As noted earlier, analysis of the data focuses on that portion of the trajectory where diving behavior is fully

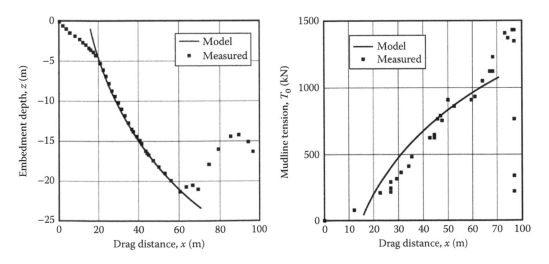

Figure 9.32 Model analysis for South Timbalier Block 295 DENNLA MK2.

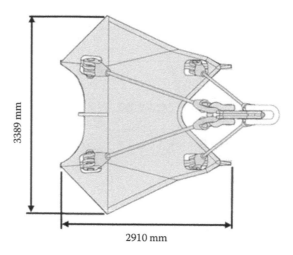

Figure 9.33 Vryhof Stevmanta anchor data for South Timbalier Block 295 test.

established. After 60 m of dragging, the anchor behavior fundamentally changes, possibly owing to the triggering of the shear pin. Since the behavior of interest is the drag embedment mode, the back-analysis of the data does not extend beyond 60 m. The operative bearing factor and load angle inferred from this data set are $N_{e0} = 5.8$ and $\theta_{e0} = 58°$.

A second anchor investigated in this JIP was a 32 kN Vryhof Stevmanta (Figure 9.33). Figure 9.34 shows the trajectory and mudline tension history for the Stevmanta installation. In this instance, continuous measurements of anchor position are not available, and only a single position measurement was recorded at the end of anchor installation. Based on the pattern of line tension mobilization versus drag distance, diving of the anchor appears to have commenced after approximately 15–20 m of drag. Trial and error selection of an operative bearing factor and load angle combination that matches both the line tension history and the final embedment depth yields an estimated bearing factor $N_{e0} = 4.7$ and load angle $\theta_{e0} = 67°$.

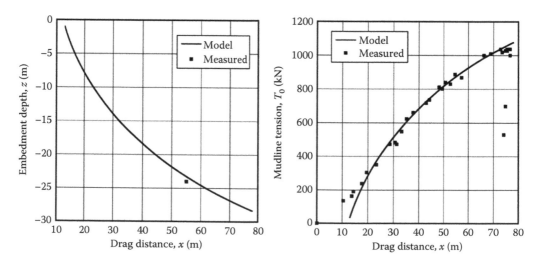

Figure 9.34 Model analysis for South Timbalier Block 295 Stevmanta.

Weight = 392 kN

Length	Dimension
A (mm)	7690
B (mm)	5768
C (mm)	2196
D (mm)	5729
E (mm)	2722
F (mm)	9655

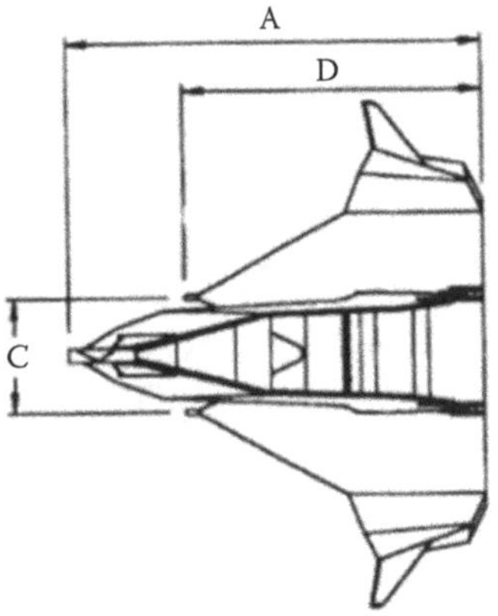

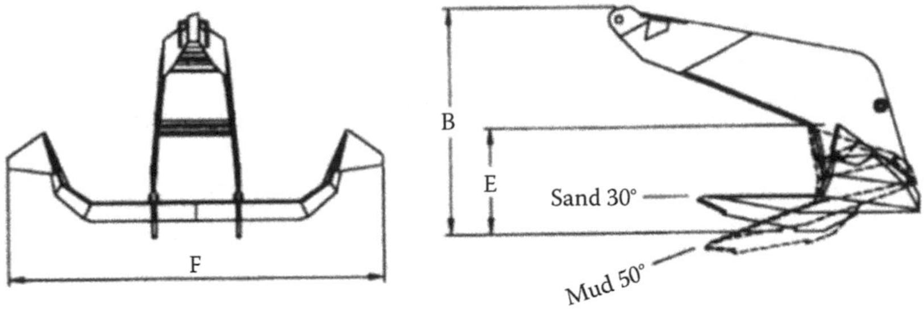

Figure 9.35 Bruce FFTS anchor data for Lihua 11-1.

9.6.3 Liuhua 11-1 field

The tests considered here were conducted in the Liuhua 11-1 field located 70 km southeast of Hong Kong, in the South China Sea (Foxton, 1996). The water depth is approximately 300 m. A contractor to Amoco conducted the tests in 1996. The Bruce FFTS MK4 anchors (Figure 9.35) were tested in soft clay using a catenary mooring system. Continuous measurements of position and line tension were not made at this site. Rather, single observations for drag distance, embedment depth and mudline tension were recorded for 11 separate anchor installations for a floating production system. The mooring system used catenary lines with chain forerunners. Soil conditions are described as slightly over-consolidated soft clay. Figure 9.36 shows that the model fits to the spread of measured positions and mudline tensions reported for this site. In this case, the operative bearing factor and load angle combination producing the best fit is $N_{e0} = 9$ and load angle $\theta_{e0} = 51°$.

9.6.4 Offshore Brazil P-13

This study was conducted at the P-13 site offshore Brazil by a contractor to Petrobras in a range of water depth of approximately 588–643 m (Foxton, 1997). Bruce DENNLA MK3 anchors (Figure 9.37) were installed in a slightly over-consolidated soft clay. The anchors were part of a taught leg mooring system. The type of forerunner was not reported but, since the anchors were part of a taut leg system, a wire forerunner is assumed. The study involved continuous measurements of position, but no measurements of line tension. Therefore, a full calibration study could not be conducted along the lines of the other case studies reported here. In this case, the calibration parameters inferred for the similarly configured but smaller

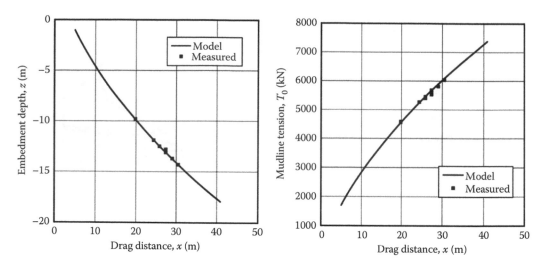

Figure 9.36 Model analysis for Lihua 11-1.

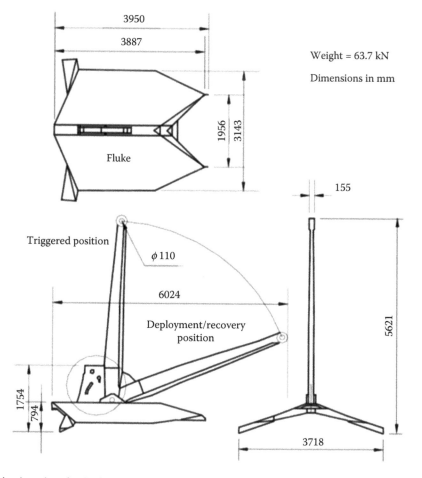

Figure 9.37 Anchor data for P-13 in offshore Brazil.

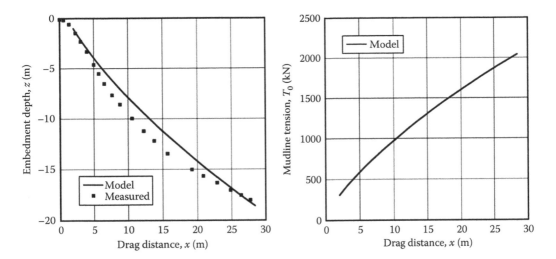

Figure 9.38 Model analysis for P-13 in offshore Brazil.

DENNLA MK2 at the South Timbalier site (discussed earlier) were applied to this case; that is, $N_{e0} = 5.8$ and load angle $\theta_{e0} = 57°$. Figure 9.38 shows the final position of the anchor calculated from the model to fortuitously match the measured final position, although the model calculations tend to under-predict penetration at intermediate drag distances.

9.6.5 Offshore Brazil Campos Basin P-27

Petrobras in 1998 conducted a large-scale anchor test at the Voadaor site in Campos Basin, P-27 in a water depth ranging from 510 to 570 m. The test employed Stevmanta anchors (Figure 9.39) deployed in a 12-line taut mooring system. The anchors were VLAs with shackles that broke when the anchor line reached the expected load. The test did not record continuous position and tension measurements. However, the drag distance and depth at which the shackles broke were recorded. The break load of the shackles provided data on anchor load capacity. The breaks occurred at drag distances of 35–60 m, with corresponding penetration depths being between 20 and 25 m. The back-analysis procedure (Figure 9.40) produced an estimated bearing factor $N_{e0} = 4.7$ and load angle $\theta_{e0} = 57°$. This bearing factor is somewhat greater than the bearing factor reported above for the South Timbalier Stevmanta, and the load angle is less. This may be attributed to a scale effect, since the Campos Basin anchors were twice the size at South Timbalier. Alternatively, the difference can be a consequence of the relatively poor resolution of the Campos Basin data.

9.7 VERTICALLY LOADED ANCHORS

In a VLA installation, the anchor is installed in a drag embedment mode with a fixed shank (Figure 9.41). Following rupture of the shear pin or other release mechanism the shank swivels freely such that the shank aligns with the pad-eye angle θ_a. The location of the shank hinge relative to the center of rotational resistance determines the new bearing factor and kinematic properties of the anchor. As depicted in Figure 9.41, the anchor now acts like a "short-shank" anchor, with the line of action of the pad-eye tension being controlled

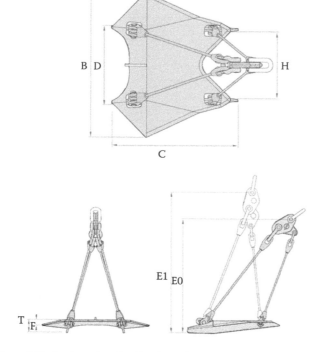

Weight = 102 kN	
B(mm)	4657
C (mm)	4410
D (mm)	2882
E_0 (mm)	4557
E_1 (mm)	4995
F (mm)	255
H (mm)	2162
T (mm)	948

Figure 9.39 Anchor data for Voador P-27 in offshore Brazil.

by the location of the hinge relative to the center of rotational resistance of the anchor. The VLA design must ensure that the hinge is placed forward of the center of rotational resistance such that the load–fluke angle θ_{e0} in the VLA configuration exceeds that in the DEA mode. While a larger θ_{e0} produces a larger bearing factor N_e for the anchor, a normal load orientation ($\theta_{e0} = 90°$), this load orientation has the undesirable effect of generating large anchor motions normal to the fluke; that is, large R_{nt} values. An optimal load angle

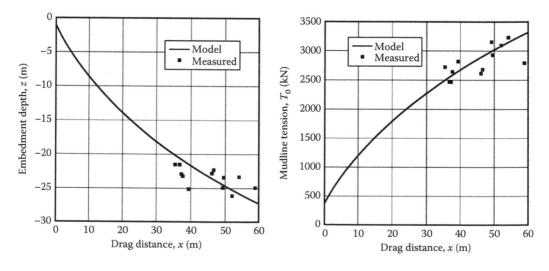

Figure 9.40 Model analysis for Voador P-27 in offshore Brazil.

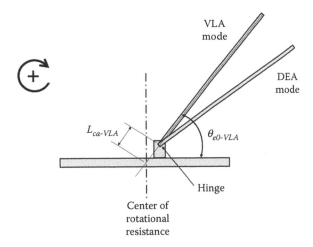

Figure 9.41 Schematic representation of vertically loaded anchor (VLA).

in the neighborhood of $\theta_{e0} = 75°$ provides a good combination of a high bearing factor N_{e0} together with tolerable normal motions R_{nt}.

9.7.1 Shank release

With the shank released, the anchor now has a new, larger, bearing factor N_e. The consequent increase in pad-eye tension T_a leads the anchor line to assume a new configuration. Thus, the first computational steps in the analysis of anchor behavior following the release of the shank are as follows: (1) establishing the load capacity versus load angle $F - \theta_{af}$ curve for the new anchor geometry using Equations 9.5 through 9.8 and 9.12, (2) for a trial pad-eye angle θ_a compute a trial pad-eye tension T_a, and (3) repeat the previous step for successive θ_a values until the trial pad-eye tension equals the anchor capacity $T_a(\theta_a) = F(\theta_a + \theta_f)$. Since the anchor has a greater load capacity in the VLA mode than in a DEA mode, the process described above involves increasing the pad-eye tension T_a at a load state below yield; consequently, no plastic displacement or rotation occurs. Accordingly, the fluke orientation θ_f is assumed to be unchanging during this process.

As VLAs are typically intended for taut mooring systems, release of the shank will normally be accompanied by a shortening of the mooring line to produce larger mudline angle. While the approximate anchor line formula in Equation 9.23 is reasonable for small θ_0, at elevated mudlines angles the more rigorous form is more appropriate:

$$\frac{T_a}{1+\mu^2}\left[\exp\left[\mu(\theta_a - \theta_0)\right](\cos\theta_0 + \mu\sin\theta_0) - \cos\theta_a - \mu\sin\theta_a\right] = \int_0^{za} Q(z)dz = \bar{Q}z_a \quad (9.25)$$

It may be recalled from 7 that Q $(=E_n N_c b s_u)$ is the unit soil resistance acting normal to the mooring line, and μ is the ratio of tangential to normal resistance (F/Q) on the mooring line.

Subsequent loading of the anchor following the release of the shank will parallel that described earlier for installation in a DEA mode. Initially, the anchor will rotate in response to moment loading in a manner similar to the initial keying, as the line of action of T_a will temporarily not pass through the center C of the plate. Some normal and tangential translation also occurs, but the model shows the fluke rotation being relatively rapid. After the fluke aligns itself such that the line of action of T_a passes through C, with sustained loading the anchor resumes its trajectory in a zero-moment state, with new relationships for bearing factor $N_{e\text{-}VLA}$ and ratio of normal to tangential motion $R_{nt\text{-}VLA}$. Owing to the relatively high load angles involved, anchor motion normal to the fluke is much more significant in the VLA as opposed to the DEA mode. During installation in the DEA mode the mudline angle θ_0 is typically maintained at or near zero.

Also as described in the earlier discussion of trajectory calculation, after keying the bearing factor N_{e0} and relative normal motion R_{nt0} become constant, thereby simplifying the calculations. The rate of change in the pad-eye angle $d\theta_a/dz$ is central to the trajectory calculation, since the anchor simply follows the mooring line, with the line of action of T_a steadily maintained to pass through the center C of the fluke. Implicit differentiation of Equation 9.25 yields the following expression that may be used recursively throughout the trajectory calculation:

$$\frac{d\theta_a}{dz} = \frac{(1+\mu^2)}{T_a} \frac{E_n N_c b s_{ua} - k N_{e0} A_f \bar{Q} z_a / T_a}{\mu \exp[\mu(\theta_a - \theta_0)](\cos\theta_0 + \mu\sin\theta_0) + \sin\theta_a - \mu\cos\theta_a} \tag{9.26}$$

9.7.2 VLA trajectory simulation

The analytical formulation described above can be illustrated by considering a VLA installation in a normally consolidated clay with mudline strength $s_{um} = 2$ kPa and strength gradient $k = 1.6$ kPa/m. The anchor as a fluke area $A_f = 4.5$ m² with a load–fluke angle $\theta_{0e\text{-}DEA} = 60°$ (Figure 9.13) when the shank is fixed in the DEA mode, and a load–fluke angle $\theta_{0e\text{-}VLE} = 75°$ when the shank is released to the VLA mode. The mooring line is wire rope with diameter $b = 0.073$ m and an assumed ratio of tangential to normal resistance $\mu = 0.1$. Anchor performance is considered for two cases in regard to the mooring line angle at the mudline. In the first case, the mooring line is in a catenary mode $\theta_0 = 0°$, during both drag embedment and subsequent VLA loading. In the second case, the mooring line is in a catenary mode ($\theta_0 = 0°$) during drag embedment and in a taut mode with $\theta_0 = 35°$ during subsequent VLA loading.

Insight into VLA behavior is provided by considering the four stages of motion (Figure 9.42) occurring during installation and subsequent VLA loading. The first two stages are identical to those described earlier for DEA installation. Immediately after the anchor is initially set on the seabed (Stage I), the line of action of T_a does not pass through the center of rotational resistance for the anchor. Thus, a net moment act on the anchor and anchor motions are largely driven by the magnitude of this moment. Details of the calculations for setting a DEA were described in the earlier discussion of DEA trajectory calculations. Figure 9.43 shows model simulations of this stage—the sign of the fluke angle θ_f is reversed here to facilitate comparison to the pad-eye angle θ_a. The simulation shows the pad-eye angle to remain relatively constant, while the anchor fluke rotates sharply downward. The anchor will eventually rotate into an orientation at which the line of action of T_a passes through the anchor's center of rotational resistance, which is the onset of Stage II. During this stage, progressive embedment of the anchor leads to increasing curvature of the mooring line, with a concomitant increase in pad-eye angle θ_a. The anchor continuously realigns itself such that the line of action of T_a passes through the center of rotational resistance and the changes in fluke angle θ_f equal the changes in pad-eye angle θ_a at all time.

Release of the shank marks the onset of Stage III as the anchor goes into a VLA mode. Computations are identical to those for Stage I, except that the angle $\theta_{e0\text{-}VLA}$ and length

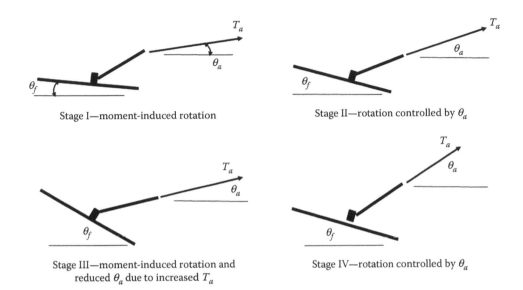

Figure 9.42 **Stages of a VLA trajectory.**

L_{ca-VLA} (Figure 9.42) corresponding to the VLA configuration should be used in the calculations, and Equation 9.25 should be used in lieu of Equation 9.23. A number of physical mechanisms drive the anchor response during Stage II. Firstly, the increased bearing capacity of the anchor in the VLA mode leads to an increased pad-eye tension T_a together with a tendency for the pad-eye angle θ_a to decrease. In the case of loading under catenary mooring conditions, $\theta_0 = 0$, the net effect is a modest reduction in θ_a during Stage III. By contrast, if the VLA is loaded in a taut line mode, $\theta_0 = 35°$, the pad-eye angle θ_a undergoes a significant increase. Secondly, the change in anchor configuration following release of the shank leads to a condition in which the line of action of T_a no longer passes through the center

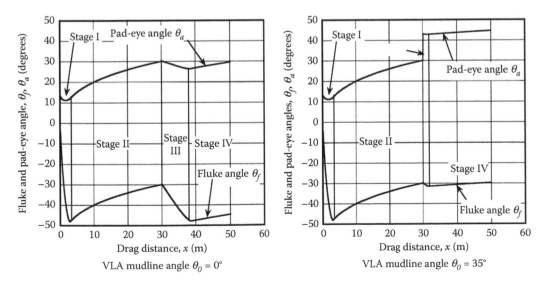

Figure 9.43 **Fluke and pad-eye angles during VLA installation and loading.**

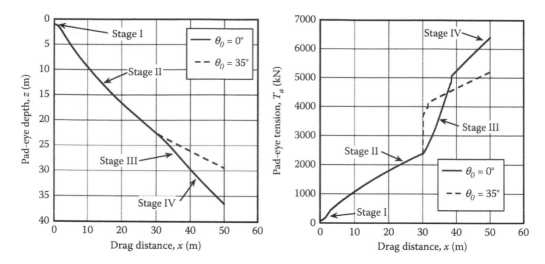

Figure 9.44 Prediction of VLA trajectory and line tension.

of rotational resistance of the anchor; thus, a moment acts on the anchor. In the case of a catenary installation, the downward rotation of the fluke is substantial (nearly 20° in this example); whereas, for a taut system, the fluke is nearly stationary. In summary, the Stage III anchor behavior can be considered as the opening of the fluke–shank angle from the DEA setting to the VLA value. In the case of catenary loading, the opening is characterized by the fluke rotating downward, while the pad-eye angle remains nearly constant. The converse occurs under taut-line loading conditions: the fluke remains more or less stationary, while the pad-eye angle increases. When the anchor rotates into a zero-moment state, Stage IV motions prevail. Trajectory calculations parallel the Stage II calculations, except Equation 9.26 should be used if high mudline angles are involved.

Figure 9.44 shows the trajectory and pad-eye tension predictions corresponding to the fluke angle history described above. In the case of catenary loading, the trajectory appears nearly linear. This may be considered remarkable in view of the dynamic nature of the fluke rotation; however, the author's experience field measurements of VLA penetration can actually show this trend. The pad-eye tension predictions show more marked behavior in response to release of the shank in the VLA mode. The increased bearing factor associate with near-normal loading on the fluke produces a distinct increase in load capacity as the shank opens. Overall, the performance of the anchor is better in the case of a catenary mooring line system. However, noting that the intent of a VLA is to resist vertical loading to secure taut moorings, the VLA performance for the taut mooring case is still more than satisfactory, with the anchor load capacity essentially doubling after the release of the shank.

9.8 STRATIFIED SOIL PROFILES

To this point, the discussion of DEAs has been restricted to soil profiles with soil strength varying in a restricted sense of uniform strength gradient. We now consider soil profiles containing stiff layers embedded in a soft clay profile. Figure 9.45 illustrates three aspects of DEA behavior that will change upon encountering a stiff layer. First, the increased soil resistance at the tip of the fluke will shift the center of rotational resistance downward, which

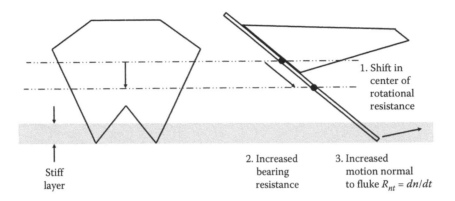

Figure 9.45 Effects of embedded stiff layer on anchor behavior.

will produce a clockwise rotation of the fluke. Second, increased bearing resistance and mooring line tension T_a—with a concomitant decrease in the pad-eye angle θ_a—will arise from partial embedment of the anchor. These first two effects are generally beneficial, since they tilt the fluke downward and increasing the tendency for a diving trajectory. Finally, the increased end resistance at the tip of the fluke will increase R_{nt}, the ratio of normal to tangential motion of the fluke. The latter effect is certainly not beneficial from the standpoint of achieving deep embedment of the anchor and, if the resistance of the stiff layer is sufficiently great—penetration ceases and the anchor simply drags at constant depth. As is evident from the figure, the teeth at the leading edge of the fluke minimize the end resistance on the fluke, thereby improving the likelihood of the anchor penetrating the stiff layer. An adaptation of the trajectory prediction model for DEAs described earlier to account for the effects illustrated in Figure 9.45 is described by Rasulo (2016). The model is being developed in a JIP on anchor performance, in a collaborative effort with Texas A&M University developing the analytical model and the University of Texas Austin performing laboratory model tests for validating and calibrating the analytical model. The laboratory model tests are currently in progress.

The soil heterogeneity considered here can be of two general forms. In the first case, a predominantly soft clay soil profile contains one or more seams of stiff clay or sand. In this case, the chief question is whether the DEA can successfully penetrate the stiff layers to achieve adequate embedment and holding capacity. Figure 9.46 shows an example of the type of problem for a DEA with fluke area $A_f = 9.7$ m^2 embedding into a soft clay with an 0.25-m thick stiff layer occurring at a depth of 7.5 m. Analyses are conducted for a series of shear strength ratios $s_{u\text{-}stiff}/s_u$, where $s_{u\text{-}stiff}$ is the strength of the embedded stiff clay layer and s_u is the soil strength in the soft clay immediately above the stiff layer. For a strength ratio $= 2$, the effect of an embedded stiff layer is nearly imperceptible. At higher strength ratios up to $s_{u\text{-}stiff}/s_u = 5$, the penetration of the anchor past the stiff layer is possible, albeit with a noticeable reduction in penetration depth and anchor holding capacity. At a high strength ratio, $s_{u\text{-}stiff}/s_u = 10$, the stiff layer halts penetration and the anchor simply plows along the top of the stiff layer under continued dragging.

Figure 9.47 show a second type of scenario, which is a two-layer system with a stiff layer underlying the soft layer. In this case, the questions are (1) under what conditions deep embedment of the anchor is possible and (2) if penetration is halted, how much holding capacity mobilizes when the anchor partially embeds into the stiff layer. In this example, the predictions show deep penetration being possible up to a strength ratio $s_{u\text{-}stiff}/s_u = 2$.

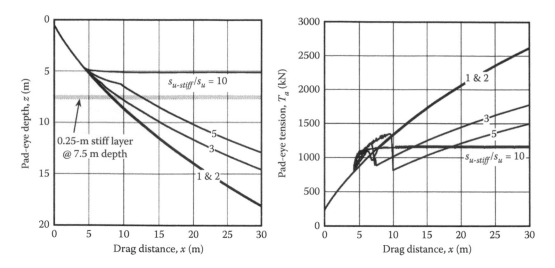

Figure 9.46 DEA performance in soil profile with thin stiff layer.

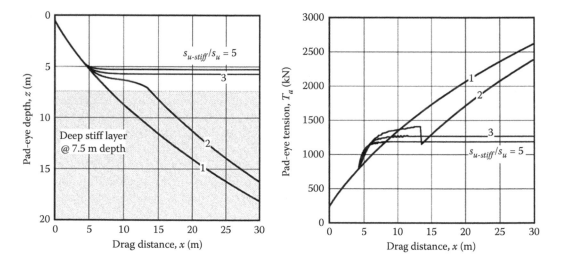

Figure 9.47 DEA performance in two-layer soil profile.

9.9 OUT-OF-PLANE LOADING

DEAs are normally designed to resist pullout loads for which the line of action of the mooring line load acts in a vertical plane containing the shank and passing through the central axis of the fluke. Situations can arise where the line of action of the load deviates from this condition, such as when one or more mooring lines fail causing the floating unit to drift off station. The discussion that follows first presents a theoretical analysis of an idealized DEA comprising a rectangular fluke and a vanishingly thin shank. The simplified analysis omits some key features of actual DEAs, but provides useful insights into how additional loading in the form of OOP shear, moments, and torsion can impact anchor performance. Second, the discussion presents findings from a series of laboratory model tests on DEAs that are more representative of those used in practice, most significantly, anchors with plate shanks having a relatively large projected area normal to the intended plane of loading.

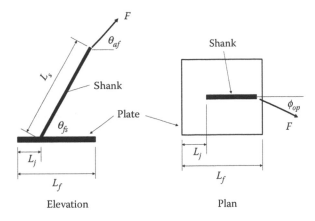

Figure 9.48 Out-of-plane loading on a plate anchor.

9.9.1 Analysis of thin shank anchors

The analysis presented here parallels that presented earlier in the development of Equation 9.12 for a plate subjected to in-plane loading, but generalized to general six DOF loading. The plate anchor has a length L_f (Figure 9.48) with a rigid shank of length L_s attached at a distance L_j from the trailing edge of the plate. The centerline of the shank forms an angle θ_{fs} with the plate. A force F applied at the pad eye is inclined at an angle θ_{af} relative to the plane of the plate. The load angle θ_{af} is measured in the plane of F, not the plane of the shank. An angle ϕ_{op} defines the OOP direction of F.

Figure 9.49 shows the coordinate system used to describe the pad-eye force F, where n is normal to the anchor plate, t_1 is parallel to the anchor plate in the plane of intended loading, and t_2 is parallel to the anchor plate perpendicular to the plane of intended loading. The orientation of the force F at the pad eye may be represented by the direction angles $(\beta_{t1}, \beta_{t2}, \beta_n)$ shown in Figure 9.49 that relate to θ_{af} and ϕ_{op} as follows:

$$\beta_{t1} = \cos^{-1}(\cos\theta_{af}\cos\phi_{op}) \tag{9.27}$$

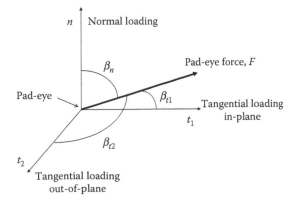

Figure 9.49 Description of 3D system of forces on fluke.

$$\beta_{t2} = \cos^{-1}(\cos\theta_{af}\sin\phi_{op}) \tag{9.28}$$

$$\beta_n = \pi/2 - \theta_{af} \tag{9.29}$$

OOP loading subjects the anchor fluke to general six DOF loading: a normal force F_n, an in-plane tangential force F_{t1}, an OOP tangential force F_{t2}, an in-plane moment M_1, an OOP moment M_2, and a torsion T. These loads are as follows:

$$F_n = F\cos\beta_n \tag{9.30}$$

$$F_{t1} = F\cos\beta_{t1} \tag{9.31}$$

$$F_{t2} = F\cos\beta_{t2} \tag{9.32}$$

$$M_1 = F[\cos\beta_n(L_s\cos\theta_{fs} + L_j - L_f/2) - \cos\beta_{t1}L_s\sin\theta_{fs}] \tag{9.33}$$

$$M_2 = F\cos\beta_{t2}L_s\sin\theta_{fs} \tag{9.34}$$

$$M_3 = F\cos\beta_{t2}(L_s\cos\theta_{fs} + L_j - L/2) \tag{9.35}$$

Applying Equations 9.30 through 9.35 to the interaction relationship Equation 4.21 yields the following expression:

$$f = \left\{\left(\frac{|c_n|N_e}{N_{nmax}}\right)^q + \left[\left(\frac{|c_{m1}|N_e}{N_{m1max}}\right)^{m1} + \left(\frac{|c_{m2}|N_e}{N_{m2max}}\right)^{m2} + \left[\left(\frac{|c_t|N_e}{N_{tmax}}\right)^n + \left(\frac{|c_{m3}|N_e}{N_{m3max}}\right)^k\right]^s\right]\right\}^{1/p} - 1 = 0 \tag{9.36}$$

where
N_e = dimensionless load capacity factor for anchor
$c_n = \cos\beta_n$
$c_s = \sqrt{\cos^2\beta_{t1} + \cos^2\beta_{t2}}$
$c_{m1} = \cos\beta_n [(L_s/L)\cos\theta_{fs} + (L_j/L) - 1/2)] - \cos\beta_{t1} (L_s/L)\sin\theta_{fs}$
$c_{m2} = \cos\beta_{t2} (L_s/L)\sin\theta_{fs}$
$c_{m3} = \cos\beta_{t2} [(L_s/L)\cos\theta_{fs} + (L_j/L) - 1/2)]$ eny

The nonlinear Equation 9.36 is then solved for N_e. Table 9.5 shows the input for an example analysis by Yang et al. (2010) for square and rectangular thin plates with a vanishingly thin shank. To illustrate the effect of OOP loading, a sweep is performed for $\phi_{op} = 0$–90°. Figure 9.50 shows the computed reduction in the anchor load capacity owing to the OOP loading effect. In this case, the maximum reduction in anchor load capacity is on the order of 30% in this example.

Table 9.5 Input parameters for out-of-plane example analysis

Description	Parameter	Square	Rectangular $L_f/w_f = 2$
Uniaxial bearing factors	Normal, N_{nmax}	12.5	12.35
	In-plane moment, N_{m1max}	1.9	1.70
	Out-of-plane moment, N_{m2max}	1.9	2.15
	Tangential, N_{tmax}	2	2
	Torsion, N_{m3max}	0.765	1.19
Interaction coefficients	Q	3.26	3.20
	m_1	1.91	2.47
	m_2	1.91	1.86
	N	2.02	1.81
	k	1.45	1.32
	s	4.80	2.84
	p	1.56	1.93

9.9.2 Experimental data

The analysis described above provides useful insight into the OOP loading effects, but does not consider the complex geometry of many actual DEAs. To further investigate the behavior of DEAs under OOP loading conditions, a program of laboratory model tests was undertaken using "generic" model anchors (Figure 9.51) having geometric configurations more representative of DEAs used in practice (Aubeny et al., 2011). In particular, the model anchors featured twin plate shanks having relatively large projected areas normal to the direction of loading, which can be critical to anchor performance under OOP loading. Two parallel test series were conducted. The first test series comprised small-scale tests conducted at the University of Texas Austin—roughly 1:30 scale—using an acrylic anchor with fluke area $A_f = 0.00669$ m^2 in a kaolin test bed. During in-plane and OOP drag installation tests, the six DOF motions

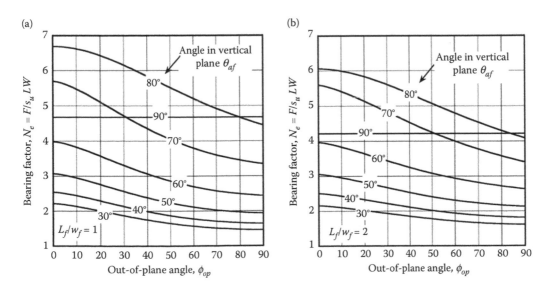

Figure 9.50 Analytical predictions of out-of-plane loading effects (a) square plate and (b) rectangular plate.

1:0 scale   1:0 scale with magnetometer

Figure 9.51 Medium- and small-scale model anchors.

of the anchor were measured with a magnetometer sensing system, permitting continuous measurements of the three-dimensional trajectory of the anchor, as well as the pitch, roll, and yaw of the fluke. Anchor mudline tension was also continuously monitored. Two wire line diameters were used in the test series; the thicker line corresponded to a normalized fluke area $A_f/b^2 = 1300$, which is broadly representative of a typical wire line anchor systems. Both uniform and linearly varying strength profiles were investigated in this test series. The uniform strength test beds had undrained shear strengths in the range of 0.5–1.2 kPa. The linearly varying strength profiles had gradients of approximately 1.5 kPa/m, which are representative of a normally consolidated clay profile. Estimated sensitivity of the kaolin is $S_t = 1.5$–2. The second test series, conducted at the Haynes Coastal Engineering Laboratory at Texas A&M University, comprised medium-scale tests—roughly 1:10 scale—using a mild steel anchor with fluke area $A_f = 0.0935$ m² in a 50–50 sand–bentonite soil mixture. The anchor was dragged with a chain that, with the instrumentation cables, had a diameter of approximately 2.5 cm. This corresponds to a normalized fluke area $A_f/b^2 = 150$, which is representative of a typical chain system. The test bed had a uniform strength profile of approximately 0.75–0.9 kPa. Anchor trajectory during dragging was measured using angular measurement transducers to track the chaser and tow line orientations (Figure 9.52) and a position sensor to measure chaser length. Accelerometers mounted on the anchor fluke provided continuous measurements of pitch and roll, and a load cell measured pad-eye tension. Both the small- and medium-scale test series investigated more than one fluke–shank angle; however, the following presentation is limited to anchors with a 50° fluke–shank angle, which is most appropriate to soft clay seabed conditions. In addition to performing drag embedment tests, both test series included "breakout" tests to measure anchor capacity under various uniaxial loading conditions, such as pure normal loading and pure in-plane shear.

The first rows of Table 9.6 show anchor bearing factors for six possible uniaxial loading modes measured in the breakout tests. The bearing factors for normal loading measured in the small- and medium-scale tests generally agree with theoretical values for a thin shank anchor (Table 9.5), indicating that a finite thickness shank does not significantly affect normal capacity. By contrast, the tangential load capacity N_{t1max} of the small-scale anchor is well above that of the medium-scale anchor—the differences may be attributed to differences in the sensitivity of the soils in the two test beds, as well as the possibility of a soil plug occurring between the twin plate shanks of the small-scale anchor. Significantly, the OOP shear-bearing factor N_{t2max} exceeds the in-plane value by nearly 50%, a likely consequence of the relatively large projected area of the shank in the OOP direction. The measured OOP

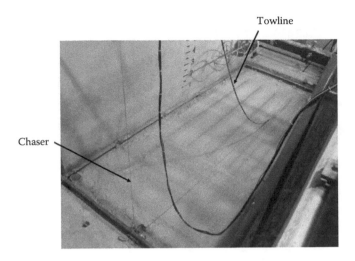

Figure 9.52 Medium-scale drag embedment test basin.

moment capacity N_{m2max} and torsional capacity N_{m3max} also exceed the corresponding theo-retical thin-shank values in Table 9.5. The general picture provided by the uniaxial measure-ments is that shank resistance of DEAs can be significant and OOP capacity can actually exceed in-plane capacity for typical DEA designs.

The drag embedment tests monitored both trajectory and anchor load capacity. Figure 9.53 shows a typical trajectory measurement for OOP loading of the small-scale anchor. In this example, the anchor was drag embedded to a depth of one fluke length in the "intended" in-plane direction of loading, after an OOP load angle of 25° was applied to the mooring line. The tests consistently indicated that anchor embedment continues under OOP loading. Figure 9.54 shows examples of the bearing factors N_e calculated during the medium-scale drag embedment tests. The lower rows of Table 9.6 show the operative bearing factors N_e measured during drag embedment under various conditions of OOP loading. The instal-lation depth in this table refers to the depth to which the anchor was embedded under in-plane conditions prior to application of an OOP load. In both the small- and medium-scale

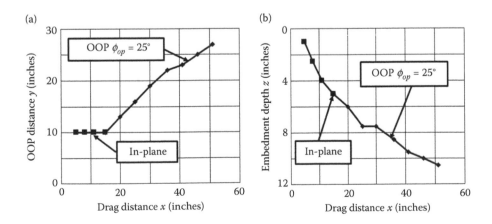

Figure 9.53 Typical trajectory measurement for out-of-plane loading tests (a) plan view and (b) elevation.

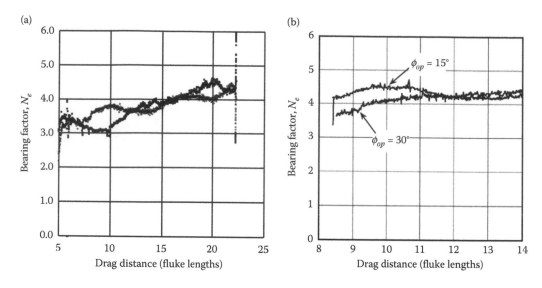

Figure 9.54 Typical records of anchor bearing factor N_e during drag embedment (a) in-plane tests and (b) out-of-plane tests.

Table 9.6 Experimental measurements of anchor bearing factors

Description	Parameter	OOP angle ϕ_{op}	Installation depth	Small 1:30 scale	Medium 1:10 scale
Uniaxial Bearing Factors	Normal N_{nmax}	–	–	10.93 ± 0.53	$11.0-11.6$
	In-plane shear N_{t1max}	–	–	4.22 ± 0.005	2.6
	OOP shear N_{t2max}	–	–	6.00 ± 0.22	–
	In-plane moment N_{m1max}	–	–	2.40 ± 0.05	–
	OOP moment N_{m2max}	–	–	2.58 ± 0.03	–
	Torsion, N_{m3max}	–	–	1.92 ± 0.06	–
Operative Bearing Factor	N_e during drag	0	–	5.82 ± 0.92	4.07 ± 0.56
		30–90°	max	6.05 ± 0.65	–
		30–90	$1L_f$	5.90 ± 0.84	–
		15	$3L_f$	–	$4.2-4.3$
		30	$1L_f$	–	4.4

tests, the OOP bearing resistance slightly exceeded the in-plane values. Thus, one should not necessarily presume that OOP loading on a DEA degrades its load capacity. Since the measured robust behavior of DEAs arises largely from shank resistance, and since many anchor models have relatively thin shanks, care should be exercised before generalizing the results shown in Table 9.6 to other anchor types.

9.10 DEAs IN SAND

Up to this point analysis of anchor performance, beyond the empirical design charts presented early in this chapter, has focused on anchor is soft clays. However, shallowly

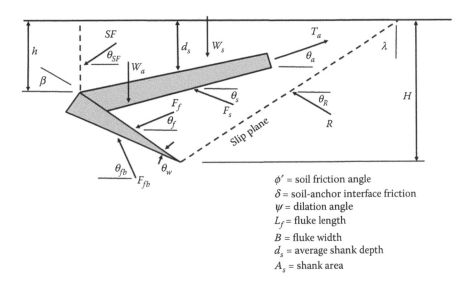

ϕ' = soil friction angle
δ = soil-anchor interface friction
ψ = dilation angle
L_f = fluke length
B = fluke width
d_s = average shank depth
A_s = shank area

Figure 9.55 Limit equilibrium analysis of anchor in sand (Adapted from Neubecker SR and Randolph MF, 1996a, *Canadian Geotechnical Journal*, 3(4), 574–558).

embedded drag anchors in sands comprise a significant component of overall usage for DEAs. Despite the utility of empirically based design charts, they offer little guidance in regard how anchor capacity is influenced by site-specific soil conditions, such as calcareous versus silica sands. To provide insights into drag anchor performance in sands, Neubecker and Randolph (1996) developed the limit equilibrium formulation illustrated in Figure 9.55. Soil parameters in the analysis include the friction angle ϕ', dilation angle ψ, effective unit weight γ', and the soil–anchor interface friction angle δ. The analysis assumes a three-dimensional failure mass defined by a slip plane oriented at an angle λ from vertical. The value of λ is treated as an unknown optimization variable, and it is varied to seek a least upper bound of ultimate anchor load capacity. The formulation assumes that the anchor fluke has a wedge geometry, defined by θ_w in Figure 9.55, as opposed to a simple plate of uniform thickness. The analysis actually involves two sequential equilibrium calculations, on for the soil failure mass and one for the anchor. Full details on the derivation of the various forces are provided in the reference; Table 9.7 provides a summary of the final form of the equations. Some pertinent points on the analysis include the following:

- The soil failure wedge has a three-dimensional geometry having a cross-sectional area A and lateral extent X, which are evaluated as shown in Table 9.7 to compute the volume and weight W_s of the soil failure mass.
- The back side of the failure mass is assumed to be bounded by a vertical plane, acting on which is a side friction (SF) force.
- The shank resistance F_s is computed from a bearing capacity equation in terms of the average depth of the shank and a bearing factor N_{qs}. Based on experiments on simple model anchors, Neubecker and Randolph (1996) measure a typical value of $N_{qs} = 35$. However, from their experimental model tests on anchors more representative of the actual geometry of commercial anchors, they deduce somewhat lower values of bearing resistance, $N_{qs} = 31$ in dense silica and calcareous sands and 25 in loose silica sand.
- The magnitude of the fluke force F_f and reaction force R on the slip plane are obtained by enforcing horizontal and vertical equilibrium. Their directions are fixed by geometric and frictional resistance considerations.

Table 9.7 Limit equilibrium model for DEAs in sand

Description	Magnitude	Direction
Weight soil wedge, W_s	$$W_s = A\gamma'\left(B + \frac{2}{3}X\right)$$ $$A = \frac{H^2 - h^2}{2\tan\beta} + \frac{H^2\tan\lambda}{2}$$ $$X = \frac{H\tan\psi}{\cos(\lambda - \psi)}$$	Vertical
Side friction, SF	$$SF = \frac{\gamma' L_f (H+h)^2(\sin\phi' - \sin\psi)}{4\cos\psi(1 - \sin\phi'\sin\psi)}$$	$\theta_{SF} = \pi/2 - \lambda + \psi$
Shank resistance, F_s	$F_s = A_s\,\gamma'\,d_s\,N_{qs}$	$\theta_s = \pi/2 - \theta_{fs} + \beta - \delta$
Top of fluke, F_f	From soil wedge equilibrium	$\theta_f = \pi/2 - \beta - \delta$
Slip surface, R	From soil wedge equilibrium	$\theta_R = \lambda + \phi - \pi/2$
Anchor weight, W_a	Anchor dependent	Vertical
Base of fluke, F_{fb}	From anchor equilibrium	$\theta_{fb} = \pi/2 - \beta + \theta_w + \delta$
Pad-eye tension, T_a	From anchor equilibrium	θ_a
Soil wedge equilibrium	$F_f\sin\theta_f + R\sin\theta_R = SF\sin\theta_{SF} + W_s + F_s\sin\theta_s$ $F_f\cos\theta_f - R\cos\theta_R = SF\cos\theta_{SF} - F_s\cos\theta_s$	
Anchor equilibrium	$T_a\sin\theta_a + F_{fb}\sin\theta_{fb} = W_a + F_f\sin\theta_f - F_s\sin\theta_s$ $T_a\cos\theta_a - F_{fb}\cos\theta_{fb} = W_a + F_f\cos\theta_f + F_s\cos\theta_s$	

Source: Neubecker SR and Randolph MF, 1996, *Canad Geotech J*, 3(4), 574–558.

- A nonzero force F_{fb} acts at the base of the fluke. Its direction is fixed by geometric and frictional resistance considerations, and its magnitude is governed by equilibrium considerations for the anchor itself. In this equilibrium calculation, the magnitudes of F_{fb} and the pad-eye tension T_a are the two unknowns to be determined.
- The equilibrium analysis can be performed for an arbitrarily specified chain angle θ_a; however, admissible combinations of T_a and θ_a are actually constrained by the mechanics of anchor lines discussed in Section 5.2 Equation 5.10. Thus, iteration is required to determine the T_a–θ_a combination that satisfies the equilibrium requirements for both the anchor and the chain.

Figure 9.56 shows an example relationship between fluke tip embedment depth H and normalized load capacity of an anchor, excluding the contribution from anchor weight W_a.

9.11 ADDITIONAL CONSIDERATIONS

During anchor penetration disturbance weakens the soil adjacent to the anchor. As discussed in Section 6.5, much of the strength reduction is associated with excess pore water pressures induced by soil remolding. After anchor installation, dissipation of excess pore pressures leads to recovery of soil strength with a concomitant increase in anchor capacity. Dunnavant and Kwan (1993) provide an assessment of the gain in load capacity owing to consolidation based on centrifuge models of Vryhof Stevpris anchors with prototype weights of 1.1 kips and 15 kips. Figure 9.57 shows an example of the measured relationship

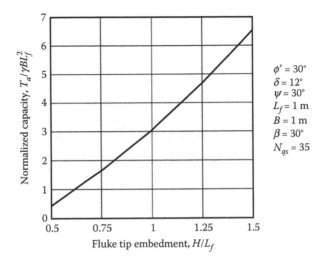

Figure 9.56 Example analysis of DEA in sand.

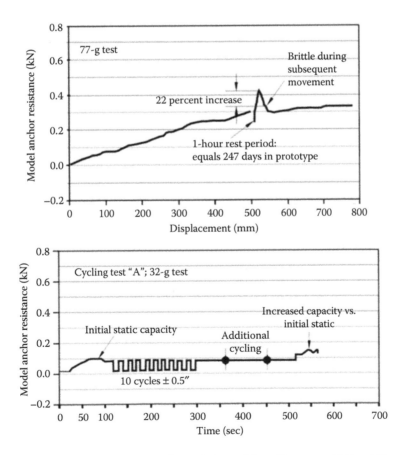

Figure 9.57 Consolidation and cyclic loading effects. (Adapted from Dunnavant TW and Kwan C-TT, 1993, *Proceedings of the 25th Annual Offshore Technology Conference*, Houston, Texas.)

between the anchor resistance and the drag distance for the model anchor. The anchor was dragged to a distance 500 mm (38.5 m prototype), allowed to set for 1 hour (247 days prototype), after which dragging was resumed. While post-set anchor resistance increased by 22%, the strength gain was rapidly lost upon continued dragging. Their results indicate that the setup process leads to brittle behavior, with the gain in load capacity owing to consolidation being quickly lost when the anchor displaces a small distance past its original position.

Dunnavant and Kwan (1993) also conducted centrifuge tests to assess the effects of cyclic loading on the load capacity of DEAs in soft clay. As shown in Figure 9.57, the load capacity of the anchor tended to increase somewhat under cyclic loading, partly owing to increased penetration of the anchor under cyclic loading.

REFERENCES

Aubeny CP and Chi C-M, 2010, Mechanics of drag embedment anchors in a soft seabed, *ASCE J Geotech Geoenviron Eng*, 136(1), 57–68.

Aubeny C, Gilbert R., Randall R, Zimmerman E, McCarthy K, Chen C-H, Drake A, Yeh P, Chi C-M, and Beemer R, 2011, *The Performance of Drag Embedment Anchors for Minerals Management Service*, Offshore Technology Research Center, College Station, TX.

Dahlberg R, 1998, Design procedures for deepwater anchors in clay, *Proceedings of the 30th Offshore Technology Conference*, OTC 8837: Houston, pp 559–567.

Dunnavant TW and Kwan C-TT, 1993, Centrifuge modeling and parametric analyses of drag anchor behavior, OTC 7202, *Proceedings of the 25th Annual Offshore Technology Conference*, Houston, Texas, pp 29–39.

Foxton P, 1996, Deepwater moorings: The Amoco Liuhua experience and beyond, *IBC Conference Mooring and Anchoring*, Aberdeen, Scotland.

Foxton P, 1997, Latest development for vertically loaded anchors, *2nd Annual Conference on Mooring and Anchoring*, Aberdeen, Scotland.

Kim BM, 2005, Upper bound analysis for drag anchors in soft clay, Doctoral Dissertation, Texas A&M University, College Station, Texas.

Murff JD, Randolph MF, Elkhatib S, Kolk HJ, Ruinen RM, Strom PJ, and Thorne CP, 2005, Vertically loaded plate anchors for deepwater applications, *Proceedings of the International Symphony on Frontiers in Offshore Geotechnics, IS-FOG05*, Perth, pp 31–48.

NCEL, 1987, *Drag Embedment Anchors for Navy Moorings*, Techdata Sheet 83-08R, Port Hueneme, California: Naval Civil Engineering Laboratory.

NAVFAC, 2011, *SP-2209-OCN Handbook for Marine Geotechnical Engineering, Naval Facilities Engineering Command*, Engineering Service Center, Port Hueneme, USA.

Neubecker SR and Randolph MF, 1995a, Profile and frictional capacity of embedded anchor chain, *J Geotech Eng Div*, ASCE, 121(11), 787–803.

Neubecker SR and Randolph MF, 1995b, The performance of embedded anchor chains systems and consequences for anchor design, *Proceedings of the 28th Offshore Technology Conference*, OTC 7712: Houston, pp 191–200.

Neubecker SR and Randolph MF, 1996, The static equilibrium of drag anchors in sand, *Canad Geotech J*, 3(4), 574–558.

Omega Marine Services International, 1990, *Joint Industry Project: Gulf of Mexico Large Scale Anchor Tests—Test Report*, Omega Marine Services International, Houston, Texas.

O'Neill M, Bransby MF, and Randolph MF, 2003, Drag anchor fluke-soil interaction in clays, *Can Geotech J*, 40, 78–94.

Rasulo M, 2016, Simplified plastic limit analysis of drag embedment anchors in layered cohesive soils, MS Thesis, Texas A&M University, College Station, 68p.

Rowe RK and Davis EH, 1982, The behaviour of anchor plates in clay, *Geotechnique*, 32(1), 9–23.

Stewart WP, 1992, Drag embedment anchor performance prediction in soft soils, *Proceedings of the 24th Offshore Technology Conference*, Houston: OTC 6970, pp 241–248.

Vivatrat V, Valent PJ, and Ponterio AA, 1982, The influence of chain friction on anchor pile design, *Proceedings of the 14th Annual Offshore Technology Conference*, Houston, Texas, OTC 4178, pp 153–163.

Vryhof, 2015, *Anchor Manual 2015 The Guide to Anchoring*, Vryhof Anchors. Capelle a/d Yssel, The Netherlands, 168p.

Yang M, Murff JM, and Aubeny CP, 2010, Undrained capacity of plate anchors under general loading, *ASCE J Geotech Geoenviron Eng*, 136(10), 1383–1393.

Yoon Y, 2002, Prediction methods for capacity of drag anchors in clayey soils, *MS Thesis*, Texas A&M University, College Station.

Chapter 10

Direct embedment plate anchors

The VLA discussed in Chapter 9 has proven to be a cost-effective and efficient anchoring system to tether floating structures to the seabed. However, uncertainties in determining the exact embedment depth and location of the anchor have limited its application particularly for permanent facilities. This chapter discusses various plate anchor systems involving direct embedment of the plate, which can reduce the uncertainty in positioning associated with DEAs and VLAs. For the purpose of discussion in this chapter, direct embedment includes any plate anchor installation method exclusive of drag embedment. The installation techniques to a large extent parallel those for piles and caissons discussed in Chapter 6: suction, dynamic, and driving. Torque installation of helical anchors is an additional alternative that is covered in this chapter.

10.1 SUCTION EMBEDDED PLATE ANCHORS

To overcome the limitations of VLAs, Dove et al. (1998) and Wilde et al. (2001) conceived and developed the SEPLA. This anchor is primarily intended for use in soft clays. The SEPLA uses a suction caisson that has known penetration depth and a closely controlled location to embed a rectangular plate anchor to the target depth. The SEPLA installation consists of the following steps: caisson self-weight and suction penetration, caisson retraction, and anchor "keying," shown schematically in Figure 10.1. First, the caisson with a plate anchor slotted vertically in its base is lowered to the seafloor and penetrated into the soil under its dead weight until the skin friction and end-bearing resistance equal the caisson's dead weight. Then, the vent valve on the top of the caisson is closed, and the water trapped inside is pumped out. The ensuing differential pressure at the top drives the caisson to the design depth. The plate anchor is then released and the water is pumped back into the caisson, causing the caisson to move upward, leaving the plate anchor in place in a vertical orientation. The caisson is retracted from the seabed and prepared to be used for the next installation. As the anchor chain attached to the plate anchor is tensioned in the design direction it cuts into soil. At the same time, the anchor line applies a load to the anchor's offset pad eye causing it to rotate or "key." It ultimately reaches the target orientation perpendicular to the direction of loading such that the maximum capacity of the anchor is mobilized.

Suction installation for SEPLAs parallels the procedures employed for conventional suction caissons, described in Section 6.1, except that the soil resistance on the plate anchor should be included in the analysis. The bearing capacity equation for tangential load capacity of a plate (Equation 4.3) is generally adequate for this purpose (Gaudin et al., 2006). The load capacity of a SEPLA is computed from the basic solutions for bearing resistance for plates described in Chapters 3 and 4. The primary issue uniquely associated with SEPLAs

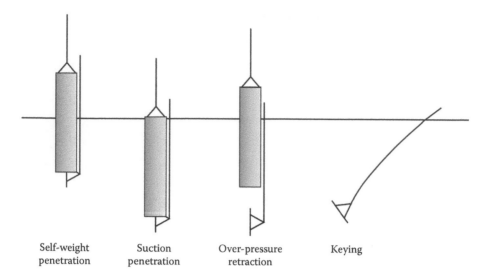

Self-weight penetration   Suction penetration   Over-pressure retraction   Keying

Figure 10.1 SEPLA installation.

is the loss of embedment occurring during the keying process. The following discussion addresses this issue.

10.1.1 System of forces on anchor

We begin by considering the system of forces acting on the anchor shown in Figure 10.2. The mooring line tension at the pad eye T_a acts at an angle θ_a measured from horizontal. The anchor has a fluke length B with a shank of length e_n attached at an offset e_s from the center of the fluke. The fluke is rectangular with width L and thickness t. At any given instant in the keying trajectory, the fluke is oriented at an angle θ_f from horizontal. An angle θ_{af} defines the angle between the mooring line force and the fluke. For the purposes of the analysis, the anchor is at a limit state with the collapse load F equaling the pad-eye tension T_a. The anchor embedded in clay has undrained shear strength s_u at the depth under consideration. Song et al. (2009) found that the anchor behavior is unaffected by the orientation for anchor embedment depths greater than three fluke lengths. This analysis assumes that anchor

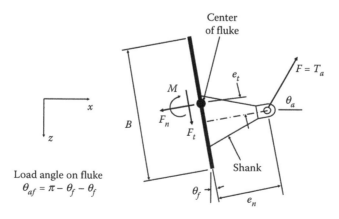

Load angle on fluke
$$\theta_{af} = \pi - \theta_f - \theta_f$$

Figure 10.2 Force system on a SEPLA.

embedment exceeds this depth, such that free surface effects may be neglected. SEPLAs used in practice commonly feature a flap intended to minimize embedment loss during keying. Studies by Yang et al. (2012) indicate that the flap does not rotate into its intended position until very late in the keying trajectory. Therefore, aside from providing additional bearing area to the anchor, the flap is not effective in reducing embedment loss during keying. The discussion that follows considers SEPLA flukes that do not incorporate a flap.

The normalized load components acting on the fluke take the following form:

$$N_n = F_n/s_u A_f = N_e \sin\theta_{af} = N_e c_1 \tag{10.1}$$

$$N_t = F_t/s_u A_f = N_e \cos\theta_{af} = N_e c_2 \tag{10.2}$$

$$N_m = \frac{M}{s_u A_f L_f} = N_m \left(\frac{L_s}{L_f} \cos\theta_{af} + \frac{e_s}{L_f} \sin\theta_{af} \right) \tag{10.3}$$

The yield locus equation for a plate (Section 4.1) is combined with Equations 10.1 through 10.3 to produce the expression shown in Equation 10.4. It may be recalled that the parameters N_{nmax}, N_{tmax}, and N_{mmax} are the bearing factors for the anchor under uniaxial normal, tangential, and rotational loading conditions, respectively. The parameters q, m, n, and p are interaction parameters defining the yield locus under combined loading (Bransby and O'Neill, 1999).

$$f = \left(\frac{c_1 N_e}{N_{nmax}} \right)^q + \left[\left(\frac{c_3 N_e}{N_{mmax}} \right)^m + \left(\frac{c_2 N_e}{N_{tmax}} \right)^n \right]^{\frac{1}{p}} - 1 = 0 \tag{10.4}$$

The operative bearing factor of the anchor N_e is then computed as the root of Equation 10.4. Wei et al. (2014) compiled a list of yield locus parameters derived from finite element studies by various investigators for various conditions (Table 10.1). The studies largely comprised two-dimensional (2D) analyses for strip anchors, but several studies considered finite aspect ratios L/B that are more representative of the actual rectangular flukes used for SEPLAs. Also considered in the various studies is the fluke thickness effect B/t and surface contact conditions. A fully bonded contact corresponds to no slippage at the soil–fluke interface; this condition nearly approximates the case of full adhesion, $\alpha = 1$. The study by Wei et al. (2014) also includes yield locus parameters that implicitly account for soil resistance on the shank of a SEPLA.

10.1.2 Anchor kinematics

Assuming an associated flow rule, the plastic potential function equals the yield locus equation, $g = f$, where f is defined by Equation 10.4. Plastic deformations can be taken as the gradient of g, as expressed by Equations 10.5 through 10.7.

$$v_n = \lambda \partial g/\partial N_n = \left(\frac{N_n}{N_{nmax}} \right)^{n-1} \frac{q}{N_{nmax}} \tag{10.5}$$

$$v_t = \lambda \partial g/\partial N_t = \left[\left(\frac{N_m}{N_{mmax}} \right)^m + \left(\frac{N_t}{N_{tmax}} \right)^n \right]^{1/p-1} \left(\frac{N_t}{N_{tmax}} \right)^{n-1} \frac{n}{N_{tmax}} \tag{10.6}$$

Table 10.1 Yield Locus parameters for plate anchors

L /B	B/t	Contact	Bearing			Interaction				Reference
			N_{mmax}	N_{tmax}	N_{mmax}	m	n	p	q	
2D	7	Fully bonded	11.87	4.29	1.49	1.26	3.72	1.09	3.16	O'Neill et al. (2003)
2D	20	$\alpha = 0.4$	11.58	1.97	1.53	1.52	5.31	1.01	2.75	Elkhatib and Randolph (2005)
2D	20	Fully bonded	11.62	3.19	1.59	1.14	4.92	1	3.39	
2D	7	$\alpha = 0.4$	11.78	3.38	1.55	2.58	3.74	1.09	1.74	
2D	7	Fully bonded	11.93	4.65	1.63	1.27	3.46	1.03	3.23	
1	20	Fully bonded	13.21	3.22	2.05	1.07	4.19	1.1	4.02	Elkhatib (2006)
2D	29	Fully bonded	14	3	2	1.1	4.2	1.1	4	Cassidy et al. (2012)
2D	29	Fully bonded	11.68	2.83	1.63	1.27	5.23	1.08	3.39	Wei et al. (2014)
2D shank	29	Fully bonded	11.70	6.28	1.63	1.24	1.80	1.06	3.33	
1.71	29	Fully bonded	12.66	2.96	1.89	1.32	5.56	1.34	3.42	
1.71Shank	29	Fully bonded	12.70	4.24	1.90	1.56	3.82	1.37	3.31	

Source: After Wei et al., 2014, Incorporating shank resistance into prediction of the keying behaviour of suction embedded plate anchors, *ASCE J Geotech Geoenviron Eng*, 141(1), 04014080-1 to 04014080-13 (electronic version).

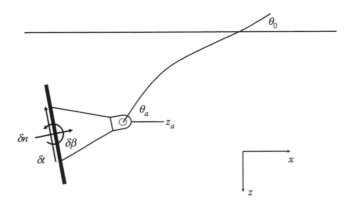

Figure 10.3 SEPLA motions.

$$\dot\beta L_f = \lambda \partial g/\partial N_m = \left[\left(\frac{N_m}{N_{mmax}}\right)^m + \left(\frac{N_t}{N_{tmax}}\right)^n\right]^{1/p-1}\left(\frac{N_m}{N_{mmax}}\right)^{m-1}\frac{m}{N_{mmax}} \tag{10.7}$$

The positive scalar muliplier λ indeterminate, but Equations 10.5 through 10.7 provide information on the relative magnitudes of plastic deformation, which is necessary for a trajectory calculation to proceed. In the drag embedment anchor trajectory calculations, the anchor motions were predominantly in the tangential direction. Therefore, it was convenient to formulate the trajectory calculation by applying an incremental displacement δt parallel to the fluke (Figure 10.3), and then computing the corresponding incremental normal displacement $\delta n = (v_n/v_t)\,\delta t$ and incremental rotation $\delta\beta = (\dot\beta B/v_t)\delta t$. This approach works to some extent in SEPLA trajectory calculations; however, as the anchor rotates into a position nearly normal to the direction of loading, tangential motions become vanishingly small. Numerical problems are avoided by identifying the dominant motion in advance of each step of the trajectory calculation. For example, if Equations 10.5 through 10.7 show normal motions to be dominant, an incremental displacement δn is applied normal to the fluke, and the corresponding tangential displacement is computed as $\delta t = (v_t/v_n)\,\delta n$ and incremental rotation is computed as $\delta\beta = (\dot\beta B/v_n)\delta n$.

10.1.3 Interaction with mooring line

The load capacity and kinematic relationships described above are formulated in terms of pad-eye load angle θ_a. In most situations, the control of the mooring line angle is actually exerted at the mudline; that is, a fixed mooring line angle at the mudline θ_0 is imposed. Thus, the anchor line equations in the soil column must be invoked to relate conditions at the mudline to those at the pad eye. The relevant anchor line equation from Section 5.2 is repeated here:

$$\frac{T_a}{1+\mu^2}\left[\exp\left[\mu(\theta_a-\theta_0)\right](\cos\theta_0+\mu\sin\theta_0)-\cos\theta_a-\mu\sin\theta_a\right]=\int_0^z Q(z)dz=\bar{Q}z \tag{10.8}$$

It may be recalled that μ is the ratio of tangential to normal soil resistance acting on the mooring line. The following sequence of steps can be executed to establish the initial values of pad-eye tension T_a and angle θ_a for trajectory calculations for a SEPLA embedded at depth z with soil shear strength s_u for a selected fixed value of θ_0:

1. Compute the bearing factor N_e of the anchor as a function of load angle θ_{af} from Equations 10.1 through 10.4.
2. Select a trial load angle θ_a and, noting that fluke is initially vertical, $\theta_f = \pi/2$, compute θ_{af}.
3. Compute the load capacity of the anchor, $F = N_e (\theta_{af}) s_u A_f$.
4. Compute the pad-eye tension T_a from Equation 10.8.
5. Repeat Steps 2 through 4 while varying θ_a until F matches T_a within an acceptable tolerance.

10.1.4 Example trajectory calculations

Using the equations described in the preceding sections, the predictions of anchor trajectory during keying can proceed as follows:

1. Establish the initial pad-eye tension T_a and angle θ_a using Equations 10.1 through 10.4 and Equation 10.8 as described in the discussion of mooring line interaction.
2. Impose an incremental displacement on the anchor—δt, δn, $\delta\beta$—as described in the above discussion of kinematic behavior.
3. Update the horizontal coordinate of the center of the fluke using $\Delta x = \delta n \sin \theta_f - \delta t \cos \theta_f$ and the vertical coordinate using $\Delta z = -(\delta n \sin \theta_f + \delta t \cos \theta_f)$, and compute the updated soil shear s_u for the new embedment depth.
4. Update the fluke angle using $\Delta\theta_f = -\delta\beta$ and the pad-eye angle θ_a from Equation 10.8.
5. Compute the new line tension T_a using the updated load–fluke angle θ_{af} and the updated soil strength.

Figures 10.4 and 10.5 show example trajectory calculations for a $B/t = 20$ anchor using the parameters developed by Elkhatib and Randolph (2005) in Table 10.1. The calculations are for a SEPLA embedded in normally consolidated clay having a linearly increasing strength profile with $s_{um} = 2$ kPa and $k = 1.60$ kPa/m. The anchor has a fluke length

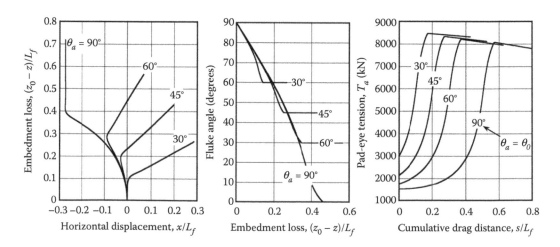

Figure 10.4 Effect of load angle on SEPLA behavior.

$B = 3$ m, width $L = 7.3$ m, and thickness $t = 0.15$ m. The shank is attached at the center of the fluke, $e_s = 0$, and the pad eye is offset a distance $L_s = 2.1$ m from the fluke. When the mooring line interaction is considered (Figure 10.5), the analysis considers a wire rope mooring line with diameter $b = 0.15$ m, with an assumed line bearing factor $N_c = 10$ and ratio of tangential to normal resistance on the line $\mu = 0.1$.

Figure 10.4 shows predicted anchor behavior for the case of no interaction with the mooring line; that is, the mooring line angle at the pad eye equals that at the mudline, $\theta_a = \theta_0$. Trajectory is plotted in terms of horizontal distance traversed by the anchor versus embedment loss, expressed in terms of fluke lengths of displacement. At small load angles, $\theta_a < 30°$, the anchor initially moves upward a small distance—in this case 0.1 fluke lengths—and then moves into the direction of applied load θ_a. At larger load angles, the anchor actually moves backward relative to the direction of the applied load—that is, in the negative x-direction in Figure 10.4—before turning into the direction of applied loading. Embedment loss increases with increasing load angle, with embedment loss in this case for pure vertical loading. The central plot in Figure 10.4 shows predicted histories of fluke angle θ_f versus embedment loss. If we take the end of the keying process as being when zero moment acts on the fluke (in the case of $e_s = 0$, this corresponds to the fluke being perpendicular to the direction of the applied pad-eye load), then this example shows the embedment loss for vertical loading to be about three times that of a 30° load angle. The right-hand plot in Figure 10.4 shows the reductions in anchor capacity—as expressed by pad-eye tension T_a—owing to the embedment loss during keying. For reference, the maximum capacity of the anchor at the original embedment depth z_0 is 8650 kN. Therefore, the load capacity reduction associated with keying ranges from 2% to 7%, depending on load angle.

Figure 10.5 shows the keying simulation for the same anchor and soil conditions, but including the chain interaction effect. The interaction effect is most significant for smaller mudline load angles θ_0, with the embedment loss for $\theta_0 = 30°$ being approximately 30% greater when the chain interaction effect is considered. The corresponding loss of load capacity T_a is also slightly greater, about 3% versus 2% computed for the no-interaction analysis. Predicted histories of pad-eye angle θ_a (center plot in Figure 10.5) show how the deviated from the mudline angle at the end of keying. For example, for a mudline load angle $\theta_0 = 30°$, the pad-eye angle approaches $\theta_a = 37°$ at the end of the keying process. Differences at the end of keying between θ_a and θ_0 diminish as θ_0 increases.

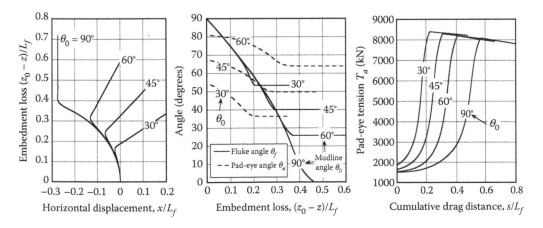

Figure 10.5 Effect of mooring line interaction on SEPLA behavior.

10.1.5 Comparison to experimental data

Gaudin et al. (2006) performed a series of centrifuge tests carried out at 145 g to evaluate SEPLA performance using a suction caisson having prototype diameter $D = 4.35$ m, length $L_f = 24.5$ m, and wall thickness $t_w = 0.58$ m. The caisson was used to install a square plate of length $L = 5.08$ m, thickness $t = 0.145$ m, and eccentricity $e_t = 0$ and $e_n = 3.33$ m. The tests were conducted in a normally consolidated kaolinite test bed having a linearly varying undrained strength with $k = 1.1$ kPa/m, with an estimated sensitivity $S_t = 2.5$. The orientation of the chain at the mudline was set at $\theta_0 = 45°$ for this test series. The chain comprised four plaited strands of wire that produced a bearing diameter $b = 0.319$ m at prototype scale. The test program investigated a number of installation-retraction scenarios, including suction versus jacked installation, pumped (overpressure) versus vented retraction, and the allowance of a set time between anchor installation and pullout. The case considered here is suction installation, pumped retraction with minimal set time (designated as test PE-ST2 the article).

In employing the plastic limit approach presented earlier for predicting loss of embedment during keying, we will utilize the O'Neill et al. (2003) bearing factors (simplified by Murff et al., 2005) defined by Equations 4.2 through 4.4 in this text in conjunction with the O'Neill et al. (2003) interaction factors listed in Table 10.1. While these simplified equations are not necessarily as accurate as the finite element solutions in Table 10.1, they do permit explicit consideration of the actual aspect ratio of the plate, $L/B = 35$. Additionally, they permit an assessment of the influence of plate roughness $\alpha = 1 / S_t$ on the keying loss predictions. In assessing the influence of sensitivity, a range $S_t = 1–2.5$ is considered. The left-hand plot in Figure 10.6 shows the predicted relationships between embedment loss and bearing factor in comparison to the measurements obtained by Gaudin et al. (2006). The plastic limit model does not compare well to measurements during the early stages of keying, with the model prediction a much steeper rise in bearing resistance N_e at this stage of keying. However, during the middle and latter stages of keying, the model appears to provide a generally realistic portrayal of actual behavior. Noteworthy is the strong influence of soil sensitivity on the predictions. A sensitivity $S_t = 1.5$ actually provides a close match to measurements. However, the match needs to be regarded as somewhat fortuitous, since the estimated sensitivity was actually 2.5. Additionally, predictions can be significantly

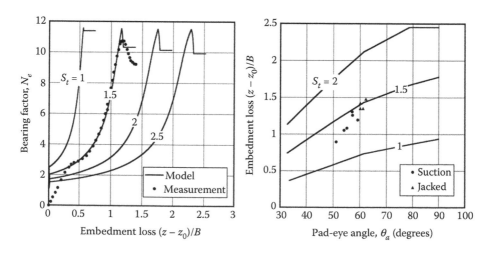

Figure 10.6 Measurements of keying embedment loss.

influenced by specific bearing factors and interaction coefficients employed in the analysis. It is noted that the keying predictions presented in the previous section (Figures 10.4 and 10.5) are based on an assumption of full bonding between soil and anchor (essentially $S_t = 1$), which results in embedment losses on the order of one half of a plate length. As is evident from Figure 10.6, embedment losses on the order of 1–1.5 of a plate length are more realistic. Gaudin et al. also demonstrate that there is a strong, nearly linear, correlation between embedment loss and pad-eye angle θ_a, as shown in the right-hand plot in Figure 10.6. In their case, all mudline load angles are $\theta_0 = 45°$, so the final pad-eye angles depend on the history of the keying trajectory. The plastic limit model predictions superimposed on the data measurements were generated by varying the mudline angle from $\theta_0 = 0$–$90°$. The trend lines for the predicted effects of pad-eye angle differ somewhat from the trend of the measured data points. This is not totally unexpected since the predicted curves were generated by varying θ_0, while the measured points were generated using a constant θ_0. Therefore, different physical mechanisms may be behind the different behaviors.

10.1.6 Parametric studies

Wang et al. (2011) undertook a series of large deformation finite element analyses to investigate the influence of a number of variables that could influence embedment loss during keying, including (in nondimensional form) the following anchor aspect ratio B/L, eccentricity e_n/L, thickness t/L, local soil strength $s_{u0}/\gamma'B$, soil rigidity E/s_u, soil strength gradient kB/s_{u0}, anchor weight $\gamma'_a t/s_{u0}$, and load angle θ_0. Their study identified the following variables that exert a significant influence on embedment loss during keying:

Anchor geometry: Square anchors experience about 40% greater ultimate loss of embedment Δz_u than strip anchors (upper left plot in Figure 10.7); therefore, predictions based on strip anchor solutions are likely to give unconservative estimates of keying embedment loss. Additionally, reduced plate thickness t leads to greater loss of embedment. They note that, while decreasing the thickness of the anchor plate is attractive from the standpoint of minimizing soil resistance during SEPLA penetration, it has the negative effect of increasing embedment loss during keying.

Soil strength: Soil rigidity and strength gradient were found to have negligible impact on keying behavior. However, the local magnitude of soil strength s_{u0} at the point of anchor embedment was found to significantly influence keying behavior, since greater soil strength produces a greater resistance to plate rotation, thereby increasing embedment loss. However, embedment loss does not increase indefinitely with increasing strength. As shown in the upper right plot Figure 10.7, a point is reached where embedment loss is essentially independent of depth. Wang et al. characterize this behavior by defining Δz_{max} as the maximum embedment loss that can occur, irrespective of soil strength. From their finite element analyses, they provide the empirical expression for Δz_{max} as a function of eccentricity e_n ($=e$ in Figure 10.7) and plate thickness t shown below:

$$\frac{\Delta z_{max}}{B} = a\left[\left(\frac{e_n}{B}\right)\left(\frac{t}{B}\right)^p\right]^q \tag{10.9a}$$

For square plates they give the fitting coefficients $a = 0.144$, $p = 0.2$, and $q = -1.15$. To account for the reduced loss of embedment associated with reduced rotational resistance at lower values of soil strength s_{u0}, they provided the following fit to the finite element solutions shown in the center plot in Figure 10.7:

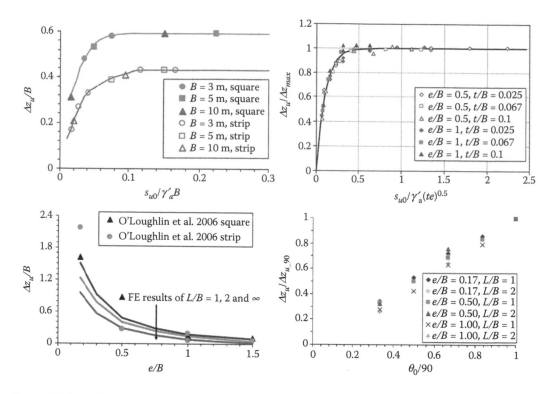

Figure 10.7 Variables affecting keying embedment loss. (After Wang D, Hu Y, and Randolph MF, 2011, *ASCE J Geotech Geoenviron Eng*, 137(12), 1244–1253.)

$$\frac{\Delta z_u}{B} = \frac{\Delta z_{max}}{B} \tanh\left[b\left(\frac{s_{u0}}{\gamma'\sqrt{te_n}}\right)^r\right] \qquad (10.9b)$$

Best fit coefficients are given as $b = 5$ and $r = 0.85$.

Load eccentricity: Eccentricity e_n significantly affects embedment loss (lower left plot in Figure 10.7), and its effect is considered in Equation 10.9. Wang et al. note that embedment loss increases dramatically when eccentricity is less than one half of the anchor width $e_n/B < 0.5$.

Chain angle: As demonstrated repeatedly in earlier sections, embedment loss increases in a nearly linear fashion with load angle θ_0. The plot in the lower right of Figure 10.7 shows that this trend holds for a wide range of anchor eccentricity and aspect ratios.

10.1.7 Performance of keying flap

The commonly used version of the SEPLA used in industry features a flap (Figure 10.8), which is intended to reduce upward movement of the anchor during keying; that is, minimize keying embedment loss. The length of the flap is typically 0.3 times the length of the fluke. The pad eye is positioned at the center of the fluke; therefore, eccentric loading conditions generally prevail on the combined fluke–flap system. The flap is configured to permit only outward (counter-clockwise in the figure) rotation, and the outward rotation is limited

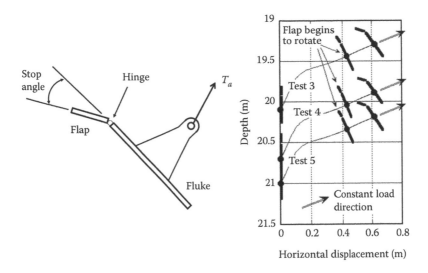

Figure 10.8 Performance of the keying flap. (After Yang M, Aubeny CP, and Murff JD, 2012, *J Geotech Geoenviron Eng*, ASCE, 138(2), 174–183.)

by a "stop angle" of 20°. Yang et al. (2012) modified the plastic limit analysis described earlier to incorporate the soil resistance mobilized by the flap. The revised formulation considers the following constraints: (1) equilibrium of the fluke in conjunction with the yield criterion, (2) equilibrium of the flap in conjunction with the yield criterion, (3) compatibility of fluke and flap displacements at the hinge, and (4) interaction with the anchor chain. The first two constraints are effectively imposed by Equation 10.4. The methods for enforcing the compatibility of displacements depend on the mode of flap motion; three modes are possible. When the flap is aligned with the fluke, the anchor can be treated simply as a rigid body with a nonzero tangential eccentricity e_t (Figure 10.2). When the flap is at the limiting stop angle, the anchor can similarly be treated as rigid, with due consideration of the angle between the two plates. When the flap is freely rotating, Equations 10.5 through 10.7 are invoked to ensure that the velocity at the top of the fluke (i.e., at the hinge) equals the velocity at the bottom of the flap. The final constraint is the interaction with the chain, which was described earlier within the context of a single plate.

The plot on the right of Figure 10.8 shows the typical results from the analysis. Significantly, flap rotation does not occur until late in the keying process, thereby indicating that the flap as currently designed is ineffective in reducing keying embedment loss. Large deformation finite element analyses by Tian et al. (2014) also produced results that lead to the questioning of the effectiveness of the design of the keying flap currently used in industry. Interestingly, they conclude that a keying flap could potentially improve the SEPLA performance, but the finite element simulations indicate that the flap motions should be constrained to rotate inward (clockwise in Figure 10.8) relative to the fluke to achieve the beneficial effect of a flap.

10.2 PILE DRIVEN PLATE ANCHORS

As shown in Figure 10.9, the concept behind pile driven plate anchors (PDPAs) is similar to that described above for SEPLAs, except that conventional pile driving techniques, rather than suction advances the pile (NAVFAC, 2011). A plate anchor is attached to the tip of

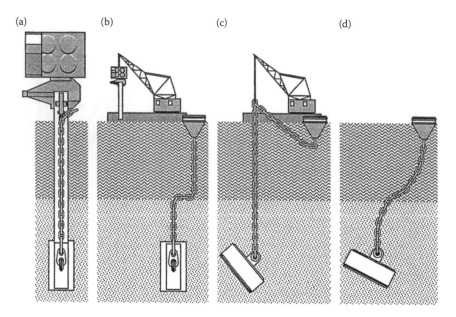

Figure 10.9 Pile driven plate anchor installation procedure (a) drive anchor, (b) recover driver and follower, (c) proof load to set anchor, and (d) anchor established. (From Forrest J, Taylor R, and Brown L, 1995, *Design Guide for Pile-driven Plate Anchors*, Technical Report TR-2039-OCN, Naval Facilities Engineering Service Center, Port Hueneme, CA, March 1995.)

a pile, termed a "follower," which can be installed by hammer driving, vibration, or jetting. After installation to the desired plate embedment depth, the follower is retracted. The plate anchor is then "keyed" to rotate from its initial vertical orientation to an orientation approximately normal to the direction of the applied mooring line load. In soft soils a keying flap is sometimes attached to the anchor. A keying flap is a hinged plate that is oriented vertically during penetration, but opens to resist uplift during keying.

In contrast to SEPLAs, which are largely restricted to soft clay soil profiles, PDPAs can be deployed in a broad range of soil profiles, including soft clay, sand, over-consolidated clays, and coral (Forrest et al., 1995). Installation is possible in sands, stiff clays, and stratified soil profiles, conditions that can render many alternative anchor systems either difficult or impossible. The PDPA plate has a relatively simple geometry (Figure 10.10), with correspondingly modest fabrication costs. Like all plate anchors, the PDPAs have a high geotechnical efficiency; thus, high load capacity is achievable using relatively small, lightweight anchors. Since deep embedment of PDPAs is possible in sands and stiff clays, they can develop significant vertical load capacity in these soil profiles. In this regard, PDPAs enjoy a notable advantage over DEAs, whose shallow embedment in sands and stiff clays largely restricts their use for resisting horizontal loads associated with catenary mooring systems. A chief potential disadvantage of PDPAs is the relatively high cost and complexity of offshore required by the pile driving operation to advance the anchor to its target depth. Nevertheless, in situations involving (1) a vertical component of loading on the anchors and (2) complex soil stratigraphy, PDPAs can be one of the few anchor types that are technically feasible.

Figure 10.11 shows that the guidelines for estimating plate anchor load capacity in clays (Forrest et al., 1995) generally recommend a uniaxial normal loading bearing factor $N_{n max} = 12(= \bar{N}_c)$ for anchor embedments greater than $h/B = 2-4$. At shallower embedment depths, a downward

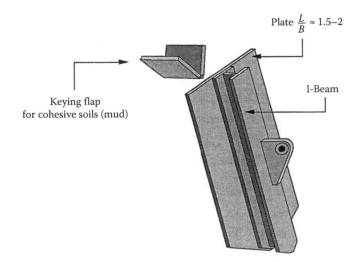

Figure 10.10 Pile driven plate anchor detail. (From Forrest J, Taylor R, and Brown L, 1995, *Design Guide for Pile-driven Plate Anchors*, Technical Report TR-2039-OCN, Naval Facilities Engineering Service Center, Port Hueneme, CA, March 1995.)

adjustment is made to the bearing factor to account for the free surface effect. This guidance is consistent with the solutions for plate anchors embedded in clay presented in Chapters 3 and 4 of this book, as well as the guidance described above (Table 10.1) for SEPLAs.

10.2.1 Load capacity of strip anchors in sand

Uplift capacity of anchors embedded in sand utilizing limit equilibrium approaches or the method of characteristics include fundamental studies by Meyerhof and Adams (1968),

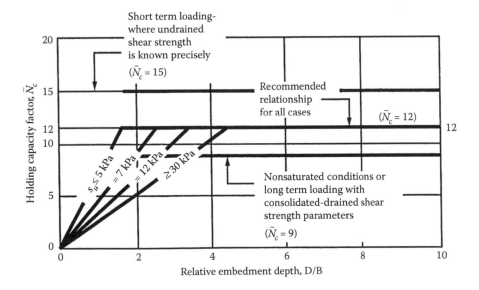

Figure 10.11 Pile driven plate anchor capacity in clay. (From Forrest J, Taylor R, and Brown L, 1995, *Design Guide for Pile-driven Plate Anchors*, Technical Report TR-2039-OCN, Naval Facilities Engineering Service Center, Port Hueneme, CA, March 1995.)

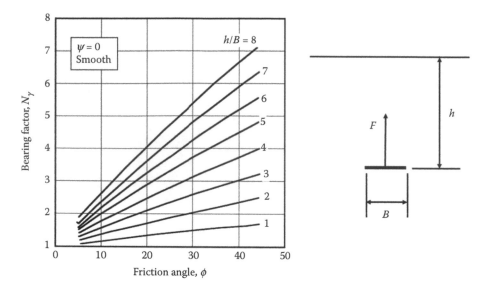

Figure 10.12 Basic bearing factor for vertically loaded plate anchor in sand. (After Rowe and Davis, 1982.)

Vesic (1971), and Merifield et al. (2006). The present discussion focuses on a finite element-based study by Rowe and Davis (1982), since subsequent assessments of their method show good overall agreement with physical measurements (Dickin, 1988; Giampa et al., 2017). As the Rowe and Davis analysis applies to strip anchors, the determination of appropriate shape factors still requires resolution. Additionally, the Rowe and Davis analysis addresses only eccentrically loaded anchors. Thus, they are generally applicable to pullout capacity assessment, but do not provide the information necessary for a keying analysis. The analyses cover two cases of centric normal loading, namely vertical loading on a horizontally oriented plate (Figure 10.12) and horizontal loading on a vertically oriented plate (Figure 10.14).

The analyses start with a base case condition of a nonassociated flow law with zero dilation angle, $\psi = 0$. The average yield pressure q_{ult} for the anchor can be expressed in terms of a bearing factor N_γ and in situ vertical effective stress $\gamma'h$ as follows:

$$q_{ult} = F/BL = \gamma'h\,N_\gamma R_\psi R_R \qquad (10.10)$$

The factors R_ψ and R_R adjust the load capacity to account for dilatancy and anchor roughness. Rowe and Davis also introduce a R_K factor to consider the effect of initial stress state but, as they conclude that it can be taken as being equal to unity, it is not considered further here. Figure 10.12 shows computed values of N_γ of as a function of friction angle ϕ and anchor embedment depth h for the case of a vertical loading on a horizontally oriented strip anchor.

The effect of dilatancy is assessed by first considering the case of an associated flow law, $\psi = \phi$. Rowe and Davis show the dilatancy multiplier for associated flow $R_{\psi a}$ is highly nonlinear with respect to variation in ϕ' as shown in Figure 10.13. However, they also show that $R_{\psi a}$ increases linearly with normalized embedment, starting with $R_{\psi a} = 1$ as h/B approaches zero. Figure 10.13 shows the dilation factor in a format that is scaled to anchor embedment depth. Finally, Rowe and Davis show that the dilatancy multiplier varies approximately linearly for intermediate dilation angles over the range $0 < \psi < \phi$.

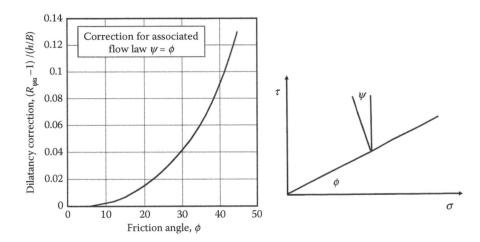

Figure 10.13 Dilatancy multiplier for vertically loaded plate anchor in sand. (After Rowe and Davis, 1982.)

Thus, the dilatancy multiplier for general nonassociated flow conditions becomes the following:

$$R_\psi = (R_{\psi a} - 1)\psi/\phi \tag{10.11}$$

For the case of a vertically loaded horizontal plate anchor, Rowe and Davis find the effect of anchor roughness to be negligible, or $R_R = 1$. To illustrate the overall analysis, we may consider the case of a horizontal plate anchor at a depth $h/B = 5$ in a sand with friction angle $\phi = 40°$ and dilatancy angle $\psi = 10°$. From Figure 10.12, the base case bearing factor is $N_\gamma = 4.5$. Figure 10.13 together with an embedment $h/B = 5$ gives a dilatancy multiplier for the case of associated flow behavior equal to $R_{\psi a} = 1.45$. Using Equation 10.11 with $\psi = 10°$ gives $R_\psi = 1.1$.

The estimate for load capacity of a horizontally loaded vertical plate anchor is depicted in Figures 10.14 and 10.15. However, anchor roughness takes on greater significance for horizontally loaded anchors at shallow embedment depths. As shown in the right-hand plot in Figure 10.14, the roughness factor R_R can exceed 1.6 as embedment depth approaches $h/B = 1$.

10.2.2 Finite length effects

The Rowe and Davis finite element studies considered only strip anchors. However, they also present experimental data from anchor vertical pullout tests in a loose sand ($D_r = 20\%$) single gravity test bed for finite length plate anchors at an embedment depth $h/B = 3$. A shape factor R_S may be introduced relating the bearing factors for strip and finite length anchors as follows:

$$R_S = \frac{N_\gamma(L/B)}{N_\gamma(L/B \to \infty)} \tag{10.12}$$

The left-hand plot in Figure 10.16 shows the trend line of their measurements in terms of the shape factor R_S. Subsequent model test data by Murray and Geddes (1987) in very dense

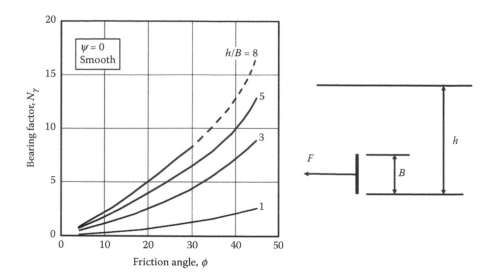

Figure 10.14 Basic bearing factor for horizontally loaded plate anchor in sand. (After Rowe and Davis, 1982.)

sand ($D_r = 85.9\%$) are superimposed on this plot. In spite of the large differences in density, the overall trend lines are similar, although for the limiting case of a square plate the shape factor for the very dense sand is approximately 25% greater than that for a loose sand. The center plot in Figure 10.16 shows the Murray and Geddes data for dense sand with embedment depths ranging from aspect ratios $h/B = 1{-}6$. An aspect ratio of 5 appears to approach a strip anchor, which is consistent with the data presented by Rowe and Davis. Shape factors for a rectangular anchor with $L/B = 2$ can exceed 2 at sufficiently large embedment depth. Most remarkable is the sharp increase in shape factor for square footings, which can exceed 4 at sufficiently large embedment depths. However, it should be kept in mind that these data apply only to very dense sands.

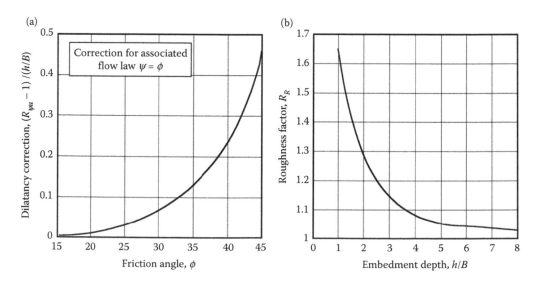

Figure 10.15 Dilatancy multiplier for horizontally loaded plate anchor in sand. (After Rowe and Davis, 1982.)

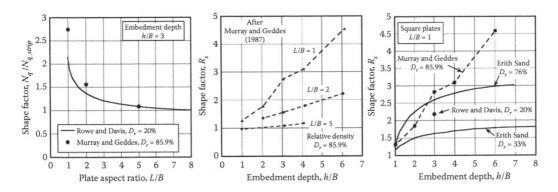

Figure 10.16 Shape effects for vertically loaded plate anchor in sand.

The right-hand plot in Figure 10.16 focuses solely on square plates. The experimental data presented by Rowe and Davis indicate that a vertically loaded square plate anchor embedded at depth $h/B = 3$ has a shape factor equal to about $R_S = 2.2$. A subsequent study by Dickin (1988) conducted centrifuge tests on vertically loaded square plate anchors embedded in sand at various embedment depths h/B. The right-hand plot in Figure 10.16 summarizes their results. Some inconsistency occurs in regard to the Rowe and Davis data for a square plate, which may be attributed to differences in the soils tested as well as the test method, that is, single gravity versus centrifuge tests. Finally, the Murray and Geddes data for very dense sand are superimposed on the plot. Although the different data sets do not provide an entirely consistent picture, a trend does appear to emerge indicating that the shape factor for square plates in sands is highly sensitive to both embedment depth and density.

10.2.3 Effect of anchor orientation

The Rowe and Davis study considered horizontally and vertically oriented plate anchors. More recent work by Yu et al. (2014) considers strip anchors inclined at intermediate angles. Their analysis is based on upper bound plastic limit analyses employing what they term a "block set" mechanism. The method constructs a kinematically admissible velocity field from two sets of rigid blocks; Figure 10.17 provides a conceptual sketch of the mechanism; the actual analyses employ many more blocks than shown in the figure. The interior angles

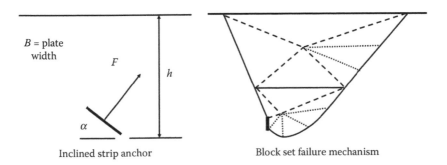

Inclined strip anchor   Block set failure mechanism

Figure 10.17 Block set failure mechanisms.

and the edge lengths of the blocks are variables that are optimized to achieve a least upper bound collapse load using a genetic algorithm combined with a pattern search method. Figure 10.18 shows the variation in bearing factor with plate inclination angle α predicted by Yu et al.

10.2.4 Chain resistance

The load capacity estimates discussed above apply to anchor loads applied at the pad eye, T_a. Estimating the corresponding line tension T_0 at the mudline requires an estimate of the additional resistance provided by the chain or wire line. For anchors embedded in clay soil profiles, the anchor line mechanics principles outlined in Chapter 5 of this book should be applicable to PDPAs. A validated analytical procedure for estimating the contribution of chain resistance to overall anchor load capacity is presently lacking. Forrest et al. (1995) provide rough guidance on chain resistance in sands with a recommendation that the pad-eye load capacity be increased by 25%, $T_0 = 1.25\,T_a$, for horizontally loaded anchors in sand. For vertical loading, the contribution of chain resistance to anchor pullout capacity should be considered negligible.

10.2.5 Keying loss of embedment

At the end of installation, a PDPA is oriented in a vertical position. Keying involves loading the anchor such that it rotates into an orientation normal to the direction of the applied pad-eye load. As is the case for SEPLAs, PDPAs experience some loss of embedment, and therefore load capacity, owing to keying of the anchor. Analytical treatment of keying embedment loss for anchors in clays should be identical to the approach presented earlier for keying of SEPLAs; however, to the author's knowledge, validated analytical methods have not been applied to estimating embedment loss of PDPAs during keying. Empirical charts have been developed, which are discussed here.

Figure 10.19 shows guidance by Forrest et al. (1995) for estimating keying embedment loss for pile driven anchors. These charts are formulated in terms of the keying arm distance, defined as the shortest distance from the pad eye to the near surface of the plate. Consistent with the experience with keying embedment losses for SEPLAs, embedment losses for PDPAs in clay are greatest for vertical loading, and the use of wire line reduces the mooring line interaction effect, which also tends to reduce keying embedment losses.

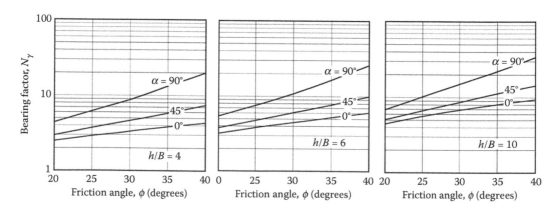

Figure 10.18 Variation in strip anchor capacity with plate orientation angle.

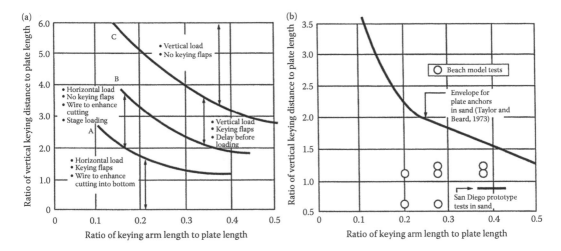

Figure 10.19 Embedment loss during pile driven plate anchor keying. (a) Anchors in clay and (b) anchors in sand. (From Forrest J, Taylor R, and Brown L, 1995, *Design Guide for Pile-driven Plate Anchors,* Technical Report TR-2039-OCN, Naval Facilities Engineering Service Center, Port Hueneme, CA, March 1995.)

The overall loss of embedment during keying of PDPAs is generally greater than that experienced during keying of SEPLAs. In comparing keying embedment losses for caissons installed by jacked versus suction installation, Gaudin et al. (2006) observed that the greater embedment loss associated with jacked installation could be due to a greater level of installation disturbance. The high keying embedment losses (relative to typical values for SEPLAs) may be similarly due to a greater level of soil disturbance occurring during pile driving.

10.3 DYNAMICALLY EMBEDDED PLATE ANCHORS

A third alternative for achieving deep embedment of plate anchors is dynamic embedment. By one approach, installation is achieved by attaching a plate to the tip of a dynamically installed pile, which serves as a follower, after which the anchor is detached from the follower and the follower is withdrawn (O'Loughlin et al., 2014). As is the case for suction embedded and PDPAs, the anchor at the end of installation is oriented vertically, so it must be keyed to orient it in a direction approximately normal to the direction of applied loading. Unlike the other plate anchors considered thus far in this chapter, the DEPLA conceived by O'Loughlin et al. is not a simple plate; rather, it comprises two circular plates attached to form a cruciform configuration. A second alternative concept for dynamic installed plate anchors is to design the plate such that it can achieve sufficient embedment without the need for a follower. Since the concept of plate penetration without a follower is in the early stages of development, the discussion that follows is directed toward the model that utilizes a follower (Figure 10.20).

Since installation entails simply suspending the anchor and releasing it over the target location, DEPLAs are expected to have relatively low installation costs (O'Loughlin et al., 2015). Additional time and equipment are required for retracting the follower, but these costs may well be offset by the high geotechnical efficiency of plate anchors in general, which lead to lower material and fabrication costs and permit smaller transport vessels.

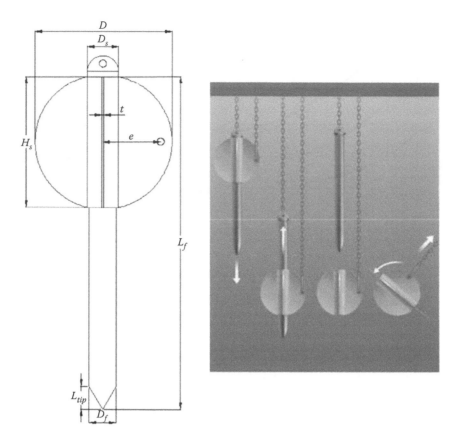

Figure 10.20 The DEPLA concept. (Reprinted from Installation and capacity of dynamically embedded plate anchors as assessed through centrifuge tests, 88, O'Loughlin CD et al., *Ocean Eng*, 204–213, Copyright 2014, with permission from Elsevier.)

In common with the dynamically installed piles discussed in Section 6.3, DEPLA installation involves somewhat greater uncertainty in installation depth relative to the other direct embedment plate anchor alternatives as well as piles and caissons. To date, the DEPLA concept developed by O'Loughlin et al. has only been considered for deployment in soft clays. The DEPLA has been thoroughly investigated through numerical studies, centrifuge tests and small-scale field trials but, to the author's knowledge, it has not been deployed on actual projects. As with all other anchor types DEPLAs have limitations—their use is limited to soft clays and, to some extent, they cannot be positioned as precisely as a number of other anchor options. Nevertheless, their efficiency and low installation cost will likely make them a competitive anchor alternative.

Prediction of DEPLA performance requires evaluation of three aspects of anchor behavior: (1) penetration depth into the seabed, (2) loss of embedment during keying, and (3) load capacity of the anchor at its installed depth. These processes have already been presented previously in this book. Impact penetration of a dynamically installed pile is covered in Section 6.3; loss of embedment owing to keying of a plate anchor originally in a vertical orientation is covered in the discussion of SEPLAs earlier in this chapter; and load capacity of a plate anchor embedded in clay is covered at length in Chapters 3 and 4. Accordingly, the discussion that follows omits much of the discussion on the basic mechanics of these processes and focuses on studies directed specifically toward DEPLAs.

10.3.1 Dynamic embedment

To gain insight into the basic mechanics of dynamic embedment, we will return to the analytical solution presented in Section 6.3.2 for dynamically installed piles. As shown in Figure 10.21, the analysis will be further simplified by considering only a single stage of penetration, in which case the governing differential equation for a penetrating pile of mass m_p is shown below:

$$m_p \ddot{z} = W_{net} B - CD^2 z \tag{10.13}$$

The net weight W_{net} of the pile is its total weight minus the weight of displaced soil during penetration. Soil resistance to penetration is assumed to be proportional to the square of pile diameter, and the average rate of increase in soil resistance during penetration is controlled by the constant C having units of force per unit volume; that is, the same units as the soil strength gradient. Thus, C may be regarded as being the product of the strength gradient k times an effective bearing factor N_e representing the combined contribution of tip resistance and side friction on all surfaces of the anchor. The analysis gives no explicit consideration of rate effects and hydrodynamic drag and relies upon a simplified characterization of soil resistance; therefore, it should be considered as a tool for conceptual understanding of the penetration process, as opposed to a rigorous predictive analysis. The differential equation is an initial value problem, with initial coordinate ($z = 0$) and impact velocity v_0 comprising the initial conditions. The solution to Equation 10.13 has the form shown below.

$$z = \frac{W_{net}}{CD^2} - \frac{W_{net}}{CD^2} \cos\sqrt{CD^2/m_p}\, t + \frac{v_0}{\sqrt{CD^2/m_p}} \sin\sqrt{CD^2/m_p}\, t$$

$$v = \frac{W_{net}/m_p}{\sqrt{CD^2/m_p}} \text{in}\sqrt{CD^2/m_p}\, t + v_0 \cos\sqrt{CD^2/m_p}\, t \tag{10.14}$$

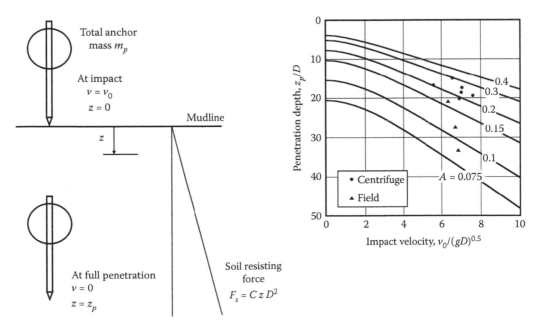

Figure 10.21 Simplified dynamic embedment model.

Penetration time T_p (in dimensionless form) is obtained by setting the velocity v to zero and solving for time.

$$T_p = \sqrt{CD^2/m_p} \; t_p = \frac{\pi}{2} + \tan^{-1}\left(\frac{W_{net}/m_p}{v_0\sqrt{CD^2/m_p}}\right) \tag{10.15}$$

Maximum penetration depth can now be predicted by substituting Equation 10.15 into 10.14. With some rearrangement of terms, the solution can more simply be expressed in terms of nondimensional parameters A, B, and V_0 as follows:

$$\frac{z_p}{D} = \frac{B}{A} - \left[\frac{B}{A} + \frac{V_0^2}{B}\right]\cos T_p$$

$$T_p = \frac{\pi}{2} + \tan^{-1}\left(\frac{B}{V_0\sqrt{A}}\right) \tag{10.16}$$

$$A = CD^3/W_p \quad B = W_{net}/W_p \quad V_0 = v_0/\sqrt{gD}$$

Three dimensionless parameters are introduced: V_0 is nondimensional impact velocity, A is the ratio of soil resistance to pile weight W_p, and B is the ratio of net to total pile weight. The dimensionless penetration time T_p in Equation 10.16 is bounded between $\pi/2$ and π; thus, the cosine term in Equation 10.16 is always negative.

At zero impact velocity the penetration depth $z_p/D = 2B/A$. In spite of the second-order velocity term in Equation 10.16, penetration depth does not quite increase as a second-order function of impact velocity, because an increase in V_0 decreases penetration time T_p, which tends to offset some of the benefit of the increased impact velocity. Plots of Equation 10.16 in Figure 10.21 illustrate this trend. At low V_0, the penetration depth versus impact velocity curves are approximately parabolic. However, in the velocity range of approximately $4 < V_0 < 10$, the curves are nearly linear.

Turning to the effect of other parameters on anchor penetration depth, the variable B is simply the ratio of net anchor weight—the weight of the anchor minus the weight of the soil it displaces—to the total anchor weight. While it has a strong influence on penetration depth, its magnitude does not vary greatly in practice; the parametric study shown here assumes $B = 0.77$. Further, aside from fabricating the anchor with an exceptionally high density material—likely a costly proposition—there is not too much that can be done from a design perspective to achieve higher B values.

The parameter A—the ratio of soil resistance to pile weight—appears in Equation 10.16 as being inversely proportional to penetration depth. For values of A greater than approximately 0.2, the predicted penetration depth is relatively insensitive to variations in A. By contrast, the plots show sharp increases in the penetration depth when A falls below 0.2. This behavior is a consequence of the roughly inverse (hyperbolic) relationship between penetration depth and normalized soil resistance. In contrast to B, the parameter A ($= CD^3/W_p$) can be manipulated to optimize anchor performance. On the one hand, a designer has limited means to reduce soil resistance, which is largely driven by the strength gradient k of the soil. On the other hand, the pile weight can be increased without increasing the pile diameter simply by increasing the follower length.

Installation tests on the DEPLA include 200-g centrifuge tests by O'Loughlin et al. (2014) in a normally consolidated kaolin test bed and field tests on reduced scale model DEPLAs

Table 10.2 DEPLA installation data

Test type	Equivalent diameter D_{eq} (m)	Anchor mass (Mg)	Impact velocity v_0 (m/sec)	Tip penetration z_p (m)	Normalized penetration z_p/D
Centrifuge	1.86	171.7	23.87	31.08	16.74
O'Loughlin et al. (2014)	1.86	171.7	30.04	32.28	17.39
	1.69	132.4	28.50	31.38	18.61
	1.50	97.6	29.17	29.18	19.42
	1.84	184.32	29.37	37.28	20.23
	2.02	212.64	29.17	30.12	14.93
Field	0.0797	0.0203	5.6	1.66	20.83
O'Loughlin et al. (2015)	0.128	0.096	7.5	3.5	27.36
	0.205	0.389	9.7	6.8	33.22

by Blake and O'Loughlin (2015). The centrifuge test program considered a number of six different configurations with varying dimensions for the follower, sleeve and plate. Blake and O'Loughlin (2015) present the results of field tests conducted in a lakebed comprising a very soft clay having strength gradient $k = 0.9$ kPa/m, which is considerably weaker than the strength profile in the kaolin test bed used in the centrifuge tests, $k = 1.3$–1.6 kPa/m. Three model anchors were used at scales 1:12, 1:7.2, and 1:4.5. Relevant data for both test series are summarized in Table 10.2. Equivalent diameter presented in the table is based on the cross-sectional area of the followers and the cruciform plate, neglecting any additional components such as the anchor sleeve, pad eye or chain.

Overlaying the test data on the analytical solution in Figure 10.21 shows the centrifuge test results to be clustered in the region between $A = 0.2$–0.3. As noted above, this just at the edge of the boundary at which penetration depth increase sharply with decreasing A. Despite the differences in details of the plate and follower geometries, penetration depths z_p/D do not vary greatly. This is consistent with the notion that, for $A > 0.2$, the penetration depth is relatively insensitive to variations in A. The field test data generally lie in the range of A between 0.1 and 0.2. To some extent, this may be due to the relatively low strength gradient at the site where the field tests were conducted. There is also more scatter in the field penetration depth measurements, which is consistent with this particular anchor–soil system being in a region where small variations in soil resistance results in large variations in penetration depth. Noting that the field tests were conducted at three different scales, sources of variability in soil resistance can include scale-dependent strain rate effects as well as differences in secondary components of the DEPLA system such as the follower and plate lines.

As emphasized earlier, the simplified solution presented here is intended solely for gaining a conceptual understanding of the dynamic penetration process. Rigorous predictions require consideration of all sources of soil bearing and frictional resistance, strain rate effects, hydrodynamic resistance, and buoyant resistance. The methodology parallel was already presented in Section 6.3 and not repeated here. Blake and O'Loughlin (2015) and O'Loughlin et al. (2013) provide details on the application of the dynamic equilibrium and energy models, originally developed for dynamically installed piles, to DEPLAs.

10.3.2 Keying loss of embedment

Nearly, all of the research related to keying behavior of DEPLAs is for the model illustrated in the sketch on the left side of Figure 10.20. In addition to the circular plate shown in the

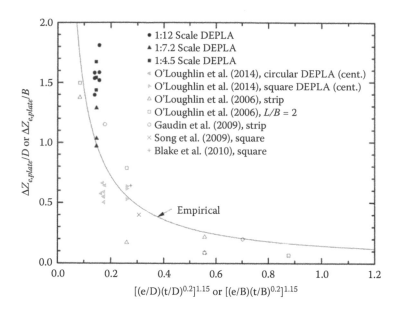

Figure 10.22 DEPLA embedment loss during keying. (From Blake AP, O'Loughlin CD, and Gaudin C, 2015, *Can Geotech J*, 52, 1270–1282, © Canadian Science Publishing or its licensors.)

figure, studies by Wang and O'Loughlin consider a diamond-shaped plate configuration. Both centrifuge (O'Loughlin et al., 2014) and field tests (Blake and O'Loughlin, 2015; Blake et al., 2015) have been performed to investigate the performance of this DEPLA model, the field tests being conducted on anchors at scales ranging from 1:12 to 1:4.5 of actual prototype dimensions. The centrifuge tests largely employed a model with follower length $L_f/D_f = 12.67$ (although lengths L_f/D_f from 8.58 to 16.92 were investigated); the models in the field tests had a similar ratio, $L_f/D_f = 12.5$. Recalling the nondimensional variables considered earlier for SEPLAs, Table 10.3 summarizes the corresponding variables for the tests conducted on DEPLAs.

In Figure 10.22, Blake et al. (2015) overlay centrifuge and field test measurements of keying embedment loss Δz_e on values obtained from various previous studies of embedment loss for keying of SEPLAs. Also shown in the plot is the empirical relation Equation 10.9a for predicting maximum embedment loss, using the DEPLA diameter D in lieu of the plate breadth B in this equation. Comparison to Equation 10.9a does not consider reductions in keying embedment losses owing to the local strength effect; however, substitution of the parameters from Table 10.2 in Equation 10.9b indicates that conditions of maximum embedment loss prevail. The embedment loss reductions measured in the centrifuge tests are actually in reasonable agreement with Equation 10.9a, with the measured $\Delta z_e/D = 0.7$–0.9 being slightly less than those predicted from the empirical equation. However, the field tests show somewhat greater embedment losses, in the range $\Delta z_e/D = 1$–1.8.

The reasons for the discrepancy between centrifuge and field test measurements are unclear at this time, so possible explanations are speculative. However, it is noted that the empirical equation and the overall interpretive framework does not consider soil sensitivity which, as discussed previously, likely has a significant influence on keying embedment loss. Since sensitivity exerts its greatest influence on tangential resistance of the plate, its effect is expected to be similar to variations in plate thickness t. Either an increase in sensitivity or a decrease in plate thickness reduce tangential resistance, thereby leading to

Table 10.3 DEPLA performance test properties

Variable	Field	Centrifuge
Eccentricity e_n/D	0.42–0.44	0.38–0.36
Thickness t/D	0.01–0.02	0.02–0.05
Soil strength $s_{u0}/\gamma'D$	4.75–7.5	0.61–1.44
Strength gradient kD/s_{u0}	0.16–0.25	0.16–0.38
Weight $\gamma'_a t/s_{u0}$	0.14–0.20	0.37–0.45

Source: Blake AP, O'Loughlin CD, and Gaudin C, 2015, *Can Geotech J*, 52, 1270–1282, © Canadian Science Publishing or its licensors.

greater upward motions accompanying plate rotations during keying. Unfortunately, for this possible explanation, the soil sensitivity at the site of the field tests reported by Blake et al. (2015) is $S_t = 2.5$, which is not dissimilar from that expected for the kaolin in the centrifuge tests. Nevertheless, the inclusion of soil sensitivity in the predictive framework for keying embedment loss may be advisable until the process is better understood. It is also noted that the current design of the DEPLA features eccentricities $e_n/D < 0.5$, which correspond to the steep portion of the embedment loss curve. Practically speaking, any predictions in this region are subject to considerable uncertainty, since minor variations in soil conditions can cause large variations in embedment loss. As noted by Blake et al., a greater eccentricity could avoid this issue, although it would entail a revision of the anchor design.

10.3.3 Load capacity

First-order assessment of the pullout capacity of a DEPLA can be based on the solutions of Martin and Randolph (2001) for a deeply embedded circular disk ($N_c = 13.11$) and Merifield et al. (2003a) for pullout of plates subjected to free surface and breakaway effects. However, these solutions can be expected to offer somewhat low estimates of bearing resistance, since they do not consider the strengthening effect of the cruciform shape of the DEPLA as designed by O'Loughlin et al. (2014). In an investigation directed specifically toward the cruciform configuration, Wang and O'Loughlin (2014) performed large deformation finite element analyses to evaluate load capacity of cruciform anchors comprised of square and circular plates, considering the effects of proximity to a free surface and potential breakaway (no suction) at the base of the anchor. In the latter case, soil self-weight contributes to capacity, which is also considered in their study. Figure 10.23 summarizes the bearing factors obtained for circular DEPLAs, which plots anchor bearing capacity factor N_c as a function of normalized embedment depth H/D. The depth H here is taken as the depth to the center of the anchor; thus, if the tip of the anchor is at a depth z_e, the value of H would be $z_e - D/2$.

O'Loughlin et al. (2014) performed a series of centrifuge tests of DEPLA dynamic installation, keying, and pullout in a normally consolidated kaolin test bed. Their test series also included tests on both circular and square plates; the measured bearing factors N_c are superimposed over the numerical solutions in Figure 10.23. The anchor embedment depths shown in the figure account for the embedment loss during keying and are presented in terms of the mid-depth H of the anchor. The measured bearing factors are seen to slightly exceed the numerical solution for the no breakaway case. One test included anchor installation by jacking (static) to assess the effects of dynamic installation. The anchor installed by static installation is seen to have a somewhat lower capacity than those installed dynamically.

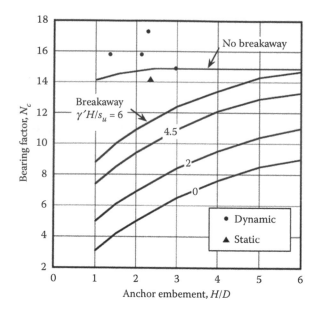

Figure 10.23 Bearing factors for circular DEPLAs.

In a subsequent study involving field tests on circular DEPLAs at reduced scales of 1:12–1:4.5 in a soft clay lakebed, Blake et al. (2015) report measured bearing capacity factors in a range $N_c = 14.3$–14.6. Details on embedment are contained in the reference, but equivalent values of H/D are generally in excess of 2. The field data thus indicate bearing factors slightly less than the numerical predictions for a no breakaway condition ($N_c = 14.9$), but well above a range associated with a breakaway condition.

10.4 HELICAL ANCHORS

Helical anchors are composed of one or more helical plates attached to a central shaft (Figure 10.24). The anchors are installed by applying torque and axial force to the shaft, which are transmitted to the helical plates. Helical anchors can be deployed in both sands and clays. In terms of typical dimensions, Elkasabgy and El Naggar (2015) categorize a typical large capacity helical pile as having a shaft diameter D_s greater than 0.178 m, and current equipment can install helical piles with shaft diameters up to 0.914 m. Table 10.4 summarizes typical helical pile/anchor dimensions reported from various field and laboratory studies. For multi-plate anchors, the ratio of plate spacing to plate diameter is largely in the range $S/D = 1$–3. The pitch of the helices also varies rather widely, with reported pitch-to-diameter ratios varying from $p/D = 0.18$–0.58. Ghaly et al. (1991) define small-, medium-, and large-pitch helical anchors as having respective p/D ratios of 0.2, 0.3, and 0.4. They also note significant sensitivity of installation torque to pitch, with large-pitch anchors requiring approximately 20% more torque for installation in sand than small-pitch anchors. Installation torque is also commonly correlated to ultimate load capacity, the correlations by Ghaly et al. (1991) being an example. Although these types of correlations are potentially useful, their role is more that of field verification than that of a predictive model. Helical foundations can resist compression, tension and lateral loads. Model tests by Prasad and Narasimha Rao (1996) on laterally loaded

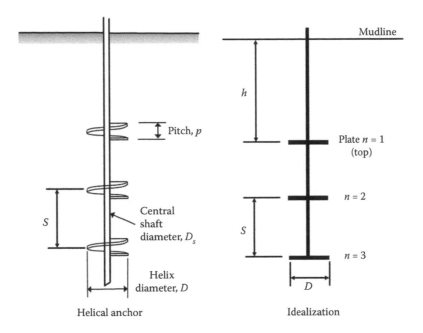

Figure 10.24 Definition sketch for multi-plate helical anchor.

helical piles showed that the helical plates increase ultimate load capacity over a single shaft pile by factors of 1.2–1.5.

10.4.1 Axial capacity in clay

Analytical estimates of pullout capacity for helical anchors in clays largely ignore the effect of pitch. In the case of single-helix anchors, the solutions for shallowly end deeply embedded circular disks presented in Section 4.1.2 and 4.1.6 are adopted. When a breakaway condition is considered, the self-weight contribution of soil to pullout resistance needs to

Table 10.4 Example helical anchor dimensions

Reference	Soil type	Test type	Shaft diameter D_s (mm)	Helix diameter D (mm)	Helix spacing S/D	Anchor depth h/D	Pitch p/D	No. of plates
Elkasabgy and El Naggar (2015)	Clay, silt, sand	Field	324	610	1.5	7.5–12.4	0.25	2
Tsuha et al. (2012)	Sand	Centri-fuge[a]	64.3	214	3	14.5	0.30	1–3
			97.7	326	3	14.1	0.21	1–3
			132	440	3	14.1	0.18	1–3
Narasimha Rao et al. (1993)	Soft clay	Lab	13.8	33	1.1–4.6	2.4–11	0.24	2–5
			33.0	75	1.3–4	2.33–7	0.33	2–4
Narasimha Rao et al. (1991)	Soft-medium clay	Lab	44	100	1.52–4.57	9.85	0.58	2–4
			60	150	1.02–3.05	0.65	0.39	2–4
			25	75	0.83–4.0	6.1–7.7	0.33	2–4

[a] Dimensions shown are prototype scale.

be incorporated into the calculation. In this case, the ultimate load capacity equation takes the following form

$$\frac{Q_u}{A} = s_u N_c$$

$$N_c = N_{c0} + \frac{\gamma' h}{s_u} \leq N_c^*$$

(10.17)

In this equation s_u is undrained shear strength, h is plate embedment depth, and N_{c0} is the bearing factor for the plate in a weightless soil. The bearing factor corresponding to the flow-around solution for a deeply embedded circular disk, N_c^*, comprises an upper bound limit on N_c. Based on lower bound and finite element studies by Merifield (2002), and (Merifield et al., 2003a), Merifield (2011) gives the following lower bound curve fit for bearing factor in a weightless soil for a smooth, shallowly embedded circular disk:

$$N_{c0} = 13.7[1 - \exp(-0.35h/D)] \leq 12.6$$

(10.18)

The upper limit on N_{c0} given by Equation 10.18 exceeds the exact solution given by Martin and Randolph (2001) for a deeply embedded smooth, circular disk, $N_c^* = 3 + 3\pi = 12.42$. Given that the Martin–Randolph value is exact, it is arguably a more appropriate limiting value for Equation 10.18, but the difference is relatively minor (1.5%).

We now consider a multiplate helical anchor system as idealized in Figure 10.24. Merifield's (2011) bases his extension of the analysis of shallowly embedded helical anchors to multi-plate systems on two premises: (1) the capacity of the uppermost plate in the system is essentially unaffected by interactions with the lower plates and (2) the capacities of the remaining lower plates are nearly independent of embedment depth h/D, but they are sensitive to plate spacing S/D. Neither premise is without exception, but finite element studies suggest that they are sufficiently valid for design purposes. In regard to the first premise, the left-hand plot in Figure 10.25 compares finite element estimates of pullout capacity for a single plate, compared to capacity estimates for the top plate in a multi-plate system for spacing ratios over a range $S/D = 1$–3. In some cases, the interference effect can reduce the uplift capacity of the upper plate by up to 4%, but overall the effect is seen to be fairly modest. Thus, Equation 10.18 provides a reasonable estimate of the load capacity of the top ($n = 1$) plate in a multi-plate anchor system.

The right-hand plot in Figure 10.25 from Merifield (2011) shows finite element predictions of load capacity of the lower plates ($n = 2$, $n = 3$) in a multi-plate helical anchor system. As noted above, the capacity mobilized by the individual plates is largely insensitive to embedment depth h/D, although the capacities are clearly seen to be sensitive to plate spacing. This observation leads Merifield to propose the concept of an "equivalent embedment depth" $(h/D)_{eq}$ expressed as a function of plate spacing. For plate spacing $S/D < 3$, Merifield proposes the following expression relating plate spacing to equivalent embedment depth:

$$(h/D)_{eq} = 1.12(S/D)^{4/3}$$

(10.19)

The embedment depth computed by Equation 10.19 is used in conjunction with Equation 10.18 to compute the bearing factor N_{c0} for any plate below the top plate; that is, for plates

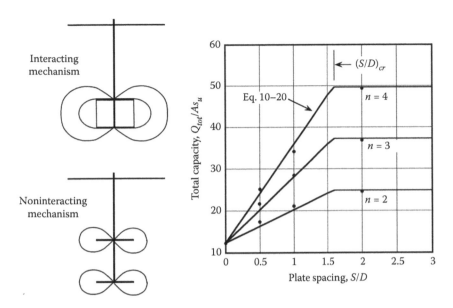

Figure 10.25 Bearing capacity of shallow multi-plate anchor in clay. (After Merifield RS, 2011, *J Geotech Geoenviron Eng*, 137(7), 704–716, with permission from ASCE.)

greater than $n = 1$. As is evident from Figure 10.25, the finite element results largely support the simplified framework proposed by Merifield, with the exception of shallow ($h/D < 3$) anchors having a wide ($S/D = 3$) plate spacing.

At sufficiently great depth, the failure mode involves localized shear around the plate or series of plates. In this case, a chief issue related to the failure mechanism for a series of plates is whether an independent failure mechanism develops for each plate or if an interaction effect occurs; the conceptual sketches on the left of Figure 10.26 illustrate the interacting and noninteracting failure mechanisms. Finite element studies by Merifield (2011) predict the combined capacity from all plates in a multi-plate anchor, Q_{u_tot}, as a function of plate spacing S/D for anchors comprising from 2 to 4 plates. Predictably, interaction effects with a concomitant reduction in total anchor capacity occurs for closely spaced plates. The finite element studies indicate that interaction effects develop for plate spacing less than a critical value of about $(S/D)_{cr} < 1.6$. The maximum capacity of a multi-plate anchor system is simply the number of plates times the single-plate capacity. Merifield (2011) proposes the equation shown below to describe this effect.

$$\frac{Q_{u\_tot}}{s_u A} = \left[(3 + 3\pi) + (n-1)\left(\frac{5\pi}{2}\right)\frac{S}{D} \right] \le n(3 + 3\pi) \qquad (10.20)$$

All axial capacity calculations described above neglect the effect of the anchor shaft. Two effects should be considered in regard to the effect of the shaft. First, the shaft provides added resistance to pullout. Estimation of the additional resistance can proceed along the lines of the side resistance calculations normally performed for axial load capacity of piles in clay. Second, the shaft reduces the effective bearing area of the helical plates. Finite element analyses by Merifield (2011) indicate that it is reasonable to assume that the shaft does not significantly affect the ultimate bearing pressure acting on the plate. Thus, the reduction in

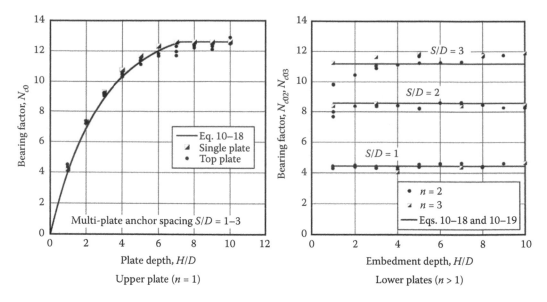

Figure 10.26 Bearing capacity of deep multi-plate anchor in clay. (After Merifield RS, 2011, *J Geotech Geoenviron Eng*, 137(7), 704–716, with permission from ASCE.)

pullout capacity owing to the shaft Q_{u_s} can be computed in terms of the reduced bearing of the plate, as shown in Equation 10.21, where D_s is the shaft diameter.

$$\frac{Q_{u\_s}}{Q_u} = 1 - \left(\frac{D_s}{D}\right)^2 \qquad (10.21)$$

The solutions presented above neglect the effects of installation disturbance, which is a potentially significant issue. Nevertheless, comparisons by Merifield (2011) of their solutions to laboratory model test measurements from Narasimha Rao et al. (1991, 1993) indicate good agreement (largely within ±10%) between calculated and measured axial load capacities.

10.4.2 Axial capacity in sand

While limit equilibrium and plastic limit models have been applied to helical anchors in sands, they have not been validated to the extent that they can be used with confidence. Pending the outcome of ongoing research known to the author on the performance of helical anchors in sand, the most reliable approach at present is arguably based on the Rowe and Davis (1982) finite element solutions for shallowly embedded anchors in sands. Details of their work were presented earlier in this chapter under the discussion of PDPAs. As was noted in that discussion, the Rowe and Davis solution actually apply to strip anchors; however, some experimental data exist to support estimates of a shape factor to relate computed capacities for strip anchors to anchors of finite length. The data in Figure 10.16 actually relate the load capacity of strip anchors to rectangular and square plates. It remains to provide an adjustment to the circular configuration characteristic of helical anchors. Such adjustment is possible through the numerical studies of Merifield et al. (2003b), which used lower bound methods and displacement-based finite element analyses to compare load

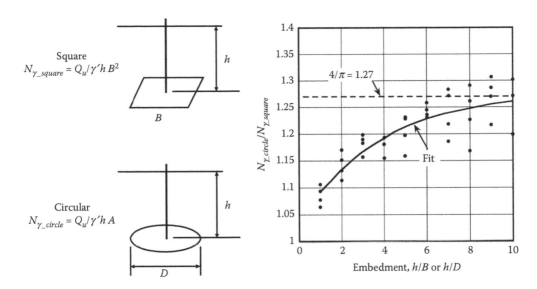

$N_{\gamma\_square} = Q_u/\gamma' h\, B^2$

Square

$N_{\gamma\_circle} = Q_u/\gamma' h\, A$

Circular

Figure 10.27 Shape effect for circular plates in sand. (After Merifield RS, Sloan SW, and Lyamin AV, 2003b, *Three-dimensional Lower Bound Solutions for Stability of Plate Anchors in Sand*, Report TR-2003-02, ISBN 1 877078 04 2, University of Southern Queensland, Australia.)

capacities of circular and square plates embedded in sand for friction angles 20°, 30°, and 40°. Figure 10.27 presents the shape factors computed from their studies. The shape factor evolves from slightly greater than unity for shallowly embedded plates to what appears to be a limiting value of 1.27 at greater embedment depths. Merifield et al. point out that the factor 1.27 is close to 4/π, or the ratio of the area of a square to that of a circle.

REFERENCES

Blake AP and O'Loughlin CD, 2015, Installation of dynamically embedded plate anchors as assessed through field tests, *Can Geotech J*, 52, 87–95.

Blake AP, O'Loughlin CD, and Gaudin C, 2010, Setup following keying of plate anchors assessed through centrifuge tests in kaolin clay, *Proceedings of the 2nd International Symposium on Frontiers in Offshore Geotechnics*, Perth, Australia, 8–10 November 2010, pp 705–710.

Blake AP, O'Loughlin CD, and Gaudin C, 2015, Capacity of dynamically embedded plate anchors as assessed through field tests, *Can Geotech J*, 52, 1270–1282, © Canadian Science Publishing or its licensors.

Bransby MF and O'Neill M, 1999, Drag anchor fluke soil interaction in clays, *Proceedings of the International Symposium on Numerical Models in Geomechanics*, Graz, Austria, pp 489–494.

Cassidy MJ, Gaudin C, Randolph, MF, Wong PC, Wang D, and Tian Y, 2012, A plasticity model to assess the keying behaviour and performance of plate anchors, *Geotechnique*, 62(9), 825–836.

Dickin EA, 1988, Uplift behaviour of horizontal anchor plates in sand, *J Geotech Eng*, ASCE 114(11), 1300–1317.

Dove P, Treu H, and Wilde B, 1998, Suction embedded plate anchor (SEPLA): A new anchoring solution for ultra-deep water mooring, *Proceedings of the DOT Conference*, New Orleans.

Elkhatib S, 2006, The behaviour of drag-in plate anchors in soft cohesive soils, PhD Thesis, The University of Western Australia, Australia.

Elkasabgy MA and El Naggar MH, 2015, Lateral performance of large-capacity helical piles, *IFCEE 2015*, American Society of Civil Engineers, Geotechnical special publication No. 256, pp 868–877.

Elkhatib S and Randolph MF, 2005, The effect of interface friction on the performance of drag-in plate anchors, *Proceedings of the International Symposium on Frontiers in Offshore Geotechnics (ISFOG)*, Perth, Australia, pp 171–177.

Forrest J, Taylor R, and Brown L, 1995, *Design Guide for Pile-driven Plate Anchors*, Technical Report TR-2039-OCN, Naval Facilities Engineering Service Center, Port Hueneme, CA, March 1995.

Gaudin C, O'Loughlin CD, Randolph MF, and Lowmass AC, 2006, Influence of the installation process on the performance of suction embedded plate anchors, *Geotechnique*, 56(6), 381–391.

Gaudin C, Tham KH, and Ouahsine S, 2009, Keying of plate anchors in NC clay under inclined loading, *Int J Offshore Polar Eng*, 19(2), 135–142.

Ghaly A, Hanna A, and Hanna M, 1991, Installation torque of screw anchors in dry sand, *Soils Found*, 31(2), 77–92.

Giampa JR, Bradshaw AS, and Schneider JA, 2017, Influence of dilation angle on drained shallow circular anchor uplift capacity, *Int J Geomech*, ASCE, 17(2), ISSN 1532–3641.

Martin CM and Randolph MF, 2001, Applications of the lower and upper bound theorems of plasticity to collapse of circular foundations, *Proceedings of the 10th International Conference of International Association for Computer Methods and Advances in Geomechanics*, Tucson, Vol. 2, pp 1417–1428.

Merifield RS, 2002, Numerical modeling of soil anchors, PhD thesis, University of Newcastle, Newcastle, Australia.

Merifield RS, 2011, Ultimate uplift capacity of multiplate helical type anchors in clay, *J Geotech Geoenviron Eng*, 137(7), 704–716.

Merifield RS, Lyamin AV, and Sloan SW, 2006, Three-dimensional lower-bound solutions for the stability of plate anchors in sand, *Geotechnique*, 56(2), 123–132.

Merifield RS, Lyamin AV, Sloan SW, and Yu HS, 2003a, Three dimensional lower bound solutions for stability of plate anchors in clay, *ASCE J Geotech Geoenviron Eng*, 129(3), 243–253.

Merifield RS, Sloan SW, and Lyamin AV, 2003b, *Three-dimensional Lower Bound Solutions for Stability of Plate Anchors in Sand*, Report TR-2003-02, ISBN 1 877078 04 2, University of Southern Queensland, Australia.

Meyerhof GG and Adams JI, 1968, Ultimate uplift capacity of foundations, *Can Geotech J*, 5(4), 225–244.

Murff JD, Randolph MF, Elkhatib S, Kolk HJ, Ruinen RM, Strom PJ, and Thorne CP, 2005, Vertically loaded plate anchors for deepwater applications, *Proceedings of the International Symposium on Frontiers in Offshore Geotechnics*, IS-FOG05, Perth, pp 31–48.

Murray EJ and Geddes JD, 1987, Uplift of anchor plates in sand, *ASCE J Geotech Eng*, 113 (3), 202–215.

Narasimha Rao S, Prasad YVSN, and Shetty MD, 1991, The behavior of model screw piles in cohesive soils, *Soils Found*, 31(2), 35–50.

Narasimha Rao S, Prasad YVSN, and Veeresh C, 1993, Behavior of embedded model screw anchors in soft clays, *Geotechnique*, 43(4), 605–614.

NAVFAC, 2011, *SP-2209-OCN Handbook for Marine Geotechnical Engineering*, Naval Facilities Engineering Command, Engineering Service Center, Port Hueneme, USA.

O'Loughlin CD, Blake A, Richardson MD, Randolph MF, and Gaudin C, 2014, Installation and capacity of dynamically embedded plate anchors as assessed through centrifuge tests, *Ocean Eng*, 88, 204–213.

O'Loughlin CD, Lowmass AC, Gaudin C, and Randolph MF, 2006, Physical modelling to assess keying characteristics of plate anchors, *Proceedings 6th International Conference on Physical Modelling in Geotechnics*, Hong Kong, China, 4–6 August 2006, Vol. 1, pp 659–666.

O'Loughlin CD, Richardson MD, Randolph MF, and Gaudin C, 2013, Penetration of dynamically installed anchors in clay, *Geotechnique*, 63 (11), 909–919.

O'Loughlin CD, White DA, and Stanier SA, 2015, Novel anchoring solutions for FLNG—Opportunities driven by scale, *Proceedings of the Offshore Technology Conference*, Houston, OTC 26032-MS, pp 1–32 (electronic format).

O'Neill MP, Bransby MF, and Randolph MF, 2003, Drag anchor fluke–soil interaction in clays, *Can Geotechl J*, 40, 78–94.

Prasad YVSN and Narasimha Rao S, 1996, Lateral capacity of helical piles in clays, *J Geotech Geoenviron Eng*, ASCE, 122(11), 938–941.

Rowe RK and Davis EH, 1982, The behaviour of anchor plates in sand, *Geotechnique*, 32(1), 24–41.

Song Z, Hu Y, O'Loughlin CD, and Randolph MF, 2009, Loss in anchor embedment during plate anchor keying in clay, *ASCE J Geotech Geoenviron Eng*, 135(10), 1475–1485.

Tian Y, Gaudin C, and Cassidy MJ, 2014, Improving anchor design with a keying flap, *ASCE J Geotech Geoenviron Eng*, 140(5), 04014009-1 to 04014009-13 (electronic version).

Tsuha CHC, Aoki N, Rault G, Thorel L, and Garnier J, 2012, Evaluation of the efficiencies of helical anchor plates in sand by centrifuge model tests, *Can Geotech J*, 49, 1102–1114.

Vesic AS, 1971, Breakout resistance of objects embedded in ocean bottom, *J Soil Mech Found Div*, 97(9), 1183–1205.

Wang D, Hu Y, and Randolph MF, 2011, Keying of rectangular plate anchors in normally consolidated clays, *ASCE J Geotech Geoenviron Eng*, 137(12), 1244–1253.

Wang D and O'Loughlin CD, 2014, Numerical study of pull-out capacities of dynamically embedded plate anchors, *Can Geotech J*, 51, 1263–1272.

Wei Q, Cassidy MJ, Tian Y, and Gaudin C, 2014, Incorporating shank resistance into prediction of the keying behaviour of suction embedded plate anchors, *ASCE J Geotech Geoenviron Eng*, 141(1), 04014080-1 to 04014080-13 (electronic version).

Wilde B, Treu H, and Fulton T, 2001, Field testing of suction embedded plate anchors, *Proceedings of the 11th ISOPE Conference, Stavanger*, International Society of Offshore and Polar Engineers, Cupertino, California, pp 544–551.

Yang M, Aubeny CP, and Murff JD, 2012, Behavior of suction embedded plate anchors during keying process, *J Geotech Geoenviron Eng*, ASCE, 138(2), 174–183.

Yu, SB, Hambleton JP, and Sloan SW, 2014, Analysis of inclined strip anchors in sand based on the block set mechanism, *Appl Mech Mater*, 553, 422–427.

Index